COLLINS
COMPLETE GUIDE TO
BRITISH INSECTS

Michael Chinery

Collins

HarperCollins Publishers Ltd.
77-85 Fulham Palace Road
London
W6 8JB

www.collins.co.uk

Collins is a registered trademark of HarperCollins Publishers Ltd.

First Published in 2005

15 14 13 12 11
10 9 8 7 6 5 4 3

A catalogue record for this book is available from the British Library.

ISBN: 978-0-00-729899-0

Edited and designed by D&N Publishing, Hungerford, Berkshire
Printed and bound in Hong Kong by Printing Express Ltd.

CONTENTS

ABOUT THIS BOOK

Insects are everywhere and, between them, the 20,000 or so British species eat just about everything. They range from minute fairy flies that grow up in the eggs of other insects to moths with wings spanning 15cm, and in the warmer parts of the world there are some very much bigger species. Whether we like them or not, we cannot avoid them, and the aim of this book is to allow the reader to put names to those insects most likely to be encountered in the British Isles and to learn just a little of their role in nature. Most of our larger and more conspicuous insects have been included, although some rarities and infrequent visitors that might be met with just once in a lifetime have been omitted. In general, insects less than 5mm long or with wings spanning less than 5mm have not been included because, even if they are noticed, they are generally difficult to identify without a microscope. Those midgets that do feature in the book merit inclusion because they are pests or because they produce conspicuous effects, such as galls or leaf mines.

The text is brief, but complements the photographs by drawing attention to the diagnostic features and those that separate the insects from superficially similar species. English names, where they exist, are given in bold type, followed by the scientific name. English names exist only for the more common and easily recognised species, so many of the smaller insects have just a scientific name. Conventionally printed in italics, the latter always consists of two parts (*see* p. 10). Some of the larger families are described in brief introductory paragraphs, but otherwise the family name is given after the scientific name. If no family name is given, it may be assumed that the insect in question belongs to the same family as the preceding species. Sizes are given for each species and, unless otherwise stated, they are average lengths and they include the wings if these are normally folded over the body at rest. Body length and wingspan are both given for dragonflies, while for butterflies and moths the measurement is the length of a single forewing as this is the easiest dimension to measure. If a species varies by more than a few millimetres, the normal upper and lower limits are given instead of an average size. Caterpillar sizes are maximum lengths.

The normal habitat is given for each species, together with the months during which the adult insects can be found, but the flight time may be shorter in the north because many species have two broods each year in the south and only one in the north. If no months are indicated, it may be assumed that the adult insects can be found throughout the year. Foodplants are given for vegetarian species, and for the butterflies and moths these are the plants on which the caterpillars feed.

Maps give the approximate distribution of each species as far as it is known, but these maps should not be regarded as complete: populations may well exist outside the indicated range of a species but have simply not been recorded. It is also important to bear in mind that an insect will not necessarily occur everywhere within the indicated range: it will normally be found only where the habitat is suitable. A 'V' in a circle next to a map indicates a migrant or vagrant species. A '?' on a map indicates an incidental species.

WHAT IS AN INSECT?

The insects make up a class of animals, the Insecta, within the phylum Arthropoda. This name means 'jointed feet' and refers to the way in which the legs are constructed from several sections. In fact, the whole body is composed of sections or segments, each enclosed in a tough outer coat that forms the exoskeleton. Flexible membranes between the segments allow the animals to move. As invertebrates, the animals have no bones or other internal skeleton. The arthropods include spiders, scorpions, crabs and centipedes as well as insects but, with well over a million known species, the insects are by far the largest group. They differ from the other arthropods in having three body regions, the head, thorax and abdomen, and three pairs of legs when adult. There is also a pair of antennae or feelers on the head, and most adult insects have two pairs of wings. No other arthropod nor even any other invertebrate animal ever has wings.

This is an insect.

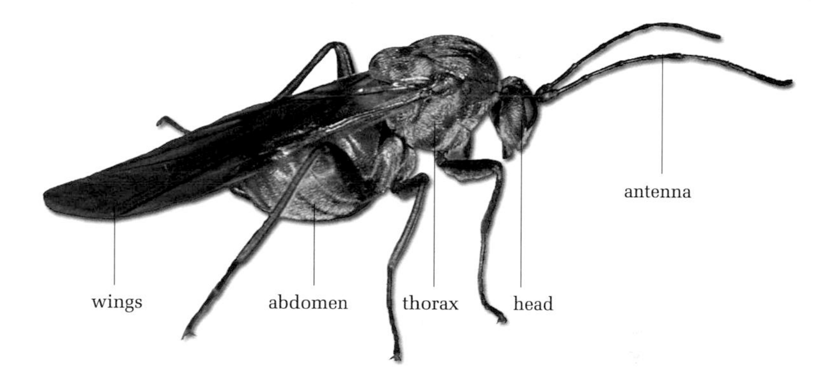

These are not insects.

A spider, four pairs of legs and only two body sections.

Woodlice, seven pairs of legs.

INSECT LIFE HISTORIES

With the exception of a few flies and certain generations of aphids, most insects reproduce by laying eggs, but the insects leaving the eggs are often very different from the adults. In particular, they have no wings. They undergo considerable changes as they grow and these changes are called metamorphosis.

An insect's tough exoskeleton is made largely of a horny substance called chitin, which is not a living material. It cannot grow with the insect, so a new covering has to be produced every now and then. When the outer coat becomes tight, the insect generates a new, looser skin underneath it. The insect then puffs itself up with air or water and the old coat, now very thin, splits open along predetermined lines of weakness. The insect wriggles out of the shattered skin and its new coat soon hardens. The air or water is then expelled, leaving room for another period of growth within the new skin. This skin-changing is called moulting or ecdysis and it occurs several times as the insect grows. Most insects undergo four or five moults, but some require as many as 25 moults before reaching maturity.

The young of dragonflies, grasshoppers, earwigs, bugs and some other insects resemble the adults in shape and are generally called nymphs, although some entomologists now prefer to call all young insects larvae. Each time it changes its skin, the nymph gets larger and more like the adult, with wing-buds appearing on the outside of the body and getting larger at each moult. At the final moult, the adult insect emerges with its wings and other organs all fully developed, although its wings are soft and crumpled as a rule and need to expand and dry before the insect can fly. The insect may need a further period of maturation before it is sexually mature. This kind of development is called partial or incomplete metamorphosis and, because the wings develop on the outside of the body, the insects are called exopterygotes.

Butterflies and moths, flies, beetles, and bees and wasps follow a different developmental pathway. The young insects look nothing like the adults, even allowing for the lack of wings, and they often eat different foods. They are called larvae and, although they get larger at each moult, they do not look any more like the adults. When a larva, such as a caterpillar, reaches its full size it sheds its skin again, but this time it reveals a pupa or chrysalis. Only now do the wings and other adult organs become apparent, although they are still not functional. The pupae of mosquitoes and some other flies are quite mobile, but most are inactive and none ever feeds. Nevertheless, there is enormous biochemical and physiological activity inside a pupa as the larval structures are broken down and converted to a thick 'soup'. Clusters of cells survive the demolition process and, fed by the 'soup', they start to grow and multiply and they eventually develop into the adult insect. The pupal skin splits in due course and the adult struggles out, sometimes after just a few days but more often several weeks or months after the pupa was formed.

Clinging to its old nymphal skin, this newly emerged dragonfly has expanded its wings, but it will not get its true colours for several days. The similarity between the nymphal and adult body is very clear.

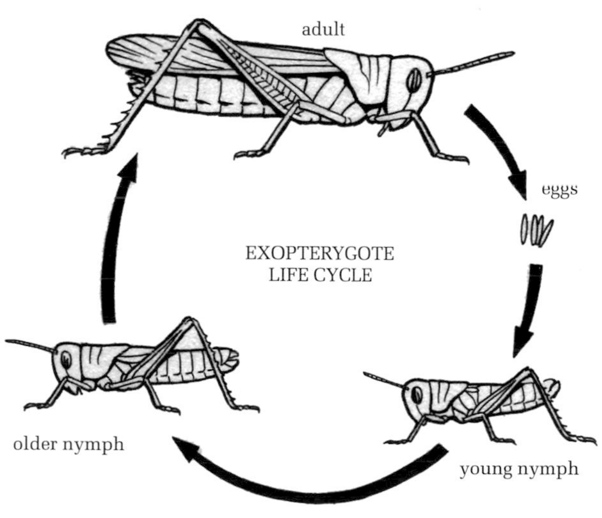

The grasshopper's life cycle is typical of the exopterygote insects, with the nymphs getting more like the adult with each moult. The wing buds, which can be seen on the back of the nymphs, get larger at each moult.

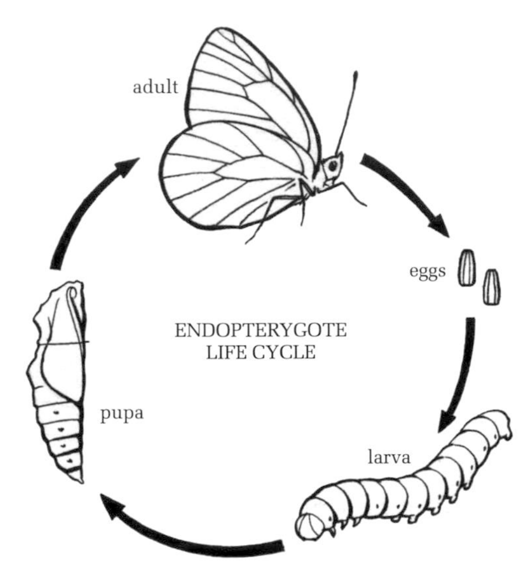

OPPOSITE PAGE: Newly-emerged insects like this Privet Hawkmoth must find room to expand their wings properly. It might take half an hour for the wings to reach their full size.

LEFT: The life cycle of a butterfly is typical of the endopterygote insects, with a pupal stage between the larva and the adult.

BELOW: The larva of an *Aphodius* beetle, seen here in the typical pose of a scarab grub, looks very different from the adult insect (*see* p. 116).

Some insects can spend several years in the pupal state. As with the exopterygotes, the wings are soft and crumpled at first, but they soon expand and harden.

This kind of development is known as complete metamorphosis and, because the wings develop inside the body, the insects are called endopterygotes. They are also known as holometabolous insects.

CLASSIFYING INSECTS

The insect world is divided into about 30 major groups called orders, although not all entomologists agree on the divisions and some recognise more orders than others. For example, the lacewings, alderflies and snakeflies have long been treated as members of a single order, the Neuroptera, but many entomologists now put them into three separate orders (*see* pp. 106–8). The orders are based on various anatomical features and most of their names are derived from the nature of the wings: Diptera (the true flies) = two wings; Trichoptera (the caddis flies) = hairy wings; and Lepidoptera (the butterflies and moths) = scaly wings. The orders are split into families, whose names all end in **-idae**. Muscidae, for example, is the family containing the house-flies and Lampyridae is the glow-worm family. Each family contains one or more genera containing a number of closely related species.

A species is an individual kind of insect or other organism and each has a scientific name that can be understood all over the world. This scientific name is always printed in italics and it consists of the name of the genus and a specific name. Thus, the scientific name of the Red Admiral butterfly is *Vanessa atalanta* and that of the Painted Lady is *Vanessa cardui*. The fact that they share the generic name *Vanessa* shows that the butterflies are closely related, something that is not apparent from their English names. A specific name ideally gives an indication of the appearance or colour of an insect, or perhaps its food or its habitat. *Vanessa* ***atalanta***, for example, is named after Atalanta, a beautiful woman in Greek mythology, while *Vanessa* ***cardui*** gets its name because its caterpillars feed mainly on thistles of the genus *Carduus*. *Musca* ***domestica*** is the House-fly and *Gryllus* ***campestris*** is the Field Cricket. Some specific names merely reflect the name of the person who first discovered the species and give no clues as to the nature of the creature. In scientific literature, the scientific name of an insect or other creature is normally followed by the name of the person who first named the species, and also the date on which the name was first published. If, as often happens, a creature's name is changed as a result of later research, the name of the original author is still used but it is enclosed in brackets. For example, the modern name of the Eyed Hawkmoth is *Smerinthus ocellata* (Linnaeus), indicating that Linnaeus first named the insect but that he gave it a different name: along with all the other hawkmoths known to him, he put it in the genus *Sphinx*. Detailed studies of the hawkmoths since the time of Linnaeus revealed big differences between many of the species and entomologists decided that they are not all closely related and should be put into several different genera. Scientific names also have to be changed if it is discovered that a species has been given two different names or two different species have been given the same name. This often happened in the past when entomologists working in different countries were not able to communicate with each other as easily as they can today.

Each order of insects found in the British Isles is briefly described in the following pages and illustrated with one or more typical species. Orders containing only tiny insects, such as fleas and thrips, are not dealt with elsewhere in the book.

ORDER THYSANURA: BRISTLETAILS

This is a small order of wingless insects characterised by a coating of shiny scales and the possession of three slender 'tails' at the rear. These 'tails' are fringed with stiff hairs and the name Thysanura literally means 'fringed tail'. The most familiar species is the Silverfish (*below*) that lives in most of our houses and eats a variety of starchy materials, including the glue of cartons and bookbindings. The Firebrat, with much longer 'tails' than the Silverfish, also lives in houses, and another six or seven species live as scavengers in the wild. *Petrobius* species, for example, are commonly found among coastal rocks.

The well-named Silverfish (*Lepisma saccharina*) is clothed with silvery scales that make it very slippery. The stiff hairs are clearly seen on its 'tails'.

The Thysanura is the only order in the subclass Apterygota. This name means 'without wings' and indicates that these insects have never had wings at any time during their evolutionary history. All other insects are included in the subclass Pterygota, meaning 'winged'. Although some of these insects, such as the lice and fleas, are wingless, they are descended from winged ancestors. The springtails and a few other groups of wingless mini-beasts were once included in the Apterygota, but are not now considered to be insects.

ORDER EPHEMEROPTERA: MAYFLIES (p. 66)

These are weak-flying insects with one or two pairs of delicate wings with numerous veins. The hindwings, when present, are very small, and the insects nearly always rest with their wings held vertically over the body.

The front legs of the male mayfly, seen here, are very long and used primarily for holding the female during copulation. The other legs can do little more than cling to supports.

The antennae are minute, but the rear end of the body bears two or three slender 'tails'. Adult mayflies do not feed and rarely live for more than a few days, hence the name Ephemeroptera, which literally means 'living for a day'. Some species enjoy no more than a few hours of adult life, although the young stages may spend a year or more growing up in ponds and streams. The youngsters all have three 'tails'. Mayflies are the only insects that moult when fully-winged: freshly-emerged adults, known as duns, are dull and hairy, but soon after emerging, sometimes within minutes, they shed their hairy coats and become shiny spinners, although they are never very colourful. The insects rarely travel far from the water in which they develop. There are about 47 species in the British Isles.

ORDER PLECOPTERA: STONEFLIES
(p. 66)

These are black, brown or yellow insects with two pairs of membranous wings, of which the hind pair is usually much broader than the front pair. The wings are folded flat over the body at rest or else rolled around it. The name Plecoptera means 'folded wings'. There is often a pair of conspicuous 'tails' or cerci at the rear. Most stoneflies breed in running water and the insects are most common in upland areas with clean water. They take up to three years to mature. The weak-flying adults are rarely found far from water. They may nibble pollen and scrape algae from trees and rocks, but most adults take no food at all. There are about 30 species in the British Isles.

Stoneflies are sometimes confused with caddis flies (*see* p. 318), but are easily distinguished by their lack of hair and by the flat or rolled position of the wings at rest. Caddis flies always rest with their wings held roof-like over the body.

ORDER ODONATA: DRAGONFLIES (pp. 68–76)

These insects have long, slender bodies and stiff wings that cannot be folded. Large eyes, with thousands of separate lenses, are very efficient at spotting moving prey, but the antennae are very short and bristle-like and often escape notice. There are two main groups: the true dragonflies of the suborder Anisoptera and the much more slender and dainty damselflies of the suborder

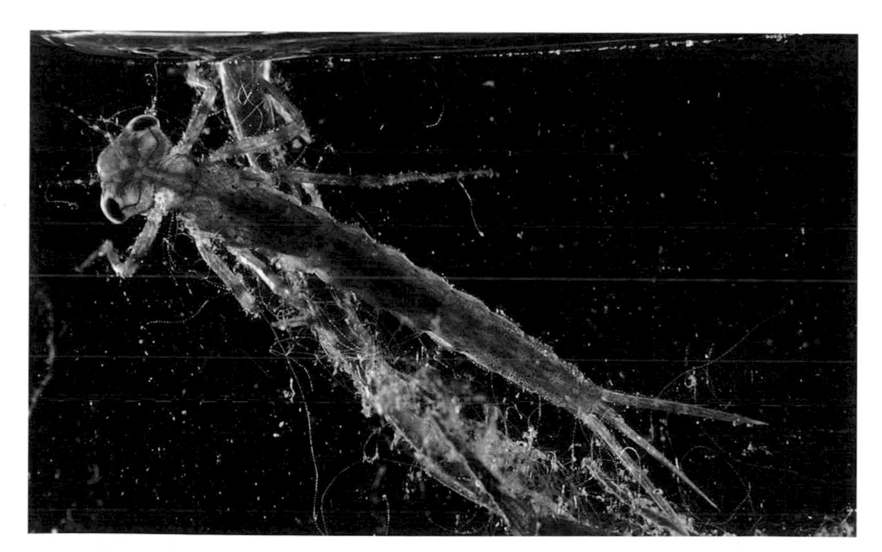

Damselfly nymphs can always be recognised by the three blade-like projections at the rear. These are the gills through which the insects breathe. True dragonflies have their gills inside the rear end of the body.

Zygoptera. The true dragonflies are mostly fast-flying, agile predators that catch other insects in mid-air by scooping them up in a 'basket' formed by the spiky legs. They rest with their wings outspread, whereas the damselflies close their wings partly or completely above the body. Damselflies also fly more slowly than the dragonflies and, unlike them, are rarely found far from water. They tend to pluck prey from the vegetation rather than catch it in flight. The prey may be eaten in flight or taken to a perch to be cut up by the toothy jaws that give the order its name. The insects are all aquatic in their early stages and fiercely predatory. Most of them live in still or slow-moving water. About 40 species are resident in the British Isles, but several others arrive as regular or infrequent visitors from the continent and some of these are showing signs of taking up residence.

LEFT: Unlike those of other insects, a dragonfly's wings are not linked and they beat independently, as is clearly shown in this picture of a damselfly in flight.

BELOW: Once their wings have hardened, the true dragonflies cannot bring them together over the body and they always rest with their wings outstretched.

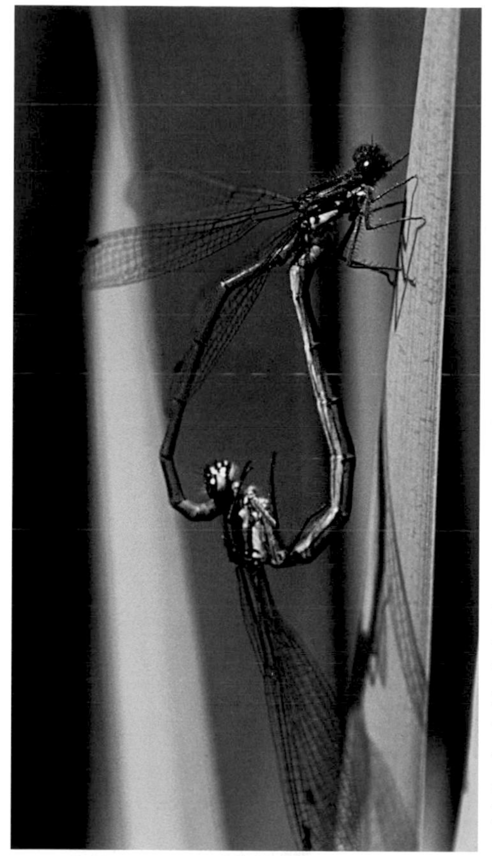

THE COPULATION WHEEL
Dragonflies and damselflies exhibit a mating routine all of their own. Although the male reproductive organs are in the normal place at the end of the abdomen, his copulatory organs are at the front of the abdomen, and before going in search of a mate he has to transfer sperm to them from the tip of his abdomen. This then leaves him free to use the claspers at the tip of his abdomen to grab a female by the neck. She then bends her own body round so that the tip of her abdomen makes contact with the male's copulatory organs to collect the sperm. This position is known as the copulation wheel and the insects can actually fly in this position. They may separate after a short while, but in some species the male continues to hold the female's neck and the pair fly in tandem, with the male pulling the female along. He may even continue to hold her while she lays her eggs in the water.

ORDER ORTHOPTERA: CRICKETS AND GRASSHOPPERS (pp. 78–82)

These are stout-bodied, bullet-shaped insects in which the hind-legs are usually modified for jumping. The order is now often called Saltatoria because of the insects' jumping abilities. There is a prominent saddle-shaped pronotum behind the broad head and typically two pairs of wings, the front pair tough and leathery and the hind ones membranous, although many species lack one or both pairs. The name Orthoptera means 'straight-winged' and refers to the way in which the wings are laid straight back along the body at rest. Grasshoppers and groundhoppers are vegetarians and have short antennae. Crickets are essentially omnivorous, although other insects predominate in the diet of many species, and they have thread-like antennae that are usually much longer than the body. Female crickets have conspicuous ovipositors.

Although flightless, the male Speckled Bush-cricket retains enough of his forewings to be able to make a soft call. The long antennae distinguish crickets from grasshoppers.

Grasshopper antennae are much shorter than the body. 'Songs' are produced when tiny pegs on the inner surface of each hind leg pass over a hard ridge on the forewing.

The insects are best known for their 'songs', which are produced by stridulation, the rubbing of one part of the body against another. Grasshoppers rub their legs against their wings, but crickets rub their wings together. It is usually only the males that make the sounds that we hear, and each species has its own 'tune' that attracts the females and deters other males. Some female grasshoppers can stridulate, but their 'songs' are very quiet and generally produced only when a male is near. There are about 30 British species, mostly living in the south.

ORDER DERMAPTERA: EARWIGS
(p. 84)

These small brown omnivorous insects are easily recognised by the tweezer-like pincers or cerci at the rear. These are more or less straight in the female but often strongly curved in the male. Although they can be used to subdue and even kill prey, they are used mainly for defence. The forewings are short and leathery and almost square. The hindwings, when present, are very soft and flimsy – Dermaptera means 'skin-wings' – and elaborately folded under the forewings, but only two of the four British species possess hind-wings. Many continental species also lack forewings. Fully-winged earwigs can fly, but rarely do so.

The pincers of this male common earwig are ready for action. The females guard and feed their young for a while and whole families can be found clustered together in crevices, especially under loose bark. White earwigs are not uncommon: they are insects that have just moulted and not regained their colour.

ORDER DICTYOPTERA: COCKROACHES
(p. 84)

These fast runners are flattened brown, black or yellowish scavengers with long spiky legs and long antennae. The head is concealed from above under the front part of the thoracic shield or pronotum. There are generally two pairs of wings, the front pair being leathery and overlapping each other. There are three native species but several aliens have become established in buildings.

The tawny cockroach is the only one of the three native species in which both sexes are fully winged and able to fly. Females of the other two species have shorter wings.

ORDER PSOCOPTERA: BOOKLICE AND BARKLICE

Also called psocids, these are all tiny, soft-bodied insects with or without wings. They have relatively large heads and long antennae. Some live indoors, where they feed on traces of moulds on books and other papers and can also cause severe damage to insect collections. These indoor species are generally wingless. Most psocids live outdoors and have four membranous wings. They feed on moulds, pollen and algae, often on tree trunks. There are about 90 species on the British list.

Psocids are commonly confused with aphids and psyllids (*see* p. 104), but they can always be distinguished by their biting jaws and more complex venation.

ORDER MALLOPHAGA: BITING LICE

These tiny insects, few of which are more than about 5mm long, are wingless parasites living on birds and mammals and using their biting jaws to feed on particles of skin, fur, feathers and some blood. The head is as wide as the body and the legs are equipped with strong claws for holding on to hairs and feathers. There are about 500 species in the British Isles, including several important poultry pests.

Biting lice live mainly on birds. Long-bodied species like those pictured here generally live among the longer feathers, but lice living among the shorter feathers of the head and neck tend to be short and plump.

ORDER ANOPLURA: SUCKING LICE

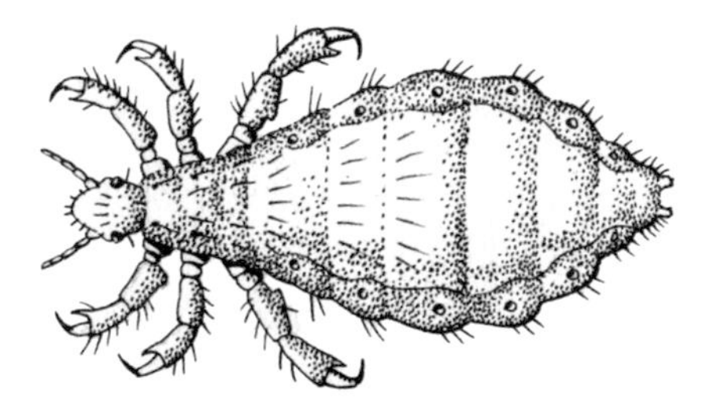

Sucking lice are best distinguished from biting lice by their very narrow heads and no clear segmentation of the thorax. Biting lice usually have distinct thoracic segments.

These are tiny wingless parasites of mammals, mostly with a very narrow head and piercing jaws designed for sucking blood. The human head louse (*Pediculus humanus*) is the most familiar of the 25 or so British species. It is common on both children and adults and easily transmitted, but its presence is no reflection on personal hygiene. It is easily picked up from seats and headrests on public transport.

ORDER HEMIPTERA: BUGS (pp. 86–104)

Many people refer to all insects as bugs, and entomologists are commonly dubbed 'bug-hunters', but this practice is to be discouraged. Strictly speaking, only members of the Hemiptera should be called bugs, and to avoid confusion they are sometimes referred to as true bugs. The order is very large and extremely varied, its members ranging from the 35mm-long water stick insect to tiny aphids and microscopic scale insects. The only thing they all have in common is a piercing and sucking beak or rostrum used for obtain-

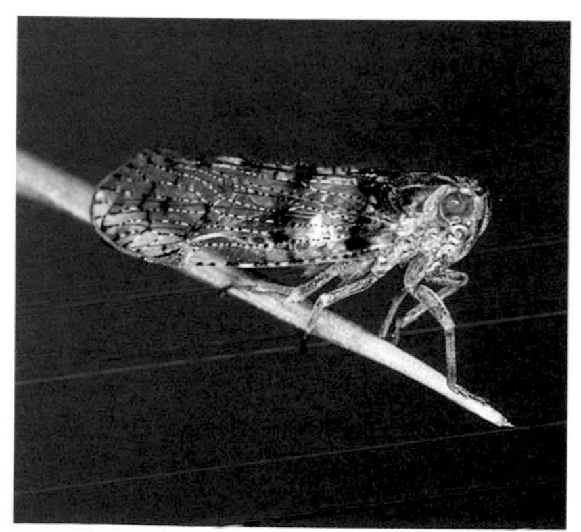

This planthopper exhibits the typical features of a homopteran bug – uniform forewings held roof-wise over the body and a beak appearing to come from the rear of the head.

This shield bug is a typical heteropteran: as well as the conspicuous membranous tips to the forewings, there is a large pronotum and a prominent, shield-shaped scutellum.

ing juices from plants or from other animals. There are typically two pairs of wings, although there are many wingless species in the order. There are two distinct suborders, Heteroptera and Homoptera, although these are regarded as separate orders by many entomologists. There are about 1,650 British species.

The Heteroptera are named for the heterogeneous nature of the forewings, each of which typically has a horny or leathery basal region and a membranous tip. Some of these bugs look as if they have only half of each wing – hence the name Hemiptera given to the whole order. The wings are laid flat over the body at rest, usually with a good deal of overlap. The scutellum is usually triangular and very conspicuous. This suborder includes both herbivorous and predatory species, with several of the latter sucking the blood of birds and mammals, including humans. All the water bugs belong to this suborder.

The Homopteran forewings, when present, are either membranous or leathery, but always of uniform texture. They are normally held roof-wise over the body at rest. The head is often very small and folded back under the thorax so that the beak appears to come from the back of the head. All are plant-feeders and they include many serious crop pests, such as the aphids and scale insects. Most of these insects are very small, with very few British species exceeding 5mm in length. Only a few are included in this book.

THRIPS: ORDER THYSANOPTERA

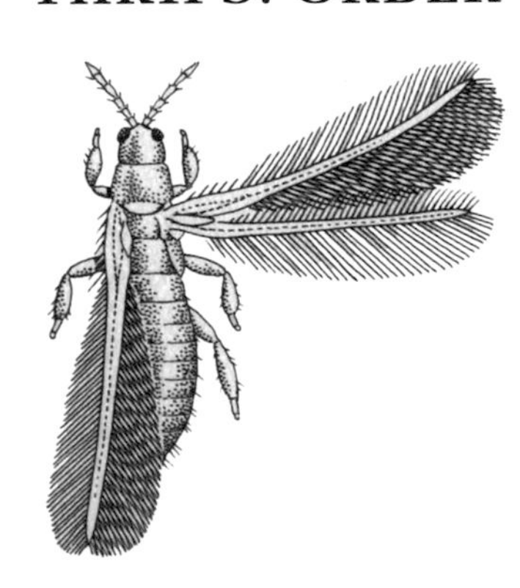

Most female thrips have minute saw-like ovipositors, with which they cut slits in plants prior to laying their eggs there.

These are tiny winged or wingless brown or black insects. The wings, when present, are like minute feathers and the name Thysanoptera means 'fringed wings'. Most are vegetarians, sucking sap from plant cells with their piercing beaks. They are abundant in flowers, but the most noticeable species are the cereal thrips, often known as thunder flies or thunder bugs because they seem to fly mainly in humid weather associated with thunderstorms. They breed in the developing ears of wheat and barley and the young females take to the air in millions as the crops ripen in the summer. They pass the winter in tiny crevices, often under loose wallpaper in houses and even inside picture frames. Most of the 150 or so British species are under 2mm long.

With the exception of the thrips, the insects in the orders described above all grow up without a pupal or chrysalis stage. The youngsters are generally called nymphs, and they are basically similar to the adults except that they have no functional wings. The wings develop gradually on the outside of the body (*see* p. 6). The following orders follow a different path of development. Their youngsters look nothing like the adult insects and are always referred to as larvae. Caterpillars and maggots are familiar larvae. They get bigger at each moult, but show no sign of wings until they are fully grown and turn into pupae. The adult body is formed during the pupal stage (*see* p. 6). The thrips are unusual because, although their wings develop on the outside of the body and they are classified with the exopterygotes (*see* p. 6), they also pass through a pupal stage.

LACEWINGS: ORDER NEUROPTERA
(pp. 106–8)

These soft-bodied, green or brown insects are named for their numerous delicate wing veins, many of which clearly fork as they reach the wing margins. The wings are held roof-wise over the body at rest, with the antennae held close together in front of the body. Despite their fragile appearance the insects are carnivorous, with both adults and larvae devouring large

This green lacewing clearly shows the resting attitude of the insects, and why they are also known as goldeneyes. The forking of the veins at the wing margins is also quite noticeable.

These green lacewing larvae are just emerging from their eggs, which are clustered together at the end of matted strands of hardened mucus. Suspended in this way, the eggs are safe from at least some of their enemies.

numbers of aphids and other small insects. The shuttle-shaped larvae are rather bristly and some species camouflage themselves by fixing the drained skins of their victims to the backs. The green lacewings of the family Chrysopidae attach their eggs to strands of mucus that harden on contact with the air so that the eggs hang down on stalks, either singly or clustered.

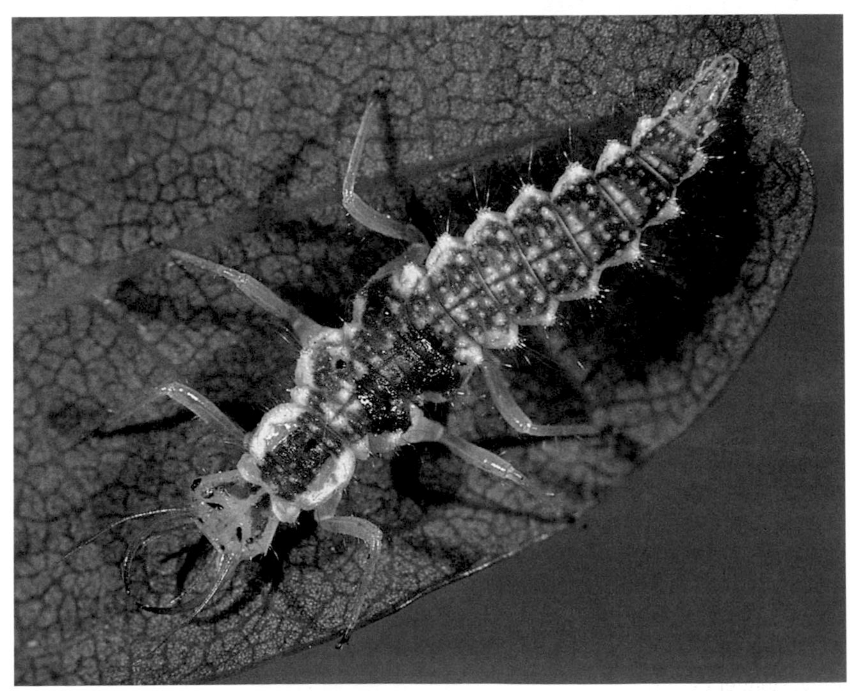

The large jaws of the lacewing larva are hollow. They pierce their victims and suck out their juices.

ALDERFLIES: ORDER MEGALOPTERA
(p. 108)

Closely related to the lacewings, and once included with them in the Neuroptera, the alder flies can be distinguished by their smoky wings and dark veins that do not fork at the wing margins. The adults are not predatory, and if they feed at all they simply nibble pollen and algae. The insects grow up in water, where the larvae feed on other invertebrates. They breathe by way of feather-like gills projecting from the sides of the abdomen. There are just three, very similar species in the British Isles.

The venation of the alderflies is very different from that of the lacewings and is one reason why entomologists now put them into different orders.

SNAKEFLIES: ORDER RAPHIDIOPTERA
(p. 108)

These insects get their name for the long 'neck-like' prothorax, on which the head can be raised well above the rest of the body. Adult snakeflies feed mainly on aphids, while the larvae feed on other insects living in decaying wood and under loose bark. There are just four, very similar species in the British Isles.

Snakeflies are essentially woodland insects and live mainly in the trees. The females each have a long, slender ovipositor, clearly seen in this picture.

SCORPION FLIES: ORDER MECOPTERA
(p. 108)

These harmless insects get their name for the upturned abdomen of most males, but the most constant and diagnostic feature of the order is the way in which the head is prolonged downwards into a stout beak, with the jaws at the tip. The adults and their caterpillar-like larvae are scavengers, with the adults feeding mainly on dead insects. Some even steal from spiders' webs. Only four species occur in the British Isles, the three very similar scorpion flies and the minute Snow Flea (*Boreus hyemalis*).

This male scorpion fly shows very clearly how the insects got their name, and also exhibits the characteristic downward-pointing beak or rostrum.

BEETLES: ORDER COLEOPTERA
(pp. 110–38)

This is by far the largest order of insects, with something in the region of 370,000 known species throughout the world, although it is not the largest order in the British Isles – that honour goes to the Diptera. Although they vary a great deal in shape, most species are easily recognised as beetles by their tough and horny forewings, known as elytra. These usually cover the whole abdomen and meet in a straight line down the centre. The name Coleoptera, meaning 'sheath-wings', refers to this feature. In many species they are conspicuously domed. The membranous hindwings are completely concealed under the elytra when the insects are not flying, although many beetles have no hindwings and a few have no wings at all. Adult beetles all have biting jaws,

Most ground beetles (*see* pp. 110–12) can be recognised as such by their overall shape, with rather oval elytra, their long antennae, and the long legs that give them a good turn of speed over the ground.

The click beetles (*see* pp. 120–2) can be recognised by their bullet-shaped bodies with the head largely concealed under the thorax. Relatively short legs indicate that they are not rapid movers.

with which they tackle just about every kind of solid food. Some, notably the aphid-eating ladybirds, are economically valuable, but the order also includes many notorious pests that attack growing and stored crops, fabrics and timber. Beetle larvae vary just as much as the adults in appearance and feeding habits: there are fast-running predators, sluggish herbivores and completely legless grubs living completely concealed inside their food-plants. Over 4,000 species live in the British Isles.

Although not as fast as the adult insect, this ground beetle larva is just as predatory, hunting other small creatures under stones and in the soil.

FLIES: ORDER DIPTERA (pp. 298–316)

Diptera means 'two-winged' and refers to the fact that these insects have just two wings, although a few species, notably some of the parasitic forms, are completely wingless. The hindwings are represented by two tiny pin-shaped structures called halteres. Best seen in the crane-flies (*see* p. 298), they aid balance during flight. All flies are liquid-feeders and they utilise a very wide range of food materials. Many have piercing mouths used for sucking blood and other body fluids, but most simply mop up exposed liquids, ranging from nectar to the foul fluids surrounding dung and rotting flesh. Some hover-flies can crush pollen grains. There are about 5,200 species in the British Isles, ranging from minute midges to sturdy horse-flies and gangly crane-flies with wings spanning up to 60mm. Some species are very colourful and mimicry is well developed in several families (*see* p. 50).

Fly larvae are legless maggots, some with biting jaws but others with virtually no jaws at all, simply sucking up liquids from their surroundings. Between them, they eat almost anything, from fungi to rotten wood, decaying flesh, and dung. Many larvae are leaf-miners and others induce gall-formation (*see* p. 38).

Pupation in many species, including hover-flies, house-flies and blow-flies, takes place in a barrel-shaped puparium formed from the last larval skin.

The common names of flies have been hyphenated in this book, as in many other publications, in order to avoid confusion with insects of other orders. Thus, crane-fly, robber-fly and house-fly are all true flies, whereas dragonflies, scorpion flies, and sawflies are not. In this context, all members of the Diptera are sometimes referred to as true flies.

ABOVE: The crane-flies belong to the group of flies known as the Nematocera, meaning 'thread-horned' and referring to the relatively long thread-like antennae of most species. Mosquitoes also belong to this group, most of which are slender flies with long, spindly legs.

RIGHT: This stoutly-built insect has very little in common with the crane-fly above, but it is still a true fly. It belongs to the Brachycera, in which the antennae project from the head like tiny horns.

FLEAS: ORDER SIPHONAPTERA

These tiny wingless insects are all blood-sucking parasites living on mammals and birds. They are brown or black and strongly flattened from side to side, enabling them to scuttle easily between the hairs or feathers. Strong claws and 'combs' of stout bristles help them to grip firmly when necessary, and powerful back legs enable them to cover 30cm or more in a single leap when not on a host. Larval fleas are wormlike and live in the nests of their hosts, where they feed on debris, including the droppings of the adult fleas which contain a certain amount of undigested blood. About 60 species live in the British Isles, although not all are natives. The largest of them, the Mole Flea, is only about 6mm long.

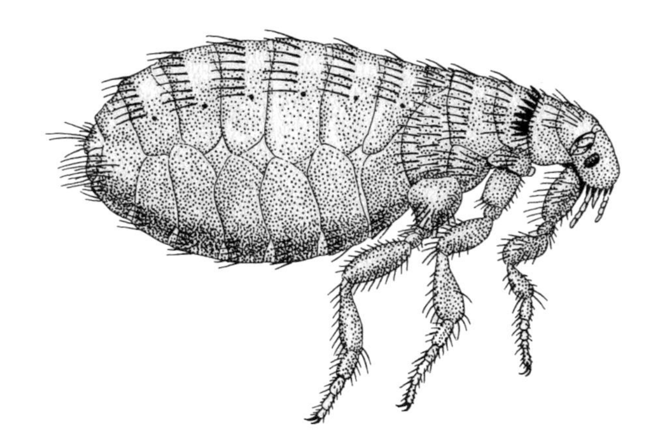

Although fleas are completely wingless, their internal anatomy suggests that they have evolved from winged ancestors.

BUTTERFLIES AND MOTHS: ORDER LEPIDOPTERA (pp. 140–296)

The name Lepidoptera means 'scale-wings' and refers to the covering of minute scales that give the wings their colours and patterns. Most species possess sucking mouthparts in the form of a slender, tubular tongue known as the proboscis and they feed mainly on nectar. When not in use, the proboscis is coiled up under the head. There is no single difference between all butterflies on the one hand and all moths on the other and the division has no scientific basis. As far as the British species are concerned, however, the form of the antennae is a good guide: all our butterflies have clubbed antennae, although the club is not always well developed, whereas the majority of moths have hairlike or feathery antennae. The day-flying burnet moths (*see* p. 160) have clubbed antennae, but these

This Dun-bar moth exhibits the typical resting position of moths, with the wings held roof-wise over the body so that the uppersides of the forewings are visible. The antennae more or less concealed.

swell gradually and not abruptly as in the butterflies. The burnets, in common with most moths, also rest with their wings folded back along the sides of the body, whereas very few butterflies can do this: butterflies often bask and feed with their wings wide open, but when truly at rest they bring their wings together above the body so that only the undersides are visible. Most moths also have a frenulum, a bristle or cluster of bristles springing from the front of the hindwing and held in place by a hook on the underside of the forewing. Its function is to link the two wings together, but none of our butterflies has such a device: their wings are held together simply by a large overlap.

Although some butterflies feed with their wings wide open, others, including the Clouded Yellow seen here, always keep their wings closed so that only the undersides are visible. When truly at rest, the antennae are pulled back between the wings. The forewings may also be partly hidden between the hindwings.

This caterpillar of the Silver-washed Fritillary exhibits the typical larval arrangement of three pairs of true legs at the front and five pairs of prolegs on the abdomen. Minute hooks on the prolegs grip the food-plants with surprising strength. The true legs are often used to manipulate the food.

The larvae of butterflies and moths are called caterpillars, and have chewing mouthparts that are very different from the tubular tongues of the adults. Most of them eat leaves, although some eat roots or tunnel into various other parts of their food-plants. Caterpillars have three pairs of true legs, which are situated on the thorax and are the forerunners of the adult legs, and up to five pairs of fleshy prolegs on the abdomen. The last pair of prolegs are called the claspers. Many moth larvae burrow into the ground before pupating, while others surround themselves with silken cocoons before turning into pupae on their food-plants or in the leaf litter and other debris on the ground. Butterfly caterpillars usually pupate naked on their food-plants, either suspended upside-down or bound vertically to the food-plant by a silken girdle in what is known as the succinct position.

Over 2,000 species are permanent residents in the British Isles, but only 56 of these are butterflies. Most of the others are rather small moths collectively known as 'micros', although they belong to many different families and some are actually more closely related to the larger moths – the 'macros', than to the other 'micros'. The division into 'macros' and 'micros' is thus an artificial one and it stems from the days of the early collectors, who concentrated on the larger species and tended to dismiss the smaller fry as not worth the effort. Just a few of the commoner or more conspicuous 'micros' are illustrated (*see* pp. 268–70).

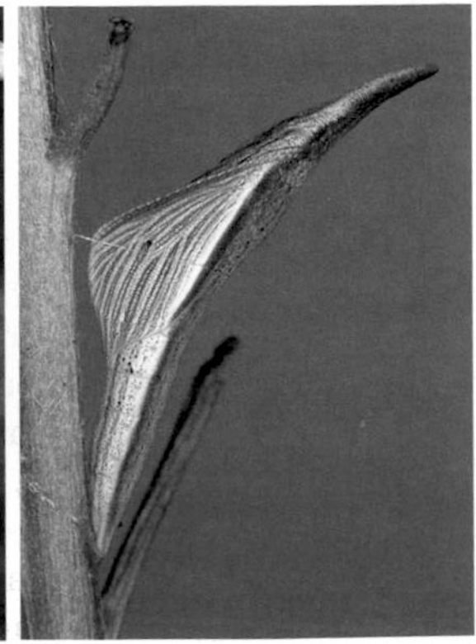
RIGHT: The chrysalis of the Silver-washed Fritillary is a typical suspended pupa.

FAR RIGHT: The chrysalis of the Orange-tip butterfly is a typical succinct pupa, held in place by a silken girdle.

CADDIS FLIES: ORDER TRICHOPTERA
(p. 318)

Trichoptera means 'hairy wings' and refers to the hairy covering on all four wings, although the hindwings are less hairy than the forewings. The wings are quite flimsy, with very few cross-veins, and flight is quite weak. At rest, the wings are held roof-wise over the body, with the long antennae held straight out to the front. Most species are nocturnal. Some species feed on nectar, but most species probably do not feed. Caddis flies nearly all grow up in ponds and streams, where many of the larvae surround themselves with portable cases made from sand or other debris. Most caddis larvae are omnivorous and some species living in running water trap food particles in silken nets spun among the vegetation. About 190 species live in the British Isles.

This caddis fly (*Agrypnia varia*) exhibits the typical resting position of the insects, with the wings held steeply roofwise over the body and the antennae close together and pointing forward.

BEES, WASPS, ANTS, AND THEIR RELATIVES: ORDER HYMENOPTERA
(pp. 320–56)

The members of this huge order vary enormously in size and shape and they include some of the smallest of all insects – the fairy flies that develop inside the eggs of various other insects. There is little to connect them all together, although typically there are two pairs of clear, membranous wings. The name Hymenoptera literally means 'membranous wings'. Venation is considerably reduced, leaving the wings with just a few large cells. There is usually a pigmented pterostigma near the tip of the forewing. The hindwings are much smaller than the forewings and are not immediately obvious in many species. Each has on its front edge a row of microscopic hooks that engage with the rear of the forewing to link the two wings firmly together. The insects all have

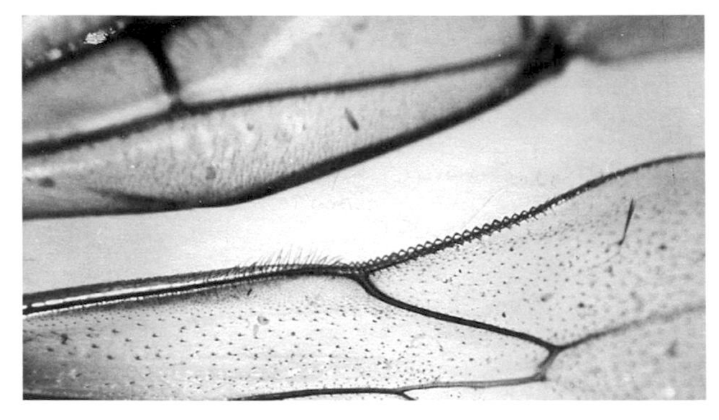

The hindwings of the Hymenoptera are usually very much smaller than the forewings and are linked to them by a series of tiny hooks. The two wings on each side then function as a single unit.

biting jaws, although most bees also have tubular tongues for sucking up nectar. The order includes vegetarians and carnivores and an enormous array of parasites (*see* p. 34), and it also includes all the social insects found in the British Isles (*see* p. 35).

There are two distinct suborders within the Hymenoptera: the Symphyta containing the sawflies and the Apocrita containing all the others. Sawflies (*see* pp. 320–4) have the thorax broadly joined to the abdomen with no obvious waist, whereas the apocritans (*see* pp. 326–56) have a marked 'wasp waist', although this is not always visible in the living insect. Sawfly larvae resemble the caterpillars of butterflies and moths, but the larvae of the apocritans have no legs because they are always surrounded by food provided by the mother or, among the social insects, other siblings.

This Hawthorn Sawfly clearly shows the large wing cells typical of the order. Not all sawflies have clubbed antennae or so much hair.

This ichneumon, one of the larger parasitoids attacking the caterpillars of butterflies and moths, clearly shows the thickened front edge of the wing typical of the family (*see* p. 330), and also the apocritan 'wasp-waist'.

Hairy coats generally distinguish bees from wasps. The pale pollen brush on which pollen is carried back to the nest is clearly visible at the base of the hind leg of this solitary bee.

PARASITIC INSECTS

It has been estimated that over a tenth of all known insect species lead parasitic lives, meaning that they live in close association with other species and take food from them without giving anything in return. About 100,000 of the known parasitic species are ichneumons (*see* pp. 330–2) and others hymenopterans, such as chalcids. Another 11,000 or so are flies, and the rest are made up mainly of lice, fleas, beetles and a few bugs. Parasitic insects attack a wide range of land and freshwater animals, from sponges to elephants, and a few lice even go to sea in the ears of seals, but other insects bear the brunt of the attack.

The parasites are either ectoparasites, living on the outside of their hosts, or endoparasites that live inside their victims. Most species live parasitically for only a part of their lives, either when young or when adult, but not both, although the lice (*see* pp. 18–19) are exceptions in being parasites throughout their lives. Insects that are parasitic as adults are almost all ectoparasites and, with the exception of the Bee Louse, which is actually a minute fly, and a few other tiny species, they all attack birds and mammals. The Bee Louse lives on honey bees, but it is more of a thief or kleptoparasite than a true parasite because it does not feed directly on its hosts. Endoparasitic species spend only their young stages in their hosts, and those that attack other insects usually affect only the young stages of their hosts. Conopid flies (*see* p. 312) are exceptions in that they parasitise adult bees and wasps, but the short adult lives of most other insects would not give the parasites time to mature.

The larvae of various flies, including the infamous warble and bot flies, grow up in vertebrate hosts and, although they may cause a good deal of distress, they do not usually kill their hosts. But those parasites that attack other insects usually *do* kill their hosts, although not until the parasites

The incredibly long ovipositor of this female *Gasteruption* species enables her to lay her eggs deep inside the nests of various solitary bees.

themselves have matured and have no further use for the hosts: in their early stages the parasites are careful to keep their hosts alive by avoiding damage to vital organs. Parasites of this kind, represented by the ichneumons and many other hymenopterans and also by the tachinid flies (*see* p. 314), are usually known as parasitoids. They include both endoparasitic and ectoparasitic species.

Most tachinid flies, characterised by their bristly bodies, parasitise the caterpillars of butterflies and moths (*see* p. 314). Some lay their eggs on or in their hosts, but others lay their eggs on the food-plants and leave the grubs to find suitable hosts. Yet others lay minute eggs that are eaten by the hosts.

SOCIAL INSECTS

Social insects are those that live in colonies, working for the benefit of the whole society and not just for themselves or their offspring. All social insects in Britain belong to the Hymenoptera and they include all our ants and numerous bees and wasps. Colony size varies enormously, but each colony is ruled by one or more fertile females, called queens, and all the other members are the offspring of these queens. Most of the population consists of sterile females called workers, and they are precisely that: they carry out most of the nest-building and food collection and they feed and tend their younger siblings, enabling the queens to concentrate on the job of laying more eggs. Males and new queens are usually reared only at certain times of the year. Wasp and bumble-bee colonies are annual affairs and only the young, mated queens survive the winter to start new colonies in the spring. Honey bee and ant colonies survive from year to year, the queens living for several years before being replaced by younger individuals.

Social insects all construct some kind of nest. Most ant nests consist of a number of simple chambers hollowed out in the soil or in rotting wood.

LEFT: This Common Wasp, identified by the anchor mark on its face, is collecting wood fibres in its jaws.

BELOW: This far-from-complete wasp nest has been broken open, revealing the neat hexagonal cells inside the protective envelope. Shiny grubs can be seen in some of the cells. Those cells that have been capped contain pupae.

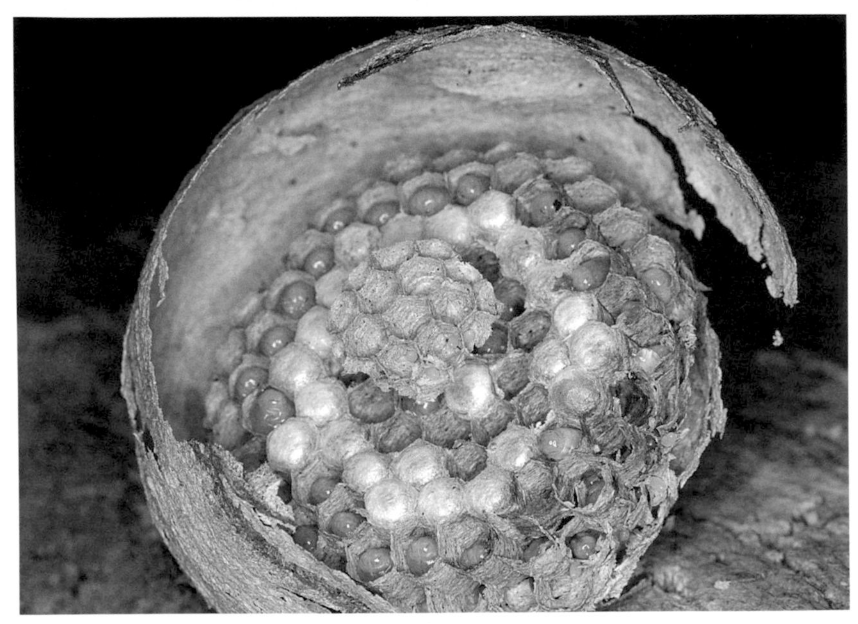

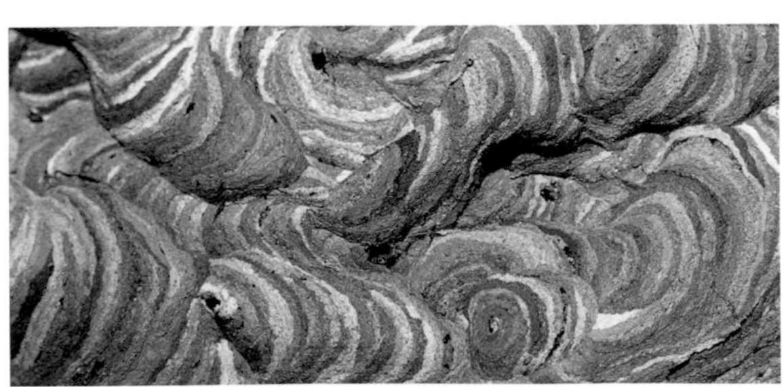

LEFT: The envelope of a Common Wasp nest consists of hundreds of shell-like lobes. Each coloured band on these lobes is put in place by a single worker, using a single load of wood pulp.

Some species, including the Yellow Meadow Ant (*see* p. 334), pile up the excavated soil and extend their tunnels and chambers into the mounds. Wood Ants (*see* p. 334) also build large mounds over their nests, but the Shining Black Ant (*Lasius fuliginosus*) goes in for true construction work. Nesting in a wide range of existing cavities, including tree stumps and hollow logs and walls, it constructs irregular chambers with chewed wood and other debris glued together with large amounts of honeydew. Commonly known as carton, this material has a durable, cardboard-like consistency.

The social wasps build elaborate nests with paper, which the insects make themselves from dead wood collected from a variety of sources. It is very easy to watch the wasps collecting wood from fences and other wooden structures, and also from the dead stems of hogweeds in the hedgerow. You can even hear the wasps at work on these plants, for the rasping actions of their jaws are amplified by the hollow stems. Having found a good source of wood, the insects return to it again and again, leaving tell-tale pale streaks that show where they have been working. The wood fibres are mixed with saliva and chewed to pulp, which is then formed into thin strips and used to construct horizontal tiers of hexagonal cells. The hornet and the common wasp select fairly rotten wood, and their nests have an overall brown or yellowish appearance and a very brittle texture. The other wasps use sound wood for paper-making, producing greyish and less fragile nests, but the building processes are, nevertheless, essentially the same for all species.

Honey bees also construct neat hexagonal cells, but these are made with wax secreted by the bees themselves and they are arranged in vertical combs. Bumble-bee nests, usually built in holes in the ground, are much less organised and consist of a number of wax boxes built on top of clumps of pollen in the centre of a ball of dried grass or moss.

The social insects are not the only nest-builders. Many solitary bees and wasps construct homes for their offspring, using a wide range of materials. The female Wool Carder Bee (*see* p. 348) lines her nest with plant hairs. Her toothy jaws, clearly seen here, are well suited to the task of gathering and combing the hairs.

GALLS

Galls are growths formed by plant tissues in response to the presence of another organism. The growth is produced by the enlargement or proliferation or both of the plant cells, and it provides both food and shelter for the gall-causing organism. The latter may be a fungus or a bacterium, but the great majority of plant galls in the British Isles are caused or induced by mites and insects. Most gall-causers are restricted to a single host species or a few closely related species and, through the action of specific chemical stimuli, each one induces the development of its own particular gall. Even very closely related species can induce totally different galls on the same plant. Some galls are clearly swollen leaves or buds, but others, such as the Robin's Pincushion Gall (*see* p. 328), appear to be completely novel structures. Although, as far as we know, the gall-causers give nothing to their hosts in return for food and shelter, and therefore behave like parasites, they do not seem to do any long-term harm to their host-plants.

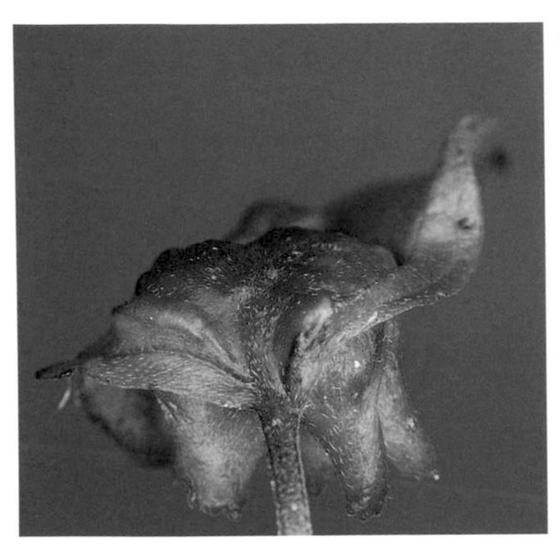

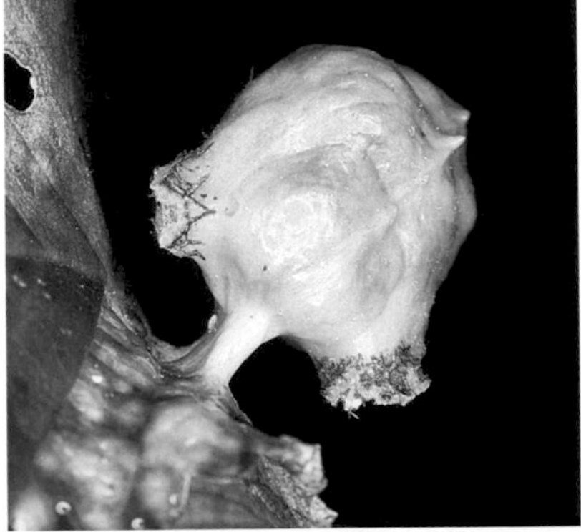

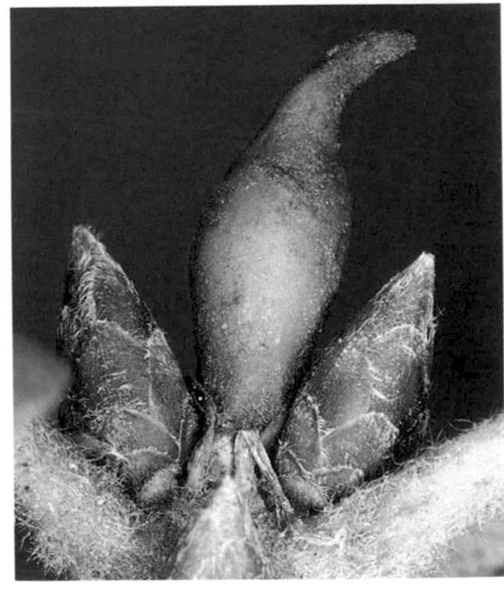

ABOVE: Rivet galls on dogwood leaves are caused by the tiny gall midge *Craneiobia corni*.

ABOVE RIGHT: The fig-gall on elm leaves is caused by the aphid *Tetraneura ulmi*. A mature gall may contain up to 50 aphids, which escape through one or more openings during the summer.

RIGHT: Lurking among the unopened oak buds, the gall of the gall wasp *Andricus solitarius* is not always easy to see. It contains a single gall wasp, which emerges in the autumn.

INSECT DEFENCES

Insects have a huge number of enemies: birds, spiders, lizards, small mammals, fish and other insects are all major predators, and these are joined by a wide range of parasites. Many of the latter are also insects, mainly flies and hymenopterans (*see* p. 34). If an insect is to stay alive long enough to reproduce – the fundamental 'purpose' of all living things – it must have some way of evading these enemies. Fleeing from danger obviously works for some insects, and some rely simply on safety in numbers, but the majority have evolved much more elaborate and fascinating defensive systems. These defences usually have a visual or chemical basis, although some insect species use sounds to deter or deceive their enemies. The abundance of insects throughout the world indicates just how well these defensive systems work.

PLAYING POSSUM

Feigning death when disturbed can sometimes result in survival because many predators are alerted by movement and fail to recognise motionless objects as food. Several moths roll over and pretend to be dead when they are alarmed, and the habit seems to be especially common in hairy species, such as the yellowtail: the dense, irritating hairs gives these moths additional protection if the predator does press home an attack. The death-feigning trick, often called playing possum, can be even more effective if the insect takes on some resemblance to its surroundings. The Water Scorpion (*see* p. 100), for example, is easily taken for a piece of dead leaf. Various weevils and other beetles draw in their limbs and resemble seeds or even bird droppings (*see* p. 41) .

The Browntail moth rolls over and plays possum when disturbed.

The simplest defensive mechanism is to avoid being noticed and a great many insects have evolved colours and patterns that enable them to blend in with their environment. Grasshoppers, for example, are generally green or brown, or a mixture of the two, and very hard to see among the grasses. Many caterpillars also blend well with their food-plants, and moths that rest on rocks and tree trunks tend to be brown and grey, matching their backgrounds so well that even experienced entomologists may have difficulty in spotting them. This simple but nevertheless effective form of camouflage is known as crypsis.

CHOOSING BACKGROUNDS

With so many different trees and other surfaces to choose from, it is obvious that the moths must be able to select their resting places with some accuracy, and various experiments have shown that they really do select the most appropriate backgrounds when given a choice. Some may actually make a conscious choice of background colour, by comparing it with their own colours, but it is likely that a genetic factor is at work in most species, in other words the moths have been programmed to seek out particular colours. It is interesting to note in this context that many autumnal moths, including the various species of sallows, have golden-brown hues that blend in with the autumn leaves. Some moths may also seek out particular scents associated with good resting sites.

Stripes and blotches and other bold markings that break up an insect's outline can make it even more difficult to see. Many moths and caterpillars go in for such disruptive coloration. Moths with prominent lines or stripes on their wings commonly rest in such a position that the markings are aligned with bark crevices, and this apparent continuation of the lines is particularly effective at concealing the outlines of the insects and making them inconspicuous to predators. But it can work only if the insects orient themselves correctly: a moth resting with its stripes running at right angles to the bark crevices would be very conspicuous however well its colours matched those of the trunk. It is possible that some species can line themselves up visually, but it seems more likely that they are programmed to respond to gravity, resting with their bodies either vertical or horizontal according to the arrangement of the stripes on their wings. The moths may also use their feet to explore the crevices and make fine adjustments to their positions.

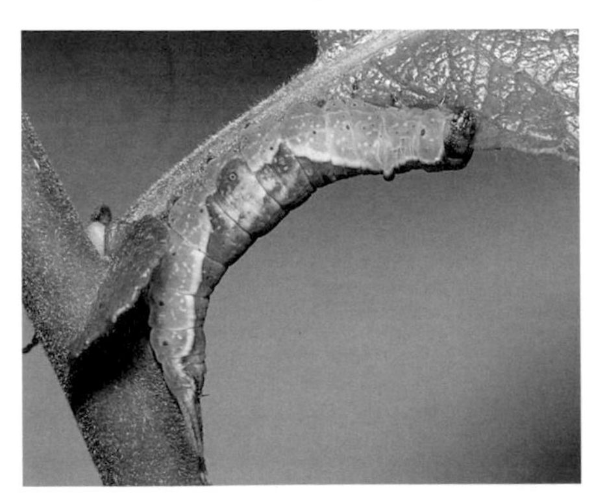

The dark saddle-like patch on the Sallow Kitten caterpillar breaks up its outline and makes it hard to pick out even from a short distance.

ABOVE: The Lilac Beauty moth is very difficult to see among the fallen leaves: its leaf-like shape and colour camouflage it and the narrow stripe breaks up its outline.

RIGHT: The Peppered Moth caterpillar holds itself at just the right angle to resemble a branch of its food-plant.

The weevil *Anthribus resinosus* falls to the ground and pulls in its legs to look just like a bird dropping when alarmed.

Protective resemblance is a rather more elaborate form of camouflage or deception, in which, instead of simply blending in with their surroundings, the insects actually resemble parts of the environment such as twigs, leaf stalks, dead leaves and even bird droppings. Predators take no notice of these objects, so the insects that resemble them have a good chance of survival, but only if they behave in the right way as well. Twig-like caterpillars and those resembling leaf stalks must hold themselves at the right angles if they are to fool the birds for long, and they must keep still! Crypsis and protective resemblance grade into each other, with many intermediate examples. The resting Mullein Moth, for example, could be said simply to blend in with its surroundings, but it could equally well be said to resemble a sliver of bark and therefore come under the heading of protective resemblance (*see* p. 238).

Pretending to be larger or fiercer than one is in reality is a good way of keeping enemies at bay and many insects have gone down this path. One of the best examples of such bluff in the British insect fauna is exhibited by the Eyed Hawkmoth (*see* p. 206). This insect is fairly well camouflaged when resting on a tree trunk, but if it *is* disturbed it opens its forewings to display a large eye-spot on each hindwing. The sudden appearance of something closely resembling a cat's eyes is enough to scare most small birds away in a great hurry. The eye-spots on the forewings of the emperor moth are equally scary. Although they are always visible, when seen from the front or from above – which is how most birds view the moth as they fly over the vegetation – the effect is again very much like that of a cat's face. The caterpillars of the Elephant Hawkmoth and the Puss Moth (*see* p. 288) both inflate their front ends and display eye-spots when disturbed, although the Puss Moth display is not entirely bluff because it can also fire an acidic spray from a gland just behind its head.

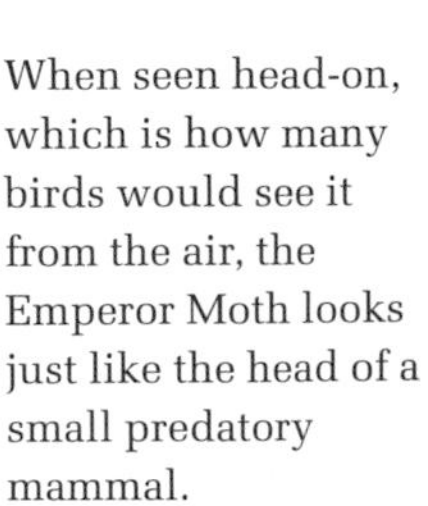

When seen head-on, which is how many birds would see it from the air, the Emperor Moth looks just like the head of a small predatory mammal.

Smaller eye-spots, such as those on the wing margins of the brown butterflies (*see* p. 154), also deceive predators but they work in a different way. If a predator does attack, it is likely to go for these false eyes instead of for the true eyes on the head, ending up with just an expendable piece of wing membrane and allowing the insect to escape.

Satyrid butterflies like this Meadow Brown are commonly seen with rips and beak marks on their wings, showing where birds have attacked the decoy eye-spots.

Chemical defences in the insect world include foul tastes, deterrent smells, repellent or even blistering sprays and other secretions, irritant hairs and powerful stings. They are usually the last line of defence and generally come into play only when the insect is actually molested. Although many insects manufacture their own poisons or toxins, most toxic species probably obtain at least some of their defensive chemicals from their food-plants. The caterpillar of the Large White butterfly, for example, absorbs pungent mustard oils from cabbage leaves and other food-plants and stores them in its body, where they remain throughout its life, protecting the pupa and adult as well as the caterpillar itself. The adult female also passes this chemical protection to her eggs.

The larvae of swallowtail butterflies are usually quite well camouflaged, but they also go in for chemical protection. When a caterpillar is alarmed, a brightly-coloured Y-shaped 'sausage', called an osmeterium, erupts from just behind its head. Its emergence is accompanied by a strong odour, which varies with the species. Easily detected by the human nose, the smells range from a quite pleasant lanolin-like scent to very sickly and downright objectionable odours. Birds generally have a poor sense of smell and are unlikely to be deterred by the scent alone, but the osmeterium also secretes a stinging fluid

At the slightest disturbance, the Swallowtail caterpillar everts the colourful and strongly-scented osmeterium just behind its head.

The bizarre, long-legged caterpillar of the Lobster Moth looks quite repulsive at all times, but when attacked it raises its head and fires formic acid at its adversary. When young, the caterpillars look remarkably like ants and birds normally leave them alone.

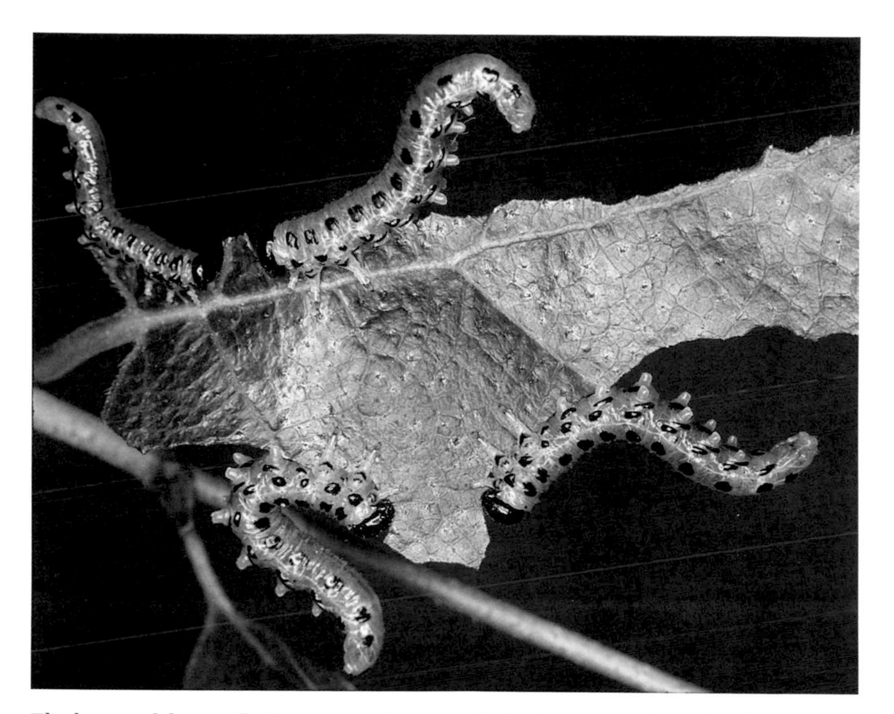

The larvae of the sawfly Croesus septentrionalis feed gregariously on birch leaves and the slightest disturbance causes them all to raise their rear ends. They release a pungent odour at the same time and few enemies are likely to press home an attack.

containing acetic and butyric acids which can make life very uncomfortable for any assailant pressing home an attack. It is possible that the sudden appearance of the colourful osmeterium alone may scare off some small birds. The odours probably serve mainly to deter parasites, which track down their hosts by smell and are totally oblivious to any form of visual camouflage, although not all parasites are going to be put off by one particular scent.

Many shield bugs, commonly known as stink bugs, protect themselves with pungent secretions exuded from various thoracic and abdominal glands. Some of these secretions smell quite sweet to the human nose and some of them can be detected from several metres away, but the insects usually taste foul and can certainly spoil a bowl of raspberries if one of them is inadvertently trapped while picking the fruit. Although stink-bug secretions may be extremely unpleasant, they are not really dangerous. But this is not true of the Spanish Fly (*see* p. 128), which is actually a shiny green beetle. Its elytra contain quantities of cantharidin which can cause severe blistering of the skin if the insect is molested. Few birds or small mammals will press home an attack on this beetle, which advertises its presence with a strong smell of mice. Several species of *Paederus* and other rove beetles have similar blistering abilities.

Predators attacking the Bombardier Beetle (*see* p. 112) are likely to get a hot reception – literally – for the beetle can respond by discharging a jet of boiling hot liquid from its rear end. Consisting largely of pungent benzoquinone, the liquid immediately disables small predators and sends birds and small mammals rapidly into reverse. The beetle obviously cannot store this volatile and corrosive liquid, so it makes it as and when required and discharges it immediately with an audible pop and a little puff of smoke. The beetle can produce up to 20 bursts in quick succession. Bombardier beetles in other parts of the world can aim their jets quite accurately by altering the position of the abdomen: some can even curve their abdomen forward between their legs and fire to the front.

The brightly-coloured, day-flying burnet moths (*see* p. 160) are rather sluggish insects and very easy to approach and catch, but they have little to fear from birds or other predators because their bodies actually contain cyanide! A slight disturbance may cause the moths to exude distasteful fluids from glands around the mouth. These fluids do not contain cyanide, but they do contain pungent compounds called pyrazines which are sufficient to repel most predators. The cyanide-containing fluids come into play only if the attack persists. They seep from joints on the legs and thorax and, although they do not contain a lot of cyanide, there is enough to cause the predator some distress and to make it back off immediately. The burnets themselves are fairly resistant to cyanide, as any collector who has tried to kill them in a cyanide bottle will know. If a predator has to peck or prod an insect before discovering that it is distasteful, it is as well for that insect to be fairly tough, and this is just what we find in the burnets: the integument is quite leathery and pliable, and if it does suffer minor damage it heals very quickly.

Loaded with cyanide and other unpleasant chemicals, and further protected by their bright colours, these burnet moths have no need to conceal themselves. No bird is likely to molest a burnet more than once.

Bold colours warn that the tiger moth is loaded with poisons, some obtained from its larval food-plants and some manufactured in its own body. Writhing movements often increase the effectiveness of the warning display, while red glands just behind the head pump out repellent fluids.

Several kinds of tiger moths (*see* p. 216) also exude nasty fluids when attacked, pumping frothy secretions from glands in the thorax or more watery ones from the feet. A dense coat of hair makes some of the tiger moths even more unpalatable.

Anyone who has handled ladybirds knows that they commonly leave a stain on the fingers if they are treated roughly. The fluid causing the stain is actually blood and the phenomenon is known as reflex bleeding. The blood oozes out through pores in the leg joints. It contains several very bitter alkaloids and also smells strongly of pyrazines. The latter are volatile compounds found in many distasteful insects, including the burnet moths mentioned above, and it is probable that they give early warning of the insects' disagreeable nature. Reflex bleeding occurs in many other beetles, notably the Bloody-nosed Beetle (*see* p. 132) which exudes blood from pores around its mouth. Leaking blood in this way is expensive in terms of an insect's body resources and it is generally used only as a last resort. Thanks to their tough outer coats and protective elytra, the beetles can stand a good deal of rough treatment and often escape without the need to discharge any blood.

The only insects possessing true stings are the bees and wasps and some ants. The sting is a much-modified ovipositor and is thus found only in females, but where the social species are concerned, this means most of the individuals. Solitary wasps use their stings to paralyse the prey that they store up for their offspring and, although some will sting if handled, they are not programmed to attack large animals. Some of their stings contain pain-inducing histamine, but they are generally quite harmless to people. Solitary bees have very feeble stings, usually no worse than a pin-prick, and use them only when mauled. They never initiate an attack and do not bother to defend their nests. The most powerful stings belong to the social bees and wasps, which sting in defence of their colonies. Honey bees and social wasps stream out from their nests at the slightest disturbance and often attack anything appearing in the flight-line close to a nest. Alarm substances released by the front-line defenders summon additional workers from the nest and few animals apart from bears and badgers are likely to press home an attack. Ants also use alarm substances to great effect. If a wood ant nest is attacked, foraging workers drop from the overhanging branches in response to the alarm substances and, although they have no stings, they bite viciously. They also squirt venom at the invader and the effect can be just as good as that of a sting.

The sting is housed in the rear of the abdomen and is visible only when in action. It consists essentially of a reservoir of venom and a set of 3 needles linked together to form a hollow tube. When the sting pierces a victim, muscles around the venom sac pump venom into the wound. Barbs on the sting of a honey bee are quite large and they prevent the sting from being withdrawn from mammalian flesh. The only way a honey bee can escape after stinging someone is to rip herself away and leave the sting behind. This is fatal for the bee, but the sacrifice of a few individuals from a colony is not a high price to pay for the security of the colony as a whole. The muscles around the venom sac continue to pump venom into the wound even after the bee has gone, and the isolated sting also continues to release alarm substances which cause more bees to join in the attack. Wasps and bumblebees have much smaller barbs, allowing them to withdraw their stings easily and use them over and over again.

Ladybirds commonly bask communally in spring and autumn. This undoubtedly gives them extra protection by enhancing the effect of their warning colours and by increasing the concentration of repellent scents around them.

The hairs of the Pale Tussock caterpillar, once known as the 'hop dog' because of its liking for hops, are extremely irritant and used to cause severe skin problems among hop-pickers.

The chemical make-up of the venom is very complex and differs from species to species. There is a mixture of numerous proteins and enzymes, usually with a certain amount of histamine which causes most of the pain and inflammation. Anti-histamine preparations can help to relieve this, but there is no real antidote to the venom. Formic acid accounts for up to 50 per cent of some ant venoms.

Many caterpillars and adult moths are protected by irritant hairs. Long hairs may just make the caterpillars look bigger or conceal their shapes, as with the larvae of The Miller (*see* p. 294), but the hairs usually provide more active protection. Some cause irritation simply because they are barbed and difficult to remove from the skin, while others are connected to poison glands in the skin and are definitely venomous. They irritate the lips, noses and eyes of any inquisitive predator that touches the insect, and some

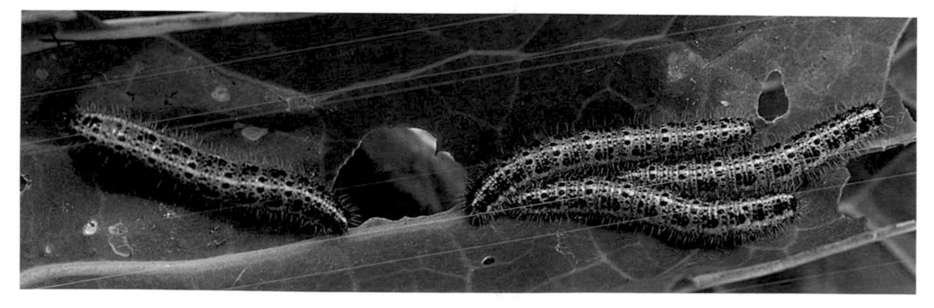

The caterpillars of the Large White butterfly feed gregariously and are protected by their warning colours and a very strong odour derived from the mustard oils in the cabbage leaves.

When alarmed, the Scarlet Tiger moth displays all the classic warning colours – black, white, yellow and black – but when it is at rest on the vegetation the spotted forewings provide surprisingly good camouflage.

people develop painful rashes after handling certain caterpillars, notably those of the Pale Tussock (*see* p. 49) and the Yellow-tail (*see* p. 290).

Chemical defences can be very effective, but replacing the chemical components uses up valuable energy and many poisonous or distasteful insects have evolved ways of warning the predators *before* they make contact and trigger off the chemical warfare, usually through the use of bright colours or bold patterns or both. Such warning coloration usually involves various combinations of black, yellow, red, and white and is well exhibited by Cinnabar Moths and their caterpillars, burnet moths, tiger moths, wasps and ladybirds. Young birds and other predators may try the insects once or twice, but quickly learn to associate the colours with foul tastes or even worse and then leave the insects alone.

Many harmless and edible insects gain protection by resembling poisonous or unpalatable species with warning colours. Many hover-flies, for example, resemble wasps in both looks and behaviour (*see* pp. 308–12). This kind of deception was first described by the 19th-century British naturalist Henry Bates and is called Batesian mimicry. Unlike protective resemblance, whose exponents are attempting to escape notice, mimics actively draw attention to themselves with the message 'look at me, I am nasty, so keep away'.

Numerous social and solitary wasps wear yellow and black and, although their patterns vary in detail, they are sufficiently alike to confuse their predators. By copying or mimicking each other, the insects all gain an advantage because a predator only has to learn one pattern before leaving all

The harmless Hornet Clearwing (*see* p. 158) moth is an excellent mimic of the Hornet (*see* p. 344), in behaviour as well as in appearance. In flight, it even buzzes like a hornet.

BELOW: With its narrow waist and long antennae, this conopid fly (*see* p. 312) mimics several kinds of solitary wasps.

Although it exhibits no membranous wings, the Wasp Beetle is sufficiently wasp-like in its coloration for birds – and people – to mistake it for some kind of solitary wasp. It is particularly wasp-like in its behaviour, scampering over the vegetation and tapping the leaves with its antennae. Resemblances do not have to be perfect for mimicry to work. Even a slight resemblance can cause a predator to hesitate, and this can be enough for the insect to escape.

these unpleasant species alone. This kind of mimicry is called Müllerian mimicry, after the German zoologist Fritz Müller who first described it, and the insects are said to form a mimicry ring. No deception is involved here because all the insects are unpalatable and they are not giving out false information, although many Batesian mimics, including hover-flies and sawflies, can also join the club and gain extra benefit by sharing the pattern.

Warning colours and other visual signals are clearly of little use at night, so some moths and other nocturnal insects make use of sound to warn or deceive their enemies. Tiger moths and some other moths emit their own sounds when they detect approaching bats. These sounds may interfere with the bats' radar and prevent them from homing in on the insects, but experimental work suggests that the moths' sounds are really the acoustic equivalent of warning colours, indicating that the moths are unpleasant. Many other moths avoid capture by simply dropping to the ground when the hear the approaching bats.

No defensive method is 100 per cent effective: if it were there would be some very hungry birds! But, as long as some insects survive long enough to reproduce, the process is effective, and it gives natural selection something to work on. If the best examples survive in each generation there will be a gradual improvement over time. This is undoubtedly how the insects have developed the wonderful examples of camouflage that we can see about us – albeit with some difficulty.

INSECT HABITATS

A journey across the British Isles, whether from east to west or from north to south, takes the traveller through a broad spectrum of scenery and an equally wide range of habitats in a relatively short distance, but few, if any, of these habitats are really natural. With the possible exception of the wilder stretches of our coastline and some of the highest mountain tops, all have been created, or certainly modified, by human activity. Because of our climate, the natural vegetation of most parts of Britain and Ireland is deciduous forest and about 8,000 years ago, after the retreat of the ice-age glaciers, almost all of the area was covered with oak and ash and a few other deciduous trees. Even the Scottish Highlands were thickly forested, although here the deciduous trees were replaced by Scots pine – and the whole of our countryside would return to forest if left to its own devices for just a few decades. This was demonstrated very dramatically in the 1950s, when myxomatosis wiped out much of the rabbit population and, in the absence of rabbit grazing, the open hillsides were quickly invaded by scrub. Roadsides that are not cut on a regular basis soon become covered with hawthorn and brambles, and even untended gardens soon turn into 'jungles'.

The destruction of the forest cover began when people started to fell the trees to make way for agriculture, and quickly accelerated when grazing animals prevented the regeneration of the trees. Combined with the complex geology of our islands, which has resulted in a wide range of land forms and soil types, forest destruction gradually led to the multiplicity of habitats or environments that we see around us today. For the naturalist, of course, this is wonderful, because it means that we can enjoy many different habitats and their varied wildlife in just a small area – and we have not lost *all* of our forest habitat.

Insects are amazingly adaptable creatures and every terrestrial and freshwater habitat has its characteristic insect populations, with each species perfectly adapted to its own little niche as a herbivore, a predator or a scavenger. Only the sea has proved difficult for the insects: although numerous flies and beetles scavenge on the strand line, very few species have managed to take to the open sea. The ability to fly enables adult insects to explore a wide range of habitats and many species can occur in a variety of environments, although their young stages are usually much more restricted. Dragonflies, for example, can be found in woods and gardens and almost anywhere else, but their nymphs are all confined to ponds and streams. Many butterflies, such as the Peacock and Brimstone, have no fixed abode and they travel far and wide in search of nectar and plants on which to lay their eggs. Migrants, such as the Painted Lady, travel hundreds or even thousands of miles in a summer. These wide-ranging insects are said to have open populations, but there are other species that rarely move more than a few hundred metres from their birth-places. About 75 per cent of the British butterfly species have such closed populations, and are therefore restricted to and very characteristic of their habitats. The figure is even higher for the more sedentary bugs and beetles.

WOODLAND INSECTS

Although today's deciduous woodlands represent only a fraction of the original wildwood, we do still have some wonderful forest habitats. Among the best known are the New Forest, Sherwood Forest, Epping Forest and Burnham Beeches, but there are also lots of smaller woodlands and many of these are being managed as nature reserves. Each forest layer, from the canopy down to the leaf mould on the ground, teems with its own assemblage of insects, making the woodland as a whole one of the richest insect habitats. Butterflies are among the most conspicuous of the insects, and the jewel in the crown of many southern oak woods is the majestic Purple Emperor (*see* p. 150). This butterfly spends much of its life floating around the canopy, occasionally descending for the somewhat dubious pleasure of drinking from dung and carrion or just sipping mineral-rich water from muddy pools. The Purple Hairstreak (*see* p. 146) also spends much of its time in the canopy, but most other woodland butterflies prefer to be nearer to the ground. Bramble blossom attracts the White Admiral (*see* p. 150) and the Silver-washed Fritillary (*see* p. 152), while the Speckled Wood (*see* p. 154) and several other species prefer to sip honeydew – the sugary secretion exuded by aphids. The caterpillars of our woodland butterflies feed mainly on the low-growing shrubs and herbaceous plants, except for those of the Purple Hairstreak, which feed on oak leaves.

The woodland butterflies are joined by numerous moths, with some woods in southern England harbouring over 300 species. Most of these are nocturnal and a good way to see them in action is to 'sugar' a few tree trunks with a

From the tops of the trees down to the dead leaves and branches on the ground, the woods are alive with insects, with sunny rides and clearings the richest places of all.

ABOVE: The Barred Sallow (LEFT) and the Centre-barred Sallow fly in the autumn and both moths blend well with the yellowing leaves.

RIGHT: The Speckled Wood is a familiar butterfly of woodland clearings, where each male defends a patch of sunlight on the ground or on low-growing vegetation.

mixture of treacle and beer. The mixture should be quite thick and it should be brushed on to the trunks at about head-height – any lower and someone is bound to lean against it! The addition of a crushed pear drop and a spoonful of rum may increase the effectiveness of the mixture. Trees on the edges of rides or clearings are best and scores of moths may come to drink, sitting around the edges of the feast and sticking their tongues into it. Sugaring can also be tried in other habitats, including gardens, but it does not always work, especially if there are good supplies of nectar in the vicinity. Another way to get to know the woodland moths is to search the tree trunks in the morning, when the insects are settling down to rest, bearing in mind that many of the moths are extremely

Goat moth caterpillars can be seen only for a very short time between emerging from the tree trunks and burrowing into the ground to pupate.

good at concealing themselves (*see* p. 39). The moth caterpillars feed on the trees and on the other woodland vegetation, including the lichens clinging to the branches. Some, including the Goat Moth and Leopard Moth caterpillars, even feed inside the trunks and branches.

Oak supports over 300 insect species in the British Isles. As well as the moth caterpillars, numerous beetles chew the leaves and equally numerous bug species, including the abundant aphids, feed on the sap. In the absence of a proper beating tray, shaking a few branches over an upturned umbrella will reveal just how many insects (and spiders) reside in the foliage of an oak tree.

The leaf litter on the woodland floor is home to a vast population of beetles. Many of them are predators, but most of the smaller species feed on the decaying leaves and the associated fungi, in company with hordes of fly grubs. A handful of leaf litter spread on a sheet of paper will yield a fascinating array of these small creatures, all of which play an important role in recycling the dead material and making it available to plant roots. Rotting tree stumps and other dead wood may house the grubs of the Stag Beetle (*see* p. 120) and several longhorn beetles (*see* pp. 128–30).

Coniferous woods have fewer tree species and a poorer ground flora than the deciduous woodlands and are consequently less rich in insect life. The tough, resinous needles of the conifers are also less nutritious than the deciduous leaves, but some insects nevertheless make their homes in the pinewoods. The caterpillars of the Pine Beauty and Bordered White moths feed on the pine needles and can cause a good deal of damage in the northern forests as well as in the coniferous plantations further south. Plantations usually contain just one or two tree species and support even less insect life than the native pinewoods, but the Pine Hawkmoth (*see* p. 206) is not uncommon in southern areas and several species of bark beetles cause damage by tunnelling under the bark. The Eyed Ladybird (*see* p. 126) is also rarely found away from pines or other conifers, where it feeds on aphids and other small pine-feeding insects. The Pine Cone Bug (*see* p. 92) pierces the needles with its sharp beak and, being very flat, it can creep between the cone scales to rest.

Scotland's much depleted Caledonian pine forest is relatively poor in insect life, partly because of the climate and partly because few insects can tackle the tough pine needles.

The caterpillar of the Pine Hawkmoth is green with white lines when young and it blends beautifully when resting among the pine needles. As it grows and becomes too big to rest among the needles, it changes colour to match the twigs.

GRASSLAND INSECTS

Grasslands of one kind or another now cover most of our land surface. Some coastal and montane grasslands may be more or less natural, but all our other grasslands have been created by humans and their grazing animals. The nature of the grassland depends partly on the underlying soil, but very largely on the way it is managed. Most of our lowland grasslands have been ploughed and sown with highly productive and nutritious grasses, often mixed with clovers and other plants. They are also regularly treated with herbicides and fertilisers. These 'improved' grasslands are certainly good for grazing animals, but the reduced number of plant species makes them less attractive to insects. In terms of wildlife, the most interesting grasslands are on the unploughed chalk and limestone slopes of the North and South Downs, the Chiltern and

Unploughed calcareous grassland, with its wide range of flowering plants, teems with butterflies and many other insects from early spring until late in the autumn, but without some kind of management it will soon revert to scrub and woodland.

Cotswold Hills, and the Yorkshire Dales. The short, springy turf was created initially by sheep-grazing and rabbits contributed to its maintenance when the Romans brought them and let them escape about 2,000 years ago. Meadow Browns, Marbled Whites, Large and Small Skippers, and other butterflies with grass-feeding caterpillars abound here, together with several blue butterflies. Grasshoppers sing throughout the summer, and on summer evenings some of these grasslands will be dotted with the lights of female Glow-worms (*see* p. 124) anxious to lure mates down from the air above. Sweeping a net through the grass at almost any time of the year will yield an amazing array of bugs, beetles and flies.

Acidic grasslands, usually found over granites and other hard rocks, and mainly in upland areas, support fewer plant species and therefore fewer animal species than the limestone pastures, but they are still quite rich when compared with the 'improved' grasslands.

ABOVE: Often called grass moors, the acidic grasslands of upland areas are relatively poor in insect life, although the wetter areas may teem with midges.

LEFT: The Common Green Grasshopper is a widely distributed species, occurring on both acidic and basic grasslands. It prefers fairly long grass and is the commonest grasshopper in upland areas.

INSECTS OF THE VERGE AND HEDGEROW

Roadside verges are excellent wildlife habitats and many stretches have been scheduled by the local naturalists' trusts as nature reserves, although the main interest is likely to be botanical rather than entomological. As well as allowing grassland wildlife to flourish in otherwise unsuitable areas, these grassy ribbons act as highways, enabling the animals to move from one area to another. Insects can move along the verges under their own steam, but they are also whisked along by the traffic and often carried in vehicles. The verge fauna is often very similar to that of other grasslands in the area, with grass-feeding butterflies the most obvious of the insects. The neighbouring hedgerows are even richer in insect life and the fauna is often very similar to that of nearby woodlands. This is especially true of the older hedgerows, some of which are actually remnants of the once-extensive forest cover. Butterflies such as the Ringlet and the Wall Brown (*see* p. 154) enjoy the nectar of brambles and other flowers and their larvae thrive on the taller grasses, while the Small Copper (*see* p. 146) breeds on the inevitable docks. Beetles and bugs abound in the dense vegetation and bumble-bees commonly nest in the bank at the bottom of the hedge. The Dark Bush-cricket (*see* p. 78) hisses his mating call by day and night during the autumn.

ABOVE: Flower-rich roadside verges like this can support hundreds of different kinds of insects in just a short stretch.

RIGHT: The Dark Bush-cricket inhabits many roadside hedgerows in the southern half of the country, nibbling all kinds of plants, especially the fruits, as well as taking other insects and spiders.

HEATHLAND INSECTS

Our heathlands developed when the forests disappeared from areas of poor sandy soils. They are most common in southern England, especially from Surrey to Dorset, and on the East Anglian Sandlings. The dominant plants are ling, which is commonly known as heather, and bell heather and, in the damper areas, cross-leaved heath. Grasses, including wavy hair grass, are usually present and in some areas they are so abundant that the habitat is called grass heath. Gorse is also common on many heathlands and birch has invaded many areas where grazing has been light. Most heathlands are dry and huge areas have been built on in recent times, making the habitat severely threatened.

ABOVE: Sandy heathland is a superb habitat for ground-nesting solitary wasps and bees, but heathland needs regular cutting or grazing – or burning – to prevent its returning to woodland.

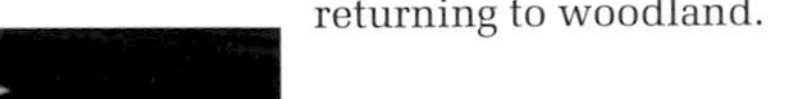

LEFT: The female Emperor Moth emits a powerful scent from glands extruded from the rear of her abdomen. The fast-flying males are attracted to her from a mile or more away and a dozen or more may swarm around her until one manages to mate with her.

Numerous bugs and beetles feed on or hunt among the heather, but most of these are rather small or inconspicuous. The most striking of the heathland insects is undoubtedly the Emperor Moth (*see* p. 220), although it is in no way restricted to heathland. The males fly rapidly over the vegetation in the spring sunshine and home in on the scent of the females with the aid of their large, feathery antennae. The females are relatively inactive and fly mainly at night. Grayling butterflies (*see* p. 154) inhabit many heathlands, especially where there is a good growth of grass, but their camouflage is excellent and they are hard to spot when resting. The female Potter Wasp (*see* p. 342) can sometimes be seen collecting fine sand for building, and a careful search of the surrounding vegetation may well reveal her beautifully fashioned, vase-shaped nest, in which she stores paralysed caterpillars for her offspring.

The upper of these two Potter Wasp nests is still open but it already contains one of the wasp's eggs and it will be sealed up as soon as the wasp has stocked it with sufficient small caterpillars to see her larva through to maturity.

Water may collect in low-lying areas of heathland and these areas become bogs full of sphagnum moss, also known as bog moss. This damp habitat is home to one of our rarest insects, the Large Marsh Grasshopper (*see* p. 82), and several uncommon and colourful dragonflies breed in the acidic pools.

INSECTS OF THE MOORS AND MOUNTAINS

The northern and western parts of the British Isles are, in general, higher than the south and east because the land is formed largely from harder rocks, including gritstones and granites. These areas are also cooler and they receive much more rain. These factors all favour the development of moorland. The vegetation is dominated by ling, bell heather and bilberry and resembles that of the lowland heaths in some ways, but the damp climate promotes the development of peat. In the wettest areas, notably in the north of Scotland and in Ireland, the ground is covered with a thick layer of peat known as blanket bog. This soggy habitat extends over huge areas and the heathers are largely replaced by sphagnum moss and an assortment of grasses and rushes.

The moors and blanket bogs support far fewer insect species than the woods and grasslands, but the wetter areas are home to the infamous biting

The heather moors of northern England are home to many bees that feast on the rich heather nectar. The caterpillars of several butterflies and moths feed on the associated grasses, with the Antler Moth (*see* p. 234) sometimes being so common that it destroys acres of grass.

midges that make life so uncomfortable in these places in the summer. Larger insects include the Large Heath, Scotch Argus and Mountain Ringlet butterflies (*see* p. 154), although the last two do not occur in Ireland. The Emperor Moth and the Fox Moth (*see* p. 162) are also widespread and often quite common on the moors. The highest peaks support little vegetation, and therefore few insects; those that do manage to survive there are generally flightless. Flying in such windy spots would probably result in being blown away.

AQUATIC INSECTS

Our freshwater habitats include various artificial ponds and canals and the more or less natural, although often much altered, lakes and rivers. All teem with insect life, both on and below the surface, although the insects of larger rivers tend to live only in the shallow water at the edge. Five fairly distinct reaches can be recognised in the archetypal river, from the headstream to the estuarine reach, although the make-up of an individual stream depends largely on the local topography and one or more reaches may be missing. The headstream, often no more than a shallow trickle running over the rocks, is home to stoneflies and some mayflies (*see* p. 66), whose nymphs cling tightly to the rocks and submerged mosses.

Further downstream, the troutbeck, bubbling rapidly over rocks and stones, also holds plenty of these crawling insects, and many caddis larvae (*see* p. 318) live there as well, but there are still not too many free-swimming insects. Still further down the stream, in the minnow reach, where the current is slower, silt or mud begins to accumulate on the bottom and there is much more aquatic vegetation. Insect life is also much more abundant. Dragonfly and damselfly nymphs crawl over the bottom and the adult insects take to the air in the summer. Mayflies, stoneflies and caddis flies are still here, but the species are different from those living higher up the river: many of the mayfly nymphs burrow in the mud. Water beetles of many kinds hunt among the submerged plants.

Some of the minnow-reach insects also occur further downstream in the lowland reach, where they are joined by many more species. Most live in the marginal shallows. The water is calmer here and moves quite slowly.

As well as the mayfly and stonefly nymphs clinging to them, the mosses of the headstream may support large numbers of midge larvae.

The fast-flowing waters of the troutbeck are home to many young mayflies, stoneflies and caddis flies, most of which cling to the stony bed to avoid being washed away.

Surface-dwelling Whirligig Beetles (*see* p. 138) and Pond Skaters (*see* p. 100) hunt here and some of our largest dragonflies patrol tirelessly overhead. The lowest reach of the river, the estuarine reach, contains brackish water and supports relatively few insects.

The still waters of our ponds and canals support an assemblage of insects not unlike that found in the lowland reach of the river, but the precise make-up of this fauna depends on the acidity of the water, which in turn is based on the underlying geology. Acidic waters generally have a less varied insect fauna than alkaline or neutral waters. Mosquitoes and an assortment of other small flies, collectively known as midges, breed in abundance in most ponds, and Water Scorpions (*see* p. 100) lurk in the mud. Most lowland ponds are

Lowland ponds are among the richest of all insect habitats. Dragonflies and damselflies abound in the air and on the vegetation, while their young stages hunt under the water alongside numerous bugs and beetles.

Upland lakes are usually cold and sparsely vegetated. Midges may breed there in abundance, and their larvae may feed a few dragonfly nymphs and some of the carnivorous stonefly and caddis fly youngsters.

fringed by reeds and other tall plants and, unless the ponds are actively managed, this vegetation gradually spreads into the centre, converting the open water to swamp or marsh and eventually to dry land. The wetlands are home to some handsome moths, many of whose caterpillars feed inside the stems of the reeds and other plants, but most famous of our wetland insects is undoubtedly the Swallowtail Butterfly, now restricted to small areas of the Norfolk Broads. Upland lakes tend to be the poorest of our freshwater habitats – cold and acidic, with rocky or stony beds, lacking vegetation – and their insect life is largely restricted to a few stoneflies and caddis flies that breed in the marginal shallows.

INSECTS OF HOUSE AND GARDEN

Parks and gardens, even in the middle of towns, support a surprisingly wide range of insect life. Butterflies and moths with open populations (*see* p. 53) visit our flower gardens to feast on the nectar without necessarily settling down for long, but numerous other species are regular inhabitants of our homes and gardens. Originally natives of the surrounding countryside, they find many of our garden plants perfectly acceptable substitutes for their natural food-plants. Many of these garden plants are, of course, merely cultivated forms of native plants and, by growing them close together in large numbers, we have positively encouraged the insects and made it easy for some of them to become pests. The white butterflies that attack our cabbages are prime examples. Even our houses are not safe from insect invaders: from the relatively harmless bluebottles (*see* p. 316) that come in to look for food to the destructive carpet beetles (*see* p. 128) and clothes moths that spend much of their lives chewing through our carpets and other fabrics. And then there are the even more destructive Furniture Beetle (*see* p. 128) and other wood-borers that attack the very structure of our homes. These insects were around long before we came on the scene and it is interesting to ponder where and how they lived before they moved in with us and became domestic pests. The wood-borers, of course, lived in dead and dying timber, of which there was an enormous amount in the forest, but what about the clothes moths and other fabric pests? Most of them would have lived in the nests of birds and small mammals, feeding on fur and feathers and other fibres, and they now find our cosy nests, full of soft fabrics, just to their liking.

A garden contains many different habitats, such as lawns and walls and trees, as well as a wide variety of plants, and can provide food and shelter for an enormous variety of insects.

Mayflies: Order Ephemeroptera
Stoneflies: Order Plecoptera

Mayflies These are flimsy insects with either 1 or 2 pairs of wings and 2 or 3 slender 'tails' at the rear of the body. They fly mostly at night and rarely stray far from water. Identification of many of the 47 British species is difficult because it necessitates examining the genitalia and the delicate venation.

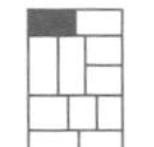

Ephemera danica Family Ephemeridae fw 15–22mm

Look for the spotted wings and the 3 'tails'. The front of the abdomen is very pale, with dark spots on the sides. Breeds in lakes and streams with plenty of silt or fine sand on the bottom; April–September.

Ephemera vulgata fw 12–20mm

The wing pattern is very similar to that of *E. danica* but the front of the abdomen is much darker. April–August, preferring still or slow-moving waters with plenty of mud on the bottom. ***E. lineata***, which breeds in large rivers, is yellowish or brick-coloured at the front of the abdomen.

Siphlonurus lacustris Family Siphlonuridae fw 15–20mm

One of several similar species with 2 'tails' and rather narrow forewings with several wavy veinlets running into the rear margin. May–September, breeding mainly in upland areas.

Potamanthus luteus Family Potamanthidae fw 15–20mm

This very local mayfly can be recognised by its yellowish colour, unspotted wings, and 3 'tails'. June–July, breeding in fast-flowing streams.

Centroptilum luteolum Family Baetidae fw 6–7mm

The front of the abdomen is pale and translucent. There are 2 'tails' and the hindwings are in the form of tiny, pointed straps. Males have huge turret-like eyes, clearly shown in the picture. April–September, breeding in stony streams and lakes. There are several similar species.

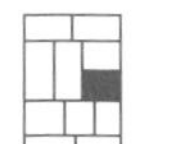

Cloeon dipterum fw 6–10mm

This species has 2 'tails' and the front edge of the forewing is yellowish, with no more than 5 small cross-veins in its outer region. There are no hindwings. April–September, breeding in ponds and ditches, and even water butts.

Stoneflies These sombre, soft-bodied insects are rarely found far from the water in which they breed. They rest with their wings either folded flat over the body or else wrapped around it. Males are often smaller than females: some have short wings and are unable to fly. There are about 30 British species.

Leuctra hippopus Family Leuctridae fw 6–10mm

This is one of several similar species, called needle-flies, because of their habit of wrapping the wings tightly around their bodies at rest. February–June, breeding in stony streams.

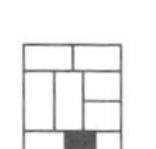

Isoperla grammatica Family Perlodidae fw 10–15mm

The only large yellow stonefly in Britain, this species is known to anglers as the large yellow sally. April–August, breeding in stony streams, especially in limestone districts. ***Chloroperla torrentium*** has forewings under 10mm long and very narrow hindwings.

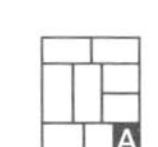

Perlodes microcephala fw 6(♂)–30mm

The female (A) can be distinguished from most other stoneflies by the cluster of small cells near the tip of the forewing. The male (B) has very short, stubby wings. March–July, breeding in stony streams. It is the only large stonefly commonly breeding in chalk streams.

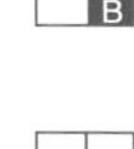

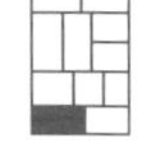

Dinocras cephalotes Family Perlidae fw 20–30mm

Pictured here with its nymphal skin, the female is one of our largest stoneflies. Look for the black pronotum contrasting with the much paler head. May–August, breeding mainly in upland streams with rocky bottoms and moss-covered boulders. ***Perla bipunctata*** has a paler body with a dark central stripe on the pronotum.

SUBORDER ZYGOPTERA

This group contains the slender-bodied damselflies – weak-flying species in which the 4 wings are all more or less alike and generally held vertically above the body at rest. The bulging eyes are well separated, giving the head and body the appearance of a tiny hammer. As with all dragonflies, the adults do not acquire their full colours until they are several days old. The shape of the pterostigma – the dark spot near the wing-tip – is a useful feature for identifying some species. There are 17 species living in the British Isles today, two more having died out here in the second half of the 20th century.

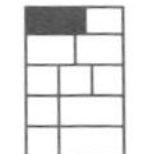

BANDED DEMOISELLE *Calopteryx splendens* Family Calopterygidae
45mm: fw 30–35mm

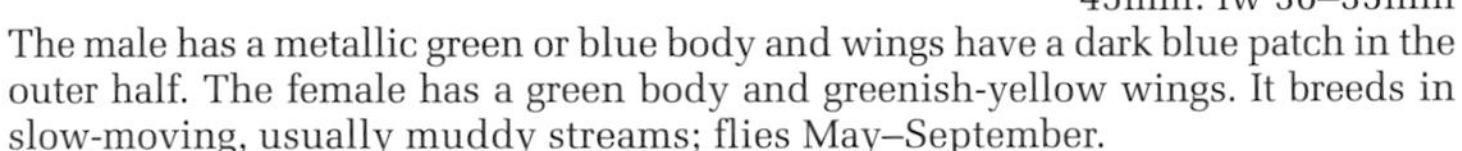

The male has a metallic green or blue body and wings have a dark blue patch in the outer half. The female has a green body and greenish-yellow wings. It breeds in slow-moving, usually muddy streams; flies May–September.

BEAUTIFUL DEMOISELLE *Calopteryx virgo* 45mm: fw 30–35mm
The male (A) has more extensive blue wing-patches than Banded Demoiselle, starting well before the middle, and the female (B) has brownish wings. It breeds in faster, clearer water than the previous species; flies May–September.

EMERALD DAMSELFLY *Lestes sponsa* Family Lestidae 38mm: fw 20–25mm
The body is metallic green in both sexes, but the mature male has a powdery blue bloom between the wings and at both ends of the abdomen. The pterostigma near the wing-tip is about 3 times longer than broad. It breeds in still water; flies June–September. Unlike most other damselflies, it rests with its wings partially open.

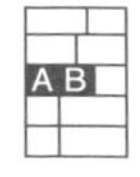

SCARCE EMERALD DAMSELFLY *Lestes dryas* *c.* 30mm: fw 20–25mm
This species is very like the Emerald Damselfly and very difficult to distinguish from it in the field, although it is a little stouter and the blue bloom on the male (A, female – B) extends no more than halfway along the 2nd abdominal segment. The pterostigma is also relatively fat – about twice as long as it is broad. Breeding in still water, this very rare damselfly flies June–September.

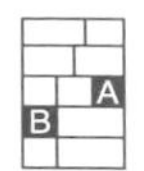

LARGE RED DAMSELFLY *Pyrrhosoma nymphula* Family Coenagrionidae
35mm: fw 20–25mm
Black legs and red stripes on the black thorax readily distinguish this from our only other red species – the Small Red Damselfly. Females (A) have more black on the abdomen than the males (B). As in all members of the Coenagrionidae, the pterostigma is diamond-shaped. It breeds in still and slow-moving water; flies April–September.

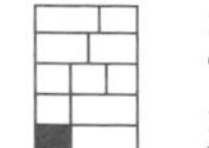

SMALL RED DAMSELFLY *Ceriagrion tenellum* *c.* 30mm: fw 15–20mm
The legs are red and there are no red stripes on the black thorax. The male abdomen is entirely red but only the front quarter of the female is red: the rest is normally black. It breeds in peat bogs and on wet heathlands; flies June–September.

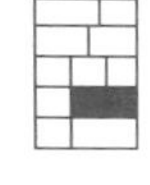

RED-EYED DAMSELFLY *Erythromma najas* *c.* 35mm: fw 20–25mm
Named for its large red eyes, this sturdy species has a black head without any pale spots. The male thorax is entirely black on the top and blue at the sides, while the female thorax has yellow sides and 2 short yellow stripes on the top. The upperside of the abdomen is largely black, although the male has a bright blue patch at the tip. It breeds in still water, where the males have a penchant for resting on floating leaves; flies May–September.

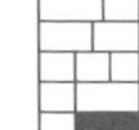

SMALL RED-EYED DAMSELFLY *Erythromma viridulum*
c. 26mm: fw 18–20mm
This species is easily confused with the Red-eyed Damselfly, although it is noticeably smaller and both sexes have yellow stripes on the top of the thorax. The sides of abdominal segments 2 and 8 in the male are entirely blue, whereas they are largely black in the Red-eyed Damselfly, and there is a black X on the top of segment 10. Females are less easy to distinguish, although always smaller than those of the previous species. Only recently recognised in Britain, it breeds in still and slow-moving water; flies June–September.

AZURE DAMSELFLY *Coenagrion puella* 30–35mm: fw 18–25mm

All male *Coenagrion* species have blue and black bodies and pale spots behind the eyes. Most can be recognised by the shape of a black mark on the 2nd abdominal segment, but the insects vary a lot and are not always easy to identify. Females are generally much darker, with little or no blue, and the markings are often obscured. The male *C. puella* has a U-shaped mark on the 2nd abdominal segment and the upperside of the penultimate segment is half blue. It breeds in still water; flies May–August.

VARIABLE DAMSELFLY *Coenagrion pulchellum* 30–35mm: fw 18–23mm

The male is very similar to the Azure Damselfly but the U-shaped mark on the 2nd abdominal segment is stalked and the penultimate segment is nearly all black. The female abdomen is normally blue and black, but the blue bands are narrower than those of the male and the black mark on the 2nd segment is more complex. It breeds in still waters and flies May–August.

NORTHERN DAMSELFLY *Coenagrion hastulatum* *c.* 30mm: fw 18–22mm

The 2nd abdominal segment of the male has a narrow black streak on each side of a black arrowhead, although the latter is very variable in appearance. The 8th and 9th segments are entirely blue. The only *Coenagrion* species in the far north, it breeds in bogs and lakes and flies May–August.

IRISH DAMSELFLY *Coenagrion lunulatum* *c.* 30mm: fw 18–20mm

The male's 2nd abdominal segment bears a black chevron or crescent and a pair of short black streaks. The blue bands in the central region of the abdomen are very narrow. This rare damselfly flies June–July and, in the British Isles, is confined to just a few boggy places in Ireland.

SOUTHERN DAMSELFLY *Coenagrion mercuriale* *c.* 30mm: fw 15–20mm

The mark on the male's 2nd abdominal segment resembles a 'ballet-dancer', and those on the two following segments resemble obelisks. This delicate species breeds in bogs and heathland streams; flies June–August.

COMMON BLUE DAMSELFLY *Enallagma cyathigerum* *c.* 32mm: fw *c.* 20mm

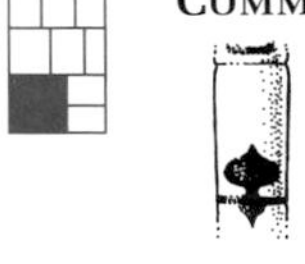

The male can usually be distinguished from *Coenagrion* species by the mushroom-like or 'ace-of-spades' mark on the 2nd abdominal segment. There is also just a single thin black line on the side of the thorax, instead of the two found in *Coenagrion*. The black and green female is easily identified by a large spine under the tip of the abdomen. Our commonest damselfly, it breeds in still and slow-moving water; flies May–October.

BLUE-TAILED DAMSELFLY *Ischnura elegans* *c.* 30mm: fw 15–20mm

Segment 8 is usually bright blue in both sexes, but otherwise the top of the abdomen is black. The pterostigma is bicoloured. Some females have segment 8 and the thorax red or purple instead of blue. The insect is abundant around still and slow-moving water May–September.

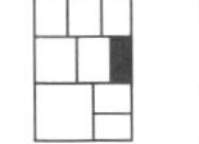

SCARCE BLUE-TAILED DAMSELFLY *Ischnura pumilio*
c. 30mm: fw 15–20mm

The male is very like that of the previous species but the blue patch covers segment 9 and just the rear portion of segment 8. The female never has a blue tail. The pterostigma is almost square and larger in the forewing than in the hindwing. It breeds in bogs and shallow pools; flies May–September.

WHITE-LEGGED DAMSELFLY *Platycnemis pennipes* Family Platycnemidae
35mm: fw *c.* 20mm

The broad, pale hind-legs separate this damselfly from all other British species. It breeds in still or slow-moving water with plenty of vegetation; flies May–August.

SUBORDER ANISOPTERA

This group contains the true dragonflies, which are generally larger and stouter than the damselflies and in which the hindwing is noticeably broader than the forewing. The insects always rest with their wings outspread. The eyes of most species meet in the middle of the head. The insects have two main methods of hunting: hawkers stay on the wing for long periods and may patrol a stretch of river or hedgerow all day, while darters remain perched for much of the time and merely fly out to snatch passing prey. There are 23 resident British species, and a few more occasional visitors from the continent.

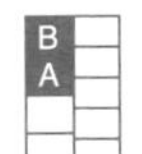

EMPEROR DRAGONFLY *Anax imperator* Family Aeshnidae
75–80mm: fw 45–52mm

The abdominal colour – bright blue in the male (A) and green and brown in the female (B) – will always identify this large hawker. It breeds in still waters with plenty of vegetation; flies May–September, with a stronger tendency to stay near water than many other hawkers.

HAIRY DRAGONFLY *Brachytron pratense* *c.* 55mm: fw 35–40mm

The hairy thorax distinguishes this from other similar species. The female lacks the broad green thoracic stripes seen here on the male, but has yellowish wing bases and the abdomen also has yellow spots instead of blue. Generally uncommon, this hawker breeds in still and slow-moving water; flies May–July, earlier than any other large dragonflies.

AZURE HAWKER *Aeshna caerulea* 60–65mm: fw *c.* 40mm

Both sexes have blue spots on the abdomen and on the last 2 segments they are more or less rectangular: all spots are smaller and duller in the female. The blue thoracic stripes are never large and often absent. It breeds in moorland pools; flies June–July.

COMMON HAWKER *Aeshna juncea* *c.* 75mm: fw 40–50mm

The abdominal spots are blue in the male and green in the female and are always rounded, even on the last 2 segments. The male has 2 narrow yellow stripes on the thorax. The front edge of each wing is distinctly yellow. It breeds in still water, with a preference for heathland and moorland pools; flies June–October.

SOUTHERN HAWKER *Aeshna cyanea* *c.* 70mm: fw 40–50mm

Both sexes have 2 broad green thoracic stripes and a conspicuous yellow triangle at the front of the abdomen. Most of the abdominal spots are green, but those on segments 8–10 of the male are blue. Unlike those of other *Aeshna* species, the spots on segments 9 and 10 are joined to form a band. This common species breeds in still water, including garden ponds; flies June–October, often roaming far from water.

BROWN HAWKER *Aeshna grandis* 70–75mm: fw 40–50mm

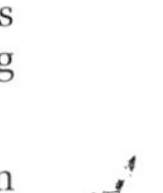

This is our only hawker with brown wing membranes. Only the male has the pale blue patches at the front of the abdomen. It breeds in still and slow-moving water; flies July–October, usually staying close to water although not uncommon in country lanes.

NORFOLK HAWKER *Aeshna isosceles* 65–70mm: fw 40–45mm

This is the only hawker with a brown body and clear wings. The wing veins are mostly black and the eyes are green. Both sexes have a conspicuous yellow triangle at the front of the abdomen, but the male lacks the blue patches seen in the Brown Hawker. Confined to the Norfolk Broads and now very rare; flies May–July.

MIGRANT HAWKER *Aeshna mixta* 60–65mm: fw *c.* 40mm

This species resembles the Common Hawker, but is less brightly coloured and can always be distinguished by the brown front edge to each wing and the small yellow triangle at the front of the abdomen. The thoracic stripes are poorly developed and sometimes absent altogether. It breeds in lakes and ponds; flies July–October, often far from water.

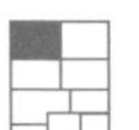

GOLD-RINGED DRAGONFLY *Cordulegaster boltoni* Family Cordulegasteridae
55–65mm: fw *c.* 45mm

Easily recognised by its gold-banded abdomen, this large hawker is the only British member of its family. It breeds in running water, mainly in upland regions; flies May–September, often hunting in sunny lanes and woodland rides far from water.

CLUB-TAILED DRAGONFLY *Gomphus vulgatissimus* Family Gomphidae
c. 50mm: fw 30–35mm

The only British member of its family, this hawker can be recognised by its widely separated green eyes and the extensive yellowish-green markings on the thorax. The abdomen, distinctly swollen at the end in the male, has a variable amount of yellow streaking and is sometimes almost completely black. It breeds mainly in slow-moving lowland rivers; flies May–June.

BLACK-TAILED SKIMMER *Orthetrum cancellatum* Family Libellulidae
c. 50mm: fw *c.* 40mm

The mature male's (A) abdomen is powdery blue with a black tip, while that of the female (B) and immature male is yellow with 2 wavy black lines down the back. The wings are completely clear and the pterostigma is black. It breeds in still water; flies May–August. It is a darter and usually perches on rocks or on the ground.

KEELED SKIMMER *Orthetrum coerulescens* 40–45mm: fw 30–35mm

This species resembles the Black-tailed Skimmer but is slimmer and the male (A) has no black tip. Females (B) and immature males lack the black abdominal pattern. The pterostigma is orange and the thorax usually bears pale stripes. It breeds mainly in peat bogs; flies June–September.

BROAD-BODIED CHASER *Libellula depressa* 40–45mm: fw 35–40mm

This species has a fairly broad abdomen, pale blue in mature males (A) and brown in females (B) and immature males. Both sexes have yellow spots along the sides of the abdomen and a dark patch at the base of each wing. It breeds in well-vegetated ponds; flies May–August. It is a darter and usually perches on reeds and bushes.

SCARCE CHASER *Libellula fulva* *c.* 45mm: fw 35–40mm

This darter resembles the Broad-bodied Chaser but is more slender (male, A; female, B). The abdomen has no yellow spots and is heavily marked with black towards the rear. Only the hindwing has a black basal patch, but all 4 wings may have brown tips. It breeds mainly in slow-moving streams with plenty of marginal vegetation; flies May–July.

FOUR-SPOTTED CHASER *Libellula quadrimaculata* 40–45mm: fw 30–40mm

This common darter is named for the dark spot about halfway along the front edge of each forewing. The male never develops a blue colour and both sexes resemble the female Broad-bodied Chaser, although there is no dark patch at the base of the forewing. The tip of the abdomen is also very black. Some specimens have a dark smudge near the pterostigma of each wing. It breeds mainly in boggy pools, but it is a great migrant and is often found far from water; flies May–August.

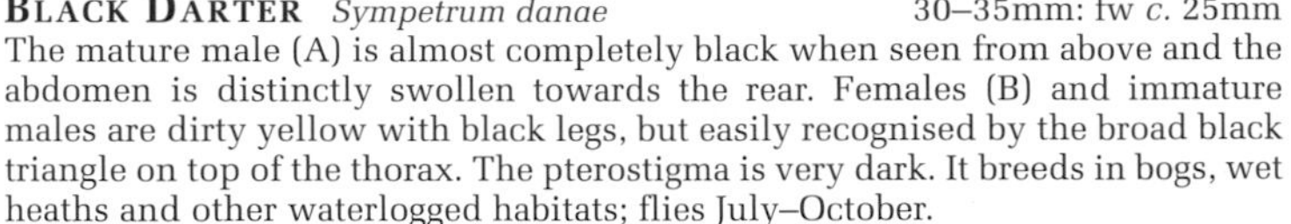

Black Darter *Sympetrum danae* 30–35mm: fw *c.* 25mm
The mature male (A) is almost completely black when seen from above and the abdomen is distinctly swollen towards the rear. Females (B) and immature males are dirty yellow with black legs, but easily recognised by the broad black triangle on top of the thorax. The pterostigma is very dark. It breeds in bogs, wet heaths and other waterlogged habitats; flies July–October.

Common Darter *Sympetrum striolatum* *c.* 40mm: fw *c.* 30mm
The male abdomen is red and slightly constricted near the front. Females and immature males are yellowish brown. In both sexes the legs are dark brown or black with a yellow stripe on the outside. This very common species breeds in still water, and flies June–October, often travelling far from the water. It perches on shrubs and hedgerows and also on the ground. Most red darters seen in England are likely to belong to this species. Individuals from the bogs and moors of Scotland and Northern Ireland have more extensive black markings on the sides of the thorax and are sometimes treated as a separate species – ***S. nigrescens*** (the **Highland Darter**). The female of this northern race could be mistaken for the Black Darter, although the latter has a black triangle on the thorax and a very dark pterostigma.

Ruddy Darter *Sympetrum sanguineum* *c.* 35mm: fw 25–30mm
The male is readily identified by its very bright red abdomen, strongly constricted near the front and bearing prominent black marks on segments 8 and 9. The female has a yellowish abdomen and resembles the female Common Darter, but differs in having completely black legs and a yellow patch at the base of the hindwing. It breeds in well-vegetated ponds; flies June–September. Many of the insects seen in the British Isles are probably immigrants from the continent.

White-faced Darter *Leucorrhinia dubia* 35–40mm: fw 25–30mm
The face is very white in both sexes. The female body has yellow markings instead of the male's red, and her wings are tinged with yellow at the base. This rare species breeds in peat bogs; flies May–July.

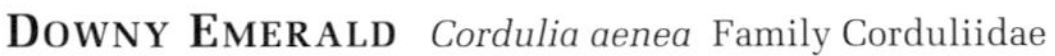

Downy Emerald *Cordulia aenea* Family Corduliidae
45–50mm: fw 30–35mm
The shiny green thorax is clothed with dense golden hair. The abdomen is darker green and that of the male is markedly constricted near the front. The wings may acquire a faint yellow tinge with age. The Brilliant Emerald is very similar, but brighter and with much longer terminal appendages. It breeds in still water, usually in wooded areas; flies May–July.

Brilliant Emerald *Somatochlora metallica* 50–55mm: fw 35–40mm
This species is very similar to the Downy Emerald but is less hairy and its brighter green abdomen is conspicuous even in flight. It has much longer terminal appendages in both sexes. It breeds in shallow still or slow-moving water; flies June–August.

Northern Emerald *Somatochlora arctica* *c.* 50mm: fw *c.* 35mm
This species is very similar to the Brilliant Emerald, but somewhat darker. The male terminal appendages are strongly curved and the female has conspicuous orange or yellow patches at the front of the abdomen. It breeds in bogs and moorland pools; flies May–August.

BUSH-CRICKETS: FAMILY TETTIGONIIDAE

Hair-like antennae, often much longer than the body, distinguish these insects from grasshoppers. The female's sturdy blade-like ovipositor is sometimes almost as long as the rest of the body. The lengths given are from the front of the head to the tip of abdomen, or to the tip of the wings if this is greater. The insects are more nocturnal than grasshoppers, although many sing by day and night.

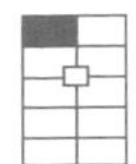

OAK BUSH-CRICKET *Meconema thalassinum* 12–17mm: ovipositor *c.* 9mm
Fully winged in both sexes. The male has a pair of long, curved cerci at the rear, while the female's ovipositor curves gently upwards. There is no song. July–November, living in a wide range of trees and often coming to lights at night.

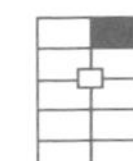

GREAT GREEN BUSH-CRICKET *Tettigonia viridissima*
40–55mm: ovipositor 18–25mm
Both sexes are fully winged, the female with ovipositor that just reaches the wing-tips. July–October in rough vegetation: the male's loud song, not unlike the sound of a sewing machine, can be heard from late afternoon until well after dark.

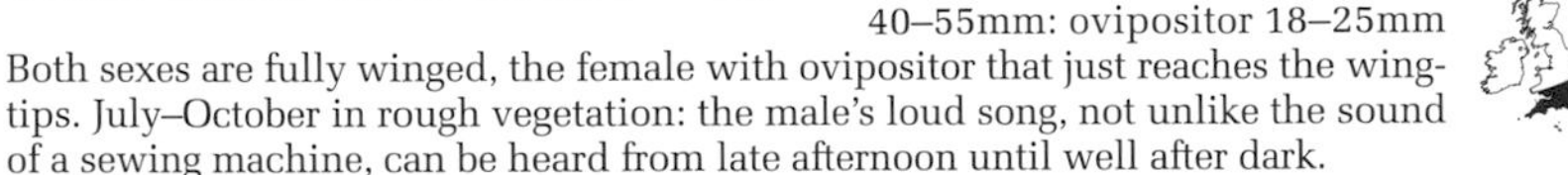

WART-BITER *Decticus verrucivorus* 30–37mm: ovipositor *c.* 20mm
Fully winged in both sexes, often with larger brown spots than shown here. The song, heard mainly by day, resembles the sound of a free-wheeling bicycle. This very rare insect occurs mainly on grassland July–September.

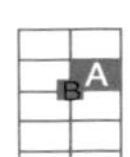

DARK BUSH-CRICKET *Pholidoptera griseoaptera*
12–20mm: ovipositor *c.* 10mm
Pale brown to almost black above and greenish yellow below. Female (B) is virtually wingless, while male (A) wings are no more than 2 pale sound-producing flaps. July–October in hedgerows and other rough vegetation. The song is a high-pitched *psst*, repeated at irregular intervals.

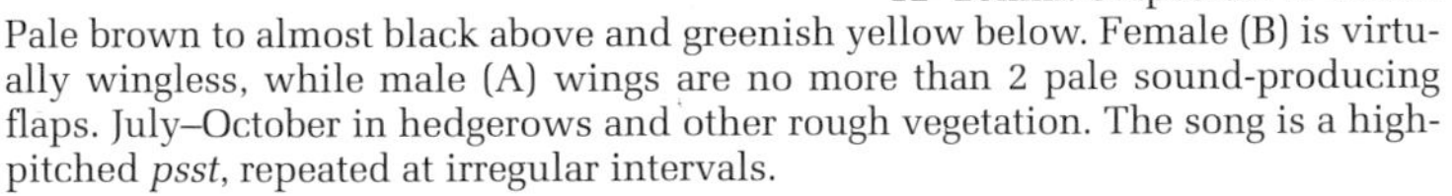

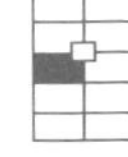

GREY BUSH-CRICKET *Platycleis albopunctata* 20–30mm: ovipositor *c.* 10mm
Mainly grey, although top of head and thorax may be brown or green. Fully winged in both sexes. The ovipositor curves sharply upwards. The insect inhabits rough vegetation July–October. The song consists of short buzzing sounds, like those made by winding a watch.

BOG BUSH-CRICKET *Metrioptera brachyptera* 10–20mm: ovipositor 8–10mm
Green or brown above and bright green below. The wings are normally short in both sexes. Look for the pale rear margin on each side of the pronotum. The song, heard mainly by day, is a series of shrill clicks, like the rapid ticking of a clock. July–November on bogs and wet heaths.

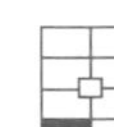

ROESEL'S BUSH-CRICKET *Metrioptera roeselii* 12–25mm: ovipositor *c.* 6mm
The pale yellow or green border all round each side of the pronotum separates this from the Bog Bush-cricket. Look also for pale spots just behind the pronotum. Both sexes are usually short-winged, although long-winged forms may occur in hot summers. The song consists of long bursts of sound resembling a dentist's drill. June–November, in lush grassland.

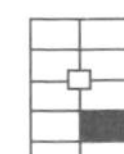

LONG-WINGED CONEHEAD *Conocephalus discolor*
15–22mm: ovipositor 9–13mm
Normally green with brown wings and a brown stripe on the head and pronotum. Both sexes are fully winged. The ovipositor is almost straight. The song is a quiet, high-pitched hiss like the sound of a knife-grinder. July–October in long grass.

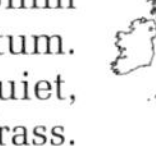

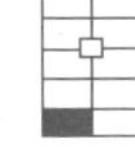

SHORT-WINGED CONEHEAD *Conocephalus dorsalis*
11–18mm: ovipositor *c.* 9mm
The wings cover only half of the abdomen and the ovipositor is strongly curved. The song resembles that of the Long-winged Conehead but rises and falls in volume. July–October, mainly in damp places and especially near the coast.

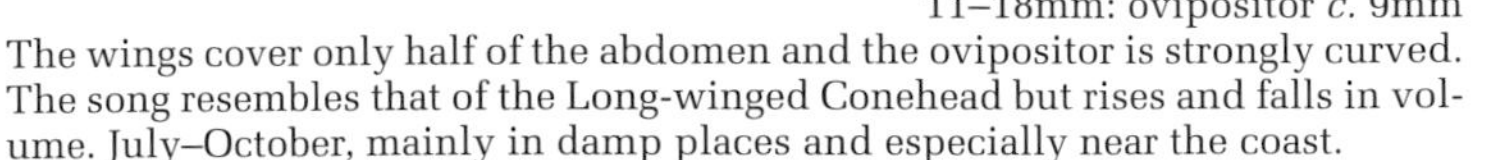

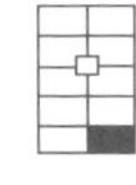

SPECKLED BUSH-CRICKET *Leptophyes punctatissima* 10–20mm
Male has tiny brown wings sitting on top of the body like a saddle. Female wings are minute flaps. Both sexes have a brown stripe along the middle of the back. The song is an almost inaudible scratching noise. July–November in rough vegetation: not uncommon in gardens.

TRUE CRICKETS: FAMILY GRYLLIDAE

These omnivorous insects resemble bush-crickets but they are more flattened and the hindwings, if present, are rolled up and extended beyond the abdomen like tails. The tarsi are only 3-segmented (4-segmented in bush-crickets) and the ovipositor is straight and needle-like. It is also possible that the crickets could be confused with cockroaches (*see* p.84), although the pronotum does not cover the head and the eyes are clearly visible. There are only 4 British species, including the very rare, wingless **Scaly Cricket** (*Mogoplistes squamiger*) known only from the Dorset coast.

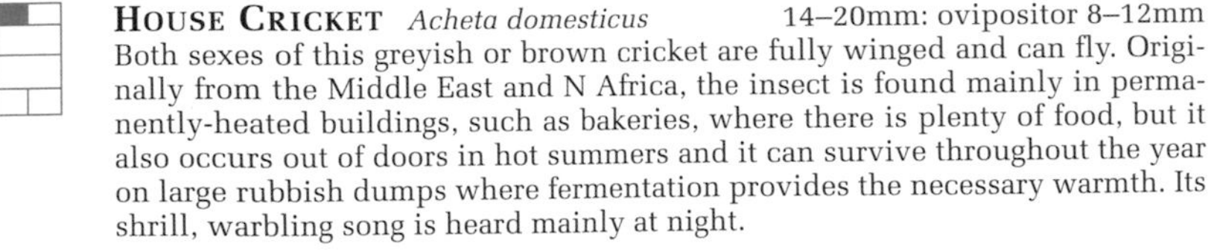

HOUSE CRICKET *Acheta domesticus* 14–20mm: ovipositor 8–12mm

Both sexes of this greyish or brown cricket are fully winged and can fly. Originally from the Middle East and N Africa, the insect is found mainly in permanently-heated buildings, such as bakeries, where there is plenty of food, but it also occurs out of doors in hot summers and it can survive throughout the year on large rubbish dumps where fermentation provides the necessary warmth. Its shrill, warbling song is heard mainly at night.

WOOD CRICKET *Nemobius sylvestris* 7–12mm: ovipositor 5–7mm

The forewings cover no more than half of the abdomen and the hindwings are completely absent. The insect lives in leaf litter in woodland rides and clearings and is adult June–November, when its soft warbling can be heard throughout the day and night.

FIELD CRICKET *Gryllus campestris* 17–25mm: ovipositor 8–12mm

This shiny black insect has no hindwings and cannot fly. It lives in burrows in short grass and adults can be found May–August, when the males sit at the mouths of their burrows and deliver their quite musical song by day and night. It is now one of our rarest insects.

MOLE CRICKET *Gryllotalpa gryllotalpa* Family Gryllotalpidae 35–45mm

The huge, spiky front legs, used for excavating its burrows, make this insect quite unmistakable. The only British member of its family, it spends most of its life chewing roots and assorted animals underground, but males come up to produce their purring and far-carrying song on warm evenings in spring and summer. Despite their size and clumsy appearance, they often fly after dark. Adults can be found throughout the year, although they are usually dormant in the winter.

GROUNDHOPPERS: FAMILY TETRIGIDAE

These insects resemble grasshoppers but the pronotum extends back to or beyond the tip of the abdomen and the forewings are reduced to small scales. The insects feed mainly on mosses and algae and require plenty of bare ground. There is no song.

COMMON GROUNDHOPPER *Tetrix undulata* 8–11mm

The colour ranges from pale brown to black according to the insect's surroundings and there is often a dark spot on each side of the pronotum. The latter has a prominent keel and does not reach beyond the tip of the abdomen. The insect is active all year in both dry and wet habitats.

SLENDER GROUNDHOPPER *Tetrix subulata* 9–14mm

The colour ranges from pale brown, often with a pinkish tinge, to grey and black. The pronotum has only a slight keel and usually reaches well beyond the end of the body. It is active all year on damp, sparsely vegetated ground. **Cepero's Groundhopper** (*T. ceperoi*), which occurs mainly along the south coast, differs slightly in the shape of the head and middle femur, but the 2 species are very difficult to separate in the field.

GRASSHOPPERS: FAMILY ACRIDIDAE

Readily distinguished from crickets by their short, stout antennae, grasshoppers are vegetarians and, true to their name, most live in grassy places. The sexes are usually alike, although the tip of the male abdomen usually turns up like the prow of a boat. The keels or ridges along the top edges of the thorax are useful guides to identification.

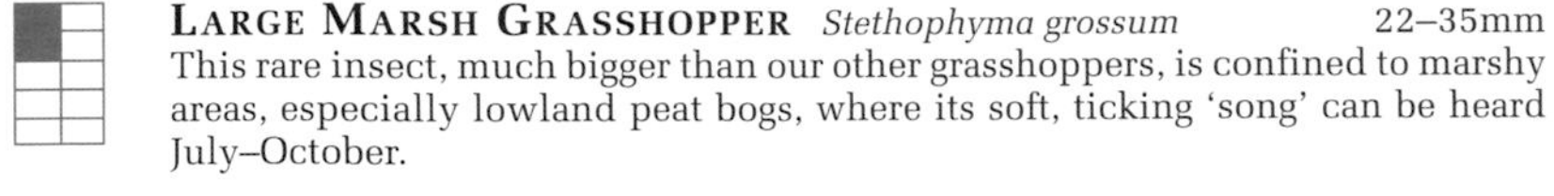

LARGE MARSH GRASSHOPPER *Stethophyma grossum* 22–35mm
This rare insect, much bigger than our other grasshoppers, is confined to marshy areas, especially lowland peat bogs, where its soft, ticking 'song' can be heard July–October.

STRIPE-WINGED GRASSHOPPER *Stenobothrus lineatus* 15–23mm
Legs and forewings may be green or brown, and the central area of the forewing has a prominent patch of parallel cross-veins. There is a white chevron near the wing-tip and the female usually has a white stripe near the front edge. The abdomen is orange only when mature. The song is a high-pitched whine, rhythmically rising and falling in volume. June–October on dry grassland.

WOODLAND GRASSHOPPER *Omocestus rufipes* 12–20mm
Usually brown with a red-tipped abdomen, but female may be green on top. The white-tipped palps are characteristic. The hissing song increases in volume for 5–10sec and then stops abruptly. Heathland and open woodland June–October.

COMMON GREEN GRASSHOPPER *Omocestus viridulus* 15–22mm
Female is always green on top, but otherwise the insect may be any combination of green, brown or grey. The forewing is often very dark at the tip. Look for the short keel on top of the head. The hissing song lasts 10–20 sec, reaching full volume at about the halfway point. July–October, mainly on lush grassland.

FIELD GRASSHOPPER *Chorthippus brunneus* 15–25mm
Usually brown, but green, grey, and reddish forms exist. Look for the sharply angled keels on top of the thorax and the dense hair below. The black marks between the keels do not reach the rear edge of the pronotum. The song is a short sequence of chirps, not unlike time-signal pips. June–October, mostly on dry grassland. The very rare **Heath Grasshopper** (*C. vagans*), confined to dry heaths in Dorset and Hampshire, is a little smaller and less hairy and the black marks between the keels reach the rear edge of the pronotum.

MEADOW GRASSHOPPER *Chorthippus parallelus* 10–22mm
A flightless species, with short forewings – especially in the female – and no hindwings. Normally green with brown wings, but completely green or brown individuals are not uncommon and some females are quite pink. Look for the almost straight keels on the thorax. The song sounds rather like a sewing machine working in 2–3 sec bursts. June–October in all kinds of grassland, but mostly damp areas.

LESSER MARSH GRASSHOPPER *Chorthippus albomarginatus* 14–21mm
This fully-winged, brown and/or green species can be recognised by the almost straight keels on the thorax. The song consists of up to 6 short chirps produced at 2-sec intervals. The insect is equally at home in dry and damp grassland, July–October.

RUFOUS GRASSHOPPER *Gomphocerippus rufus* 14–22mm
Mainly brown with slightly clubbed, white-tipped antennae. Some females are tinged with purple. When mature, the abdomen and back legs are tinged with orange. The song resembles a sewing machine working in 5-sec bursts. July–November on grassy slopes, usually on limestone.

MOTTLED GRASSHOPPER *Myrmeleotettix maculatus* 12–20mm
A very variable insect, ranging from green to dark brown or almost black, although the wings are never green. Antennae are clubbed in the male, but only slightly thickened in the female. The song is a series of up to 30 short chirps, delivered over a period of 10–15 sec and sounding like the winding of a clock. Heaths and dry grassland, June–October.

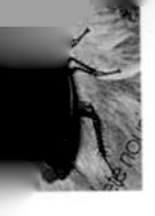

COCKROACHES: ORDER DICTYOPTERA
EARWIGS: ORDER DERMAPTERA

COCKROACHES These flattened, fast-running, omnivorous scavengers have spiky legs and long, thread-like antennae. The head is concealed under the large shield-like pronotum. Britain has just 3 native cockroaches, but several species from warmer parts of the world have now become pests in buildings. They are mainly nocturnal.

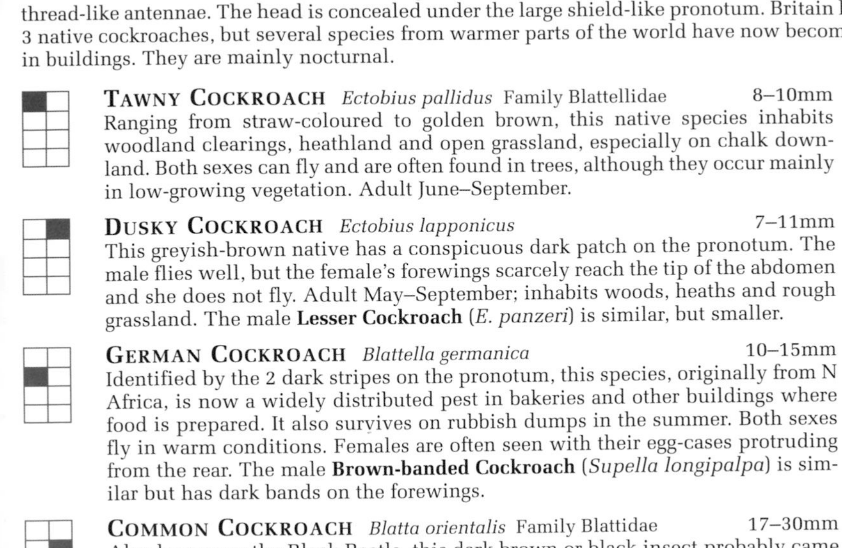

TAWNY COCKROACH *Ectobius pallidus* Family Blattellidae 8–10mm
Ranging from straw-coloured to golden brown, this native species inhabits woodland clearings, heathland and open grassland, especially on chalk downland. Both sexes can fly and are often found in trees, although they occur mainly in low-growing vegetation. Adult June–September.

DUSKY COCKROACH *Ectobius lapponicus* 7–11mm
This greyish-brown native has a conspicuous dark patch on the pronotum. The male flies well, but the female's forewings scarcely reach the tip of the abdomen and she does not fly. Adult May–September; inhabits woods, heaths and rough grassland. The male **Lesser Cockroach** (*E. panzeri*) is similar, but smaller.

GERMAN COCKROACH *Blattella germanica* 10–15mm
Identified by the 2 dark stripes on the pronotum, this species, originally from N Africa, is now a widely distributed pest in bakeries and other buildings where food is prepared. It also survives on rubbish dumps in the summer. Both sexes fly in warm conditions. Females are often seen with their egg-cases protruding from the rear. The male **Brown-banded Cockroach** (*Supella longipalpa*) is similar but has dark bands on the forewings.

COMMON COCKROACH *Blatta orientalis* Family Blattidae 17–30mm
Also known as the Black Beetle, this dark brown or black insect probably came from N Africa and now occurs in buildings all over the British Isles. It also occurs on rubbish dumps, especially in the summer. Both sexes are flightless, the female having only vestigial wings.

AMERICAN COCKROACH *Periplaneta americana* 25–45mm
Both sexes are entirely chestnut-brown and fully-winged and fly well in warm conditions. Originally from tropical Africa, it is now a cosmopolitan pest. Rarely seen out of doors in Britain, it is most common in large towns and around ports.The **Australian Cockroach** (*P. australasiae*) differs from the American Cockroach in having yellow markings on the pronotum and the front of the wings. This species is most likely to be found in large glasshouses and around ports and warehouses dealing in imported foodstuffs. It probably originated in Africa.

EARWIGS These insects are easily recognised by the tweezer-like pincers, strongly curved in males but more or less straight in females. Only 4 species are native to the British Isles, but others occasionally arrive in produce from the continent. Adults can be found throughout the year, although they are usually dormant in winter.

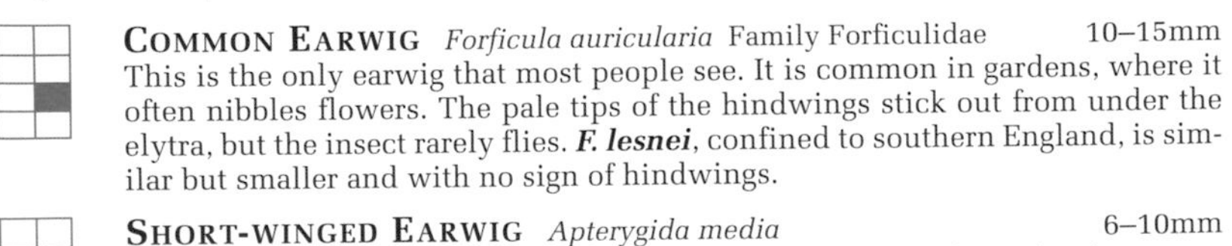

COMMON EARWIG *Forficula auricularia* Family Forficulidae 10–15mm
This is the only earwig that most people see. It is common in gardens, where it often nibbles flowers. The pale tips of the hindwings stick out from under the elytra, but the insect rarely flies. ***F. lesnei***, confined to southern England, is similar but smaller and with no sign of hindwings.

SHORT-WINGED EARWIG *Apterygida media* 6–10mm
The tiny hindwings are completely hidden by the elytra. The male pincers are more gently curved than in *Forficula* and bear 1 or 2 teeth. The insect is most often seen in flowers and seed heads in late summer and autumn.

LESSER EARWIG *Labia minor* Family Labiidae *c.* 5mm
This tiny and rather hairy earwig has a very dark head and is most often found in and around dung heaps and compost heaps. It is fully winged and often flies in warm weather. At rest, the hindwings stick out from under the elytra.

SUBORDER HETEROPTERA

The forewings of these insects, when present, are laid flat over the body at rest, usually with a good deal of overlap. Each wing normally consists of a horny or leathery basal region and a membranous tip, although many of these bugs have no more than tiny flaps for wings and some are completely wingless. The scutellum is usually triangular and very conspicuous. This suborder includes both herbivorous and predatory species. Many bugs can be confused with beetles at first, but the beetles' forewings normally meet in the midline with no overlap and they do not have membranous tips. Beetles also lack the piercing beak that characterises the bugs.

HAWTHORN SHIELD BUG *Acanthosoma haemorrhoidale*
Family Acanthosomatidae 13–15mm
The tarsi are 2-segmented in this and other members of the family, whereas other shield bugs have 3-segmented tarsi. The insect feeds on fruits and leaves of hawthorn and many other trees and shrubs. All year, although usually dormant in the winter. It shares the name 'stinkbug' with many other shield bugs because of the pungent fluid exuded when the insects are alarmed.

BIRCH SHIELD BUG *Elasmostethus interstinctus* 8–10mm
This bug differs from the Hawthorn Shield Bug in its smaller size and in having the front half of the scutellum red. It feeds mainly on birch and hazel – often on the developing catkins in the autumn. All year, although dormant in the coldest months.

JUNIPER SHIELD BUG *Elasmostethus tristriatus* *c.* 10mm
The orange spots on the forewings and the white spots at the base of the scutellum readily distinguish this bug from the 2 preceding species. Active in all but the coldest months; rarely found away from juniper bushes, where it feeds mainly on the fruits.

PARENT BUG *Elasmucha grisea* 7–9mm
This bug is named for the female's habit of sitting on her eggs and young nymphs and protecting them from parasites. The forewings are greyish brown, often tinged with purple. There is a black patch on the scutellum and the edges of the abdomen are chequered. The bug is associated mainly with birch. All year, although it hibernates in leaf litter in winter.

FAMILY PENTATOMIDAE

This is the largest family of shield bugs and nearly all its members are clearly shield-shaped. The antennae are 5-segmented, as in all shield bugs. Most other land-living heteropteran bugs have 4-segmented antennae. Unless otherwise stated, the adults can be found throughout the year, although most of them become dormant in the winter. There are 17 species in the British Isles, most of them restricted to the southern half.

TURTLE BUG *Podops inuncta* 5–6mm
The parallel-sided, tongue-like scutellum covers most of the abdomen as in the Tortoise Bug (*see* p.88), but the insect is smaller and has 2 tiny horns on the front of the pronotum. Feeds on grasses in a variety of habitats.

BISHOP'S MITRE *Aelia acuminata* 8–9mm
This bug is unlikely to be confused with any other. There are 3 ridges on the pronotum. Rough grassy places, including sand dunes. ***Sciocotis cursitans*** is more oval and has a tongue-shaped scutellum.

SLOE BUG *Dolycoris baccarum* 10–12mm
A coating of long hairs distinguishes this from several other superficially similar bugs. It feeds on the flowers and fruits of a wide range of woody and herbaceous plants and is not confined to sloe bushes. Most common in hedgerows and wood margins.

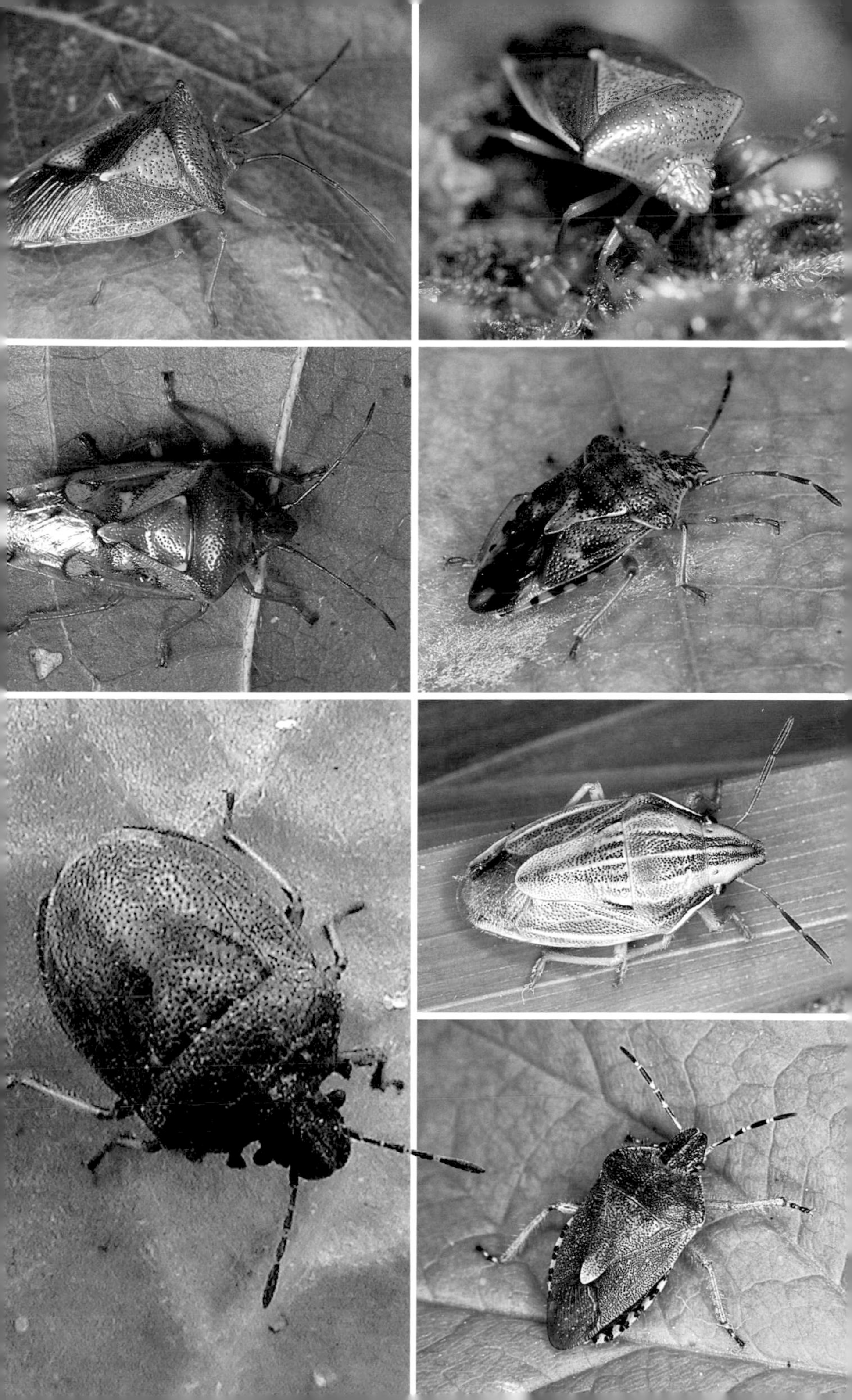

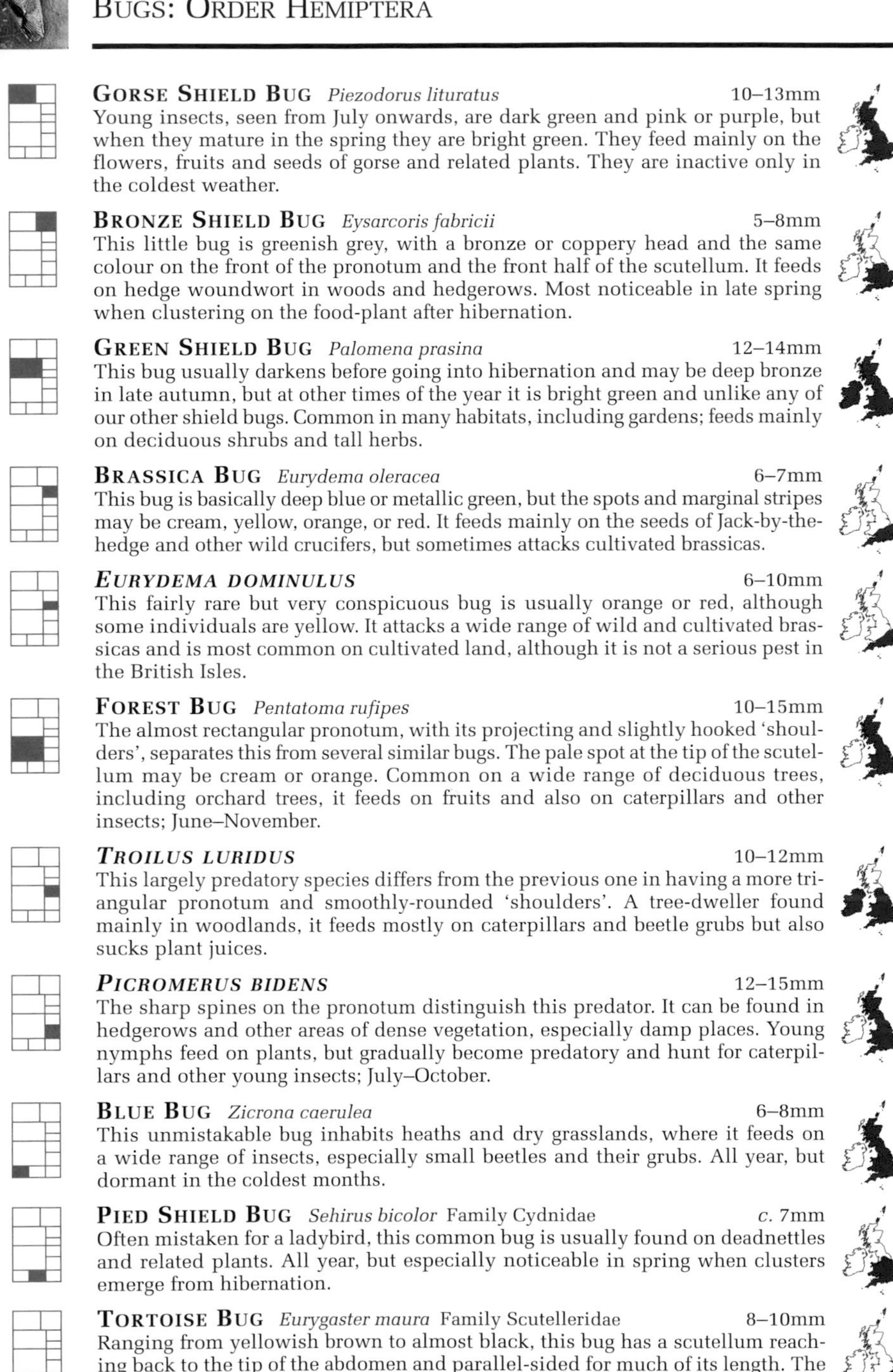

GORSE SHIELD BUG *Piezodorus lituratus* 10–13mm
Young insects, seen from July onwards, are dark green and pink or purple, but when they mature in the spring they are bright green. They feed mainly on the flowers, fruits and seeds of gorse and related plants. They are inactive only in the coldest weather.

BRONZE SHIELD BUG *Eysarcoris fabricii* 5–8mm
This little bug is greenish grey, with a bronze or coppery head and the same colour on the front of the pronotum and the front half of the scutellum. It feeds on hedge woundwort in woods and hedgerows. Most noticeable in late spring when clustering on the food-plant after hibernation.

GREEN SHIELD BUG *Palomena prasina* 12–14mm
This bug usually darkens before going into hibernation and may be deep bronze in late autumn, but at other times of the year it is bright green and unlike any of our other shield bugs. Common in many habitats, including gardens; feeds mainly on deciduous shrubs and tall herbs.

BRASSICA BUG *Eurydema oleracea* 6–7mm
This bug is basically deep blue or metallic green, but the spots and marginal stripes may be cream, yellow, orange, or red. It feeds mainly on the seeds of Jack-by-the-hedge and other wild crucifers, but sometimes attacks cultivated brassicas.

EURYDEMA DOMINULUS 6–10mm
This fairly rare but very conspicuous bug is usually orange or red, although some individuals are yellow. It attacks a wide range of wild and cultivated brassicas and is most common on cultivated land, although it is not a serious pest in the British Isles.

FOREST BUG *Pentatoma rufipes* 10–15mm
The almost rectangular pronotum, with its projecting and slightly hooked 'shoulders', separates this from several similar bugs. The pale spot at the tip of the scutellum may be cream or orange. Common on a wide range of deciduous trees, including orchard trees, it feeds on fruits and also on caterpillars and other insects; June–November.

TROILUS LURIDUS 10–12mm
This largely predatory species differs from the previous one in having a more triangular pronotum and smoothly-rounded 'shoulders'. A tree-dweller found mainly in woodlands, it feeds mostly on caterpillars and beetle grubs but also sucks plant juices.

PICROMERUS BIDENS 12–15mm
The sharp spines on the pronotum distinguish this predator. It can be found in hedgerows and other areas of dense vegetation, especially damp places. Young nymphs feed on plants, but gradually become predatory and hunt for caterpillars and other young insects; July–October.

BLUE BUG *Zicrona caerulea* 6–8mm
This unmistakable bug inhabits heaths and dry grasslands, where it feeds on a wide range of insects, especially small beetles and their grubs. All year, but dormant in the coldest months.

PIED SHIELD BUG *Sehirus bicolor* Family Cydnidae *c.* 7mm
Often mistaken for a ladybird, this common bug is usually found on deadnettles and related plants. All year, but especially noticeable in spring when clusters emerge from hibernation.

TORTOISE BUG *Eurygaster maura* Family Scutelleridae 8–10mm
Ranging from yellowish brown to almost black, this bug has a scutellum reaching back to the tip of the abdomen and parallel-sided for much of its length. The 2nd antennal segment is much longer than the 3rd. It feeds mainly on grasses, usually in dry places. All year, but dormant in winter. ***Eurygaster testudinaria***, usually found in damper places, has a scutellum gradually narrowing towards the rear and the 2nd antennal segment is hardly longer than the 3rd segment.

CO*REUS MARGINATUS* Family Coreidae *c.* 15mm

This is one of several superficially similar species, but can be distinguished by its larger size and also the presence of 2 tiny horns between its antennae. Feeding mainly on the fruits and seeds of docks and their relatives, it also attacks rose-buds and soft fruit. It is most often noticed in autumn and spring, when it congregates just before and after hibernation.

SYROMASTES RHOMBEUS 9–11mm

Easily recognised by its diamond-shaped abdomen extending well beyond the edges of the folded wings, this bug occurs on heaths and dry grasslands and also in open woodland if the soil is well-drained. It feeds on sandworts and related plants. All year, but dormant in winter.

ENOPLOPS SCAPHA 10–12mm

The oval abdomen with boldly chequered margins extends sideways well beyond the folded wings. Lives mainly in coastal areas, where it feeds chiefly on scentless mayweed and other composites. All year, but dormant in winter.

CORIOMERIS DENTICULATUS 7–8mm

Although superficially like *Enoplops*, this bug can be recognised by its stout, hairy antennae and triangular pronotum, which is also rather hairy. It lives in well-drained open habitats, where it feeds on various leguminous plants. All year, but dormant in winter.

CERALEPTUS LIVIDUS *c.* 10mm

This bug looks very much like the previous species but is much less hairy and has a conspicuous brown line along each side of the head and pronotum, whereas *Coriomeris* has a white-edged pronotum. Feeds on clovers and related plants in dry habitats. All year, but dormant in winter.

ALYDUS CALCARATUS Family Alydidae 10–12mm

This very dark and somewhat bristly bug is hard to spot when resting on vegetation, but when it takes flight it exposes a bright red patch on its abdomen. Found mainly on heathland, it feeds on a variety of plants and also takes some animal food. All year, but dormant in winter.

RHOPALUS SUBRUFUS Family Rhopalidae 7–8mm

This is one of 3 very similar bugs, but can usually be recognised by the light and dark banding on the margins of the abdomen. The forewings are largely membranous and the tip of the scutellum is lightly indented so that it appears to end in 2 tiny points. Feeds on a variety of plants in woodland clearings and other scrubby places. All year, but dormant in winter.

MYRMUS MIRIFORMIS 7–10mm

This bug is either brown or green, with a red stripe down the abdomen. The wings are usually very short with conspicuous red veins, but long-winged forms are not uncommon. The bug feeds on grasses, especially on the developing seeds, and is found in all kinds of dry and damp grassland; June–August.

CHOROSOMA SCHILLINGI *c.* 15mm

The wings of this very slender, straw-coloured or greenish bug usually finish about halfway along the abdomen. Found among tall grasses, mainly in coastal areas but also on the grassy heaths of Breckland; August–September.

SPEAR THISTLE LACEBUG *Tingis cardui* Family Tingidae 3–4mm

Although very small, this bug, one of several very similar species, is readily found on the flower heads of spear and marsh thistles. A large flower head may support 12 or more bugs. The lacebugs are named for the lace-like nature of the forewings and pronotum, the latter extending back to cover the scutellum. Most of the year, but dormant in winter. The very similar ***T. ampliata*** lives on creeping thistles. ***Stephanitis rhododendri***, 3–4mm long, has broad, lacy wings and long, slender antennae. It infests rhododendrons almost everywhere and causes white spots to develop on the leaves; June–October.

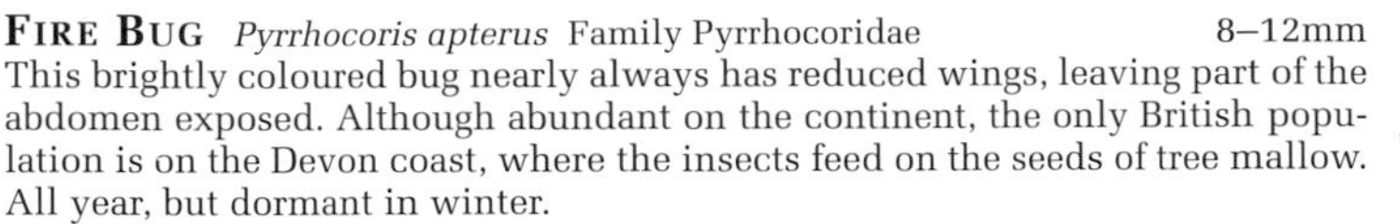

FIRE BUG *Pyrrhocoris apterus* Family Pyrrhocoridae 8–12mm
This brightly coloured bug nearly always has reduced wings, leaving part of the abdomen exposed. Although abundant on the continent, the only British population is on the Devon coast, where the insects feed on the seeds of tree mallow. All year, but dormant in winter.

NETTLE GROUND BUG *Heterogaster urticae* Family Lygaeidae 6–7mm
The banded legs serve to distinguish this rather hairy bug from several similarly coloured ground bugs. It feeds on stinging nettles and is especially noticeable in late spring when large numbers may congregate and mate on the plants. All year, but dormant in winter.

EUROPEAN CHINCHBUG *Ischnodemus sabuleti* *c.* 5mm
This very flat bug exists in 2 main forms: with very short wings (A) and with wings as long as the body (B). Rare individuals have wings of intermediate length. The nymphal abdomen is largely red. The insect is often abundant throughout the year in grassland, especially damp areas. The short-winged forms generally congregate on the lower parts of the grass blades, while the long-winged individuals tend to gather on the flowerheads.

RHYPAROCHROMUS PINI 7–8mm
The front half of the pronotum is black and the rear half is greyish brown, and the black scutellum is shorter than the visible junction between the forewings. Inhabitant of pinewoods and heathlands, it feeds mainly on seeds. Most of the year, but dormant in winter.

PINE CONE BUG *Gastrodes grossipes* 5–7mm
The bright rust-coloured wings and contrasting black scutellum make this bug easy to recognise. It is very common on the lower branches of Scot's pines, taking sap from the needles or, more often, from ripening seeds. It can often be found in fallen cones, where it usually spends the winter. The less brightly coloured **Spruce Cone Bug** (*G. abietum*) lives on spruce.

NEIDES TIPULARIUS Family Berytidae 10–12mm
This is one of several long-legged species known as stilt bugs. The legs are all of similar thickness and the antennae are slightly clubbed. It lives in long grass and other dense vegetation in fairly dry places and feeds on a wide variety of plants. All year, but dormant in winter.

ASSASSIN BUGS: FAMILY REDUVIIDAE

Named for their predatory habits, these bugs have a distinct neck, brought about by the narrowing of the head behind the eyes, and a sturdy, 3-segmented beak. When handled, the insects can screech quite loudly by rubbing the tip of the beak along a ridged groove under the thorax. There are only 5 resident British species, all feeding on other insects and sometimes spiders.

HEATH ASSASSIN BUG *Coranus subapterus* 9–12mm
This rather bristly bug generally has very short wings, although long-winged individuals are not uncommon in the north. It can be found on heaths and sand dunes July–October. It could be mistaken for some of the damsel bugs, but the latter are much less bristly. In addition, in the Heath Assassin the 1st antennal segment is the longest, whereas in the damsel bugs the 2nd antennal segment is always longer than the 1st.

FLY BUG *Reduvius personatus* 15–18mm
This rather bristly brown or black bug lives mainly in and around buildings. It flies at night and is often attracted to lights; May–September. Like all assassin bugs, it feeds mainly on other insects. Its beak can pierce human skin if it is handled.

DAMSEL BUGS: FAMILY NABIDAE

Despite their name, these bugs are voracious predators, resembling the assassin bugs but differing in having a 4-segmented beak. The 2nd antennal segment is always the longest. All 12 of the British species are some shade of brown and they feed on a wide range of other insects.

COMMON DAMSEL BUG *Nabis rugosus* 6–8mm
The wings usually just reach the tip of the abdomen, although individuals with longer or shorter wings occasionally occur. One of several very similar species, it lives in grassy places nearly everywhere. All year, but dormant in winter.

MARSH DAMSEL BUG *Dolichonabis limbatus* 7–9mm
Almost always with very short wings, this bug usually has a black ⊥ mark on the pronotum. It lives in damp grassland and other lush vegetation, July–November.

TREE DAMSEL BUG *Himacerus apterus* 8–12mm
Our only tree-dwelling damsel bug, this insect usually has short wings, although the forewings may reach the tip of the abdomen. It lives mainly in deciduous trees and feeds on mites as well as on small insects; July–October. The ground-dwelling **Ant Damsel Bug** (*H. mirmicoides*) is very similar but its antennae are much shorter than the thorax and abdomen.

CAPSID OR MIRID BUGS: FAMILY MIRIDAE

The forewings in this family, when present, are fairly soft and each has a prominent triangular area, known as the cuneus, just in front of the membrane. It often differs in colour from the rest of the wing. Unlike that of the superficially similar flower bugs (*see* p. 98), the membrane generally has 1 or 2 cells at the base. Most mirids are herbivorous, with fruits and seeds figuring largely in their diet. There are over 200 species in the British Isles.

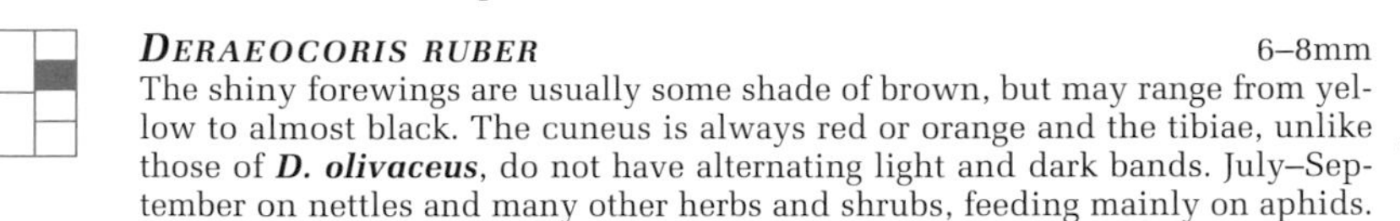

DERAEOCORIS RUBER 6–8mm
The shiny forewings are usually some shade of brown, but may range from yellow to almost black. The cuneus is always red or orange and the tibiae, unlike those of ***D. olivaceus***, do not have alternating light and dark bands. July–September on nettles and many other herbs and shrubs, feeding mainly on aphids.

PILOPHORUS CINNAMOPTERUS 4–5mm
The pale bands on the forewings are formed by silvery hairs, and the area of the forewing behind the 2nd band is quite shiny. July–October on pines and other conifers. The bug is often mistaken for an ant as it scuttles around in the trees looking for its aphid prey. ***P. perplexus*** has a less shiny rear end and occurs on deciduous trees, especially oaks.

DRYOPHILOCORIS FLAVOQUADRIMACULATUS 6–7mm
This boldly marked bug can be recognised by the strong upward bulging of the rear of its black pronotum. Common on oak trees May–July, it feeds on the young leaves and catkins and also attacks other small insects and their eggs. ***Globiceps cruciatus*** is similarly coloured but the pale areas at the front of the wings are less extensive. It lives mainly on low-growing vegetation.

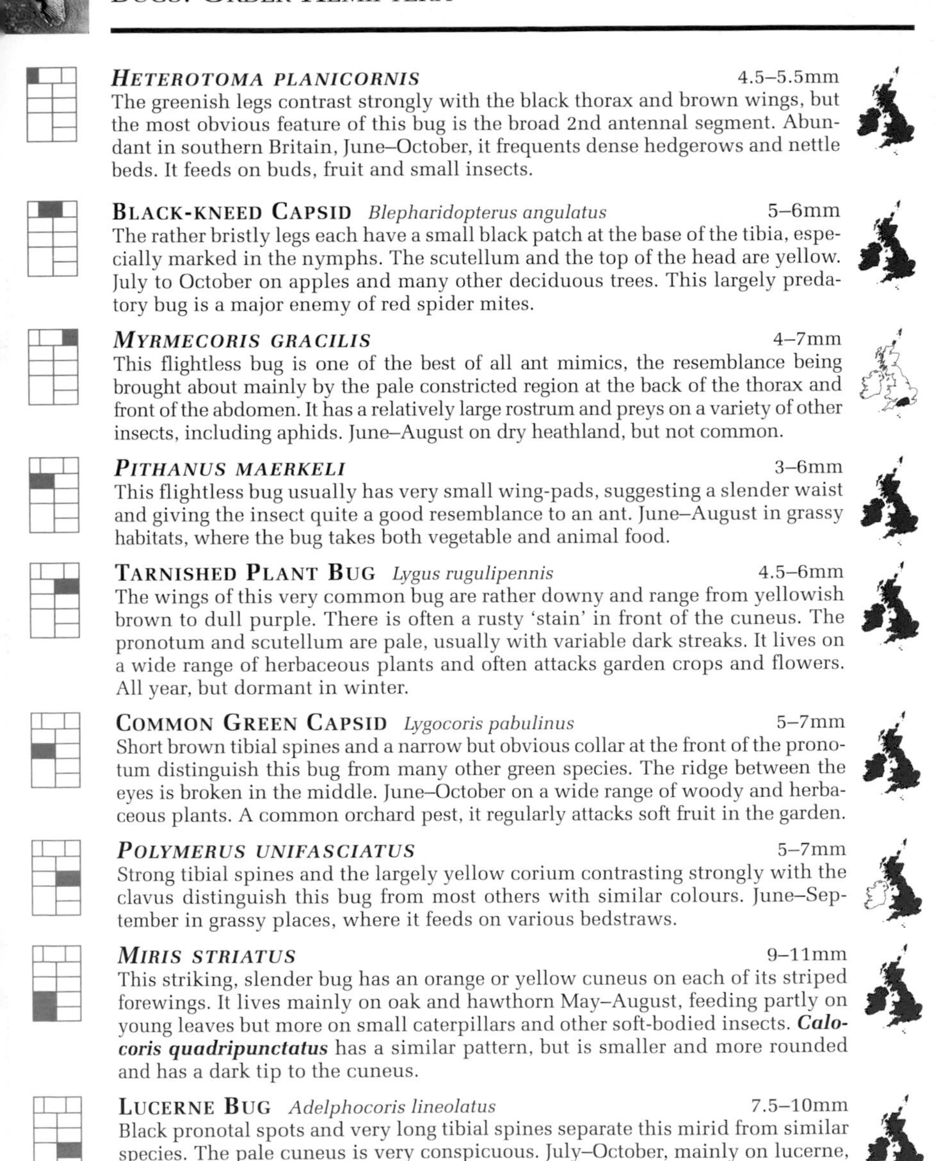

HETEROTOMA PLANICORNIS 4.5–5.5mm
The greenish legs contrast strongly with the black thorax and brown wings, but the most obvious feature of this bug is the broad 2nd antennal segment. Abundant in southern Britain, June–October, it frequents dense hedgerows and nettle beds. It feeds on buds, fruit and small insects.

BLACK-KNEED CAPSID *Blepharidopterus angulatus* 5–6mm
The rather bristly legs each have a small black patch at the base of the tibia, especially marked in the nymphs. The scutellum and the top of the head are yellow. July to October on apples and many other deciduous trees. This largely predatory bug is a major enemy of red spider mites.

MYRMECORIS GRACILIS 4–7mm
This flightless bug is one of the best of all ant mimics, the resemblance being brought about mainly by the pale constricted region at the back of the thorax and front of the abdomen. It has a relatively large rostrum and preys on a variety of other insects, including aphids. June–August on dry heathland, but not common.

PITHANUS MAERKELI 3–6mm
This flightless bug usually has very small wing-pads, suggesting a slender waist and giving the insect quite a good resemblance to an ant. June–August in grassy habitats, where the bug takes both vegetable and animal food.

TARNISHED PLANT BUG *Lygus rugulipennis* 4.5–6mm
The wings of this very common bug are rather downy and range from yellowish brown to dull purple. There is often a rusty 'stain' in front of the cuneus. The pronotum and scutellum are pale, usually with variable dark streaks. It lives on a wide range of herbaceous plants and often attacks garden crops and flowers. All year, but dormant in winter.

COMMON GREEN CAPSID *Lygocoris pabulinus* 5–7mm
Short brown tibial spines and a narrow but obvious collar at the front of the pronotum distinguish this bug from many other green species. The ridge between the eyes is broken in the middle. June–October on a wide range of woody and herbaceous plants. A common orchard pest, it regularly attacks soft fruit in the garden.

POLYMERUS UNIFASCIATUS 5–7mm
Strong tibial spines and the largely yellow corium contrasting strongly with the clavus distinguish this bug from most others with similar colours. June–September in grassy places, where it feeds on various bedstraws.

MIRIS STRIATUS 9–11mm
This striking, slender bug has an orange or yellow cuneus on each of its striped forewings. It lives mainly on oak and hawthorn May–August, feeding partly on young leaves but more on small caterpillars and other soft-bodied insects. ***Calocoris quadripunctatus*** has a similar pattern, but is smaller and more rounded and has a dark tip to the cuneus.

LUCERNE BUG *Adelphocoris lineolatus* 7.5–10mm
Black pronotal spots and very long tibial spines separate this mirid from similar species. The pale cuneus is very conspicuous. July–October, mainly on lucerne, clover and other leguminous plants in both wet and dry grassland. It is sometimes a pest of garden flowers.

STENOTUS BINOTATUS 5.5 –7mm
This bug closely resembles the Lucerne Bug but is a little smaller and somewhat paler and the cuneus contrasts less with the rest of the forewing. It feeds on the flowers and ripening seeds of various grasses. June–September in rough grassland, including roadside verges.

CALOCORIS ROSEOMACULATUS 6–8mm

The rosy patches on the forewings and the dark central line on the pale green scutellum readily identify this bug. June–October on dry grassland, where it feeds on flower heads of numerous plants, including bird's-foot trefoil, salad burnet and various composites.

POTATO CAPSID *Calocoris norwegicus* 6–8mm

This common species is one of several superficially similar green bugs, the orange tinge being apparent only in mature males. The black spots at the front of the scutellum are not always visible. June–October in hedgerows and other areas of dense vegetation, feeding on a wide range of herbaceous plants. It sometimes damages potatoes and other garden and field crops.

PHYTOCORIS VARIPES 5–7mm

The ground colour varies from pinkish brown to chestnut, but the insect can be distinguished from most other bugs by its very long, banded back legs, which enable it to leap about. The middle legs are not banded. June to October in rough grassy places, where it feeds on the flowers and developing fruits of many plants.

PHYTOCORIS TILIAE 6–7mm

This bug is pale grey or greenish with black mottling, including dark edges to the pronotum. There is a conspicuous pale patch just in front of the cuneus. Common June–October on a wide range of deciduous trees, including apples, it feeds mainly on small insects and mites.

CAPSUS ATER 5–6mm

The oval body and dark wings, and the much swollen 2nd antennal segment, distinguish this common species from most other bugs. The head and pronotum may be black or rust-red. June–September in grassy places almost everywhere. It feeds mainly on the lower stems of grasses.

CAPSODES GOTHICUS 5–7mm

Although the amount of orange on the wings varies a little, this hairy bug can be recognised by its coloration and by its broad head with bulging eyes. It differs from other *Capsodes* species in not having orange or yellow stripes on the pronotum. June–August in lush, grassy places.

MEADOW PLANT BUG *Leptopterna dolabrata* 7–10mm

Males (A) are fully winged, the forewings starting off yellow but gradually becoming brick-red. Females (B) are usually short-winged, the translucent forewings exposing the yellow and black abdomen. Both sexes have very hairy legs and antennae and emit a pungent odour when disturbed. Feeding on grass, it is abundant June–September in lush grassland of all kinds. In drier grassland it is replaced by the very similar ***L. ferrugata***.

NOTOSTIRA ELONGATA 6–9mm

The mature male (A) is black with greenish edges to the forewings and pronotum. Females of the summer generation are entirely green (B); autumn females are light brown but acquire a green abdomen in spring. The tibiae and the 1st antennal segment are very hairy in both sexes. Common in grassy places: males June–October and females all year, although dormant in winter.

COMMON FLOWER BUG *Anthocoris nemorum* Family Anthocoridae 3–4.5mm

The black pronotum and completely shiny forewings distinguish this from several related species. Abundant on leaves and flowers, feeding on aphids and other soft-bodied prey for much of the year but hibernating under bark or in debris in winter.

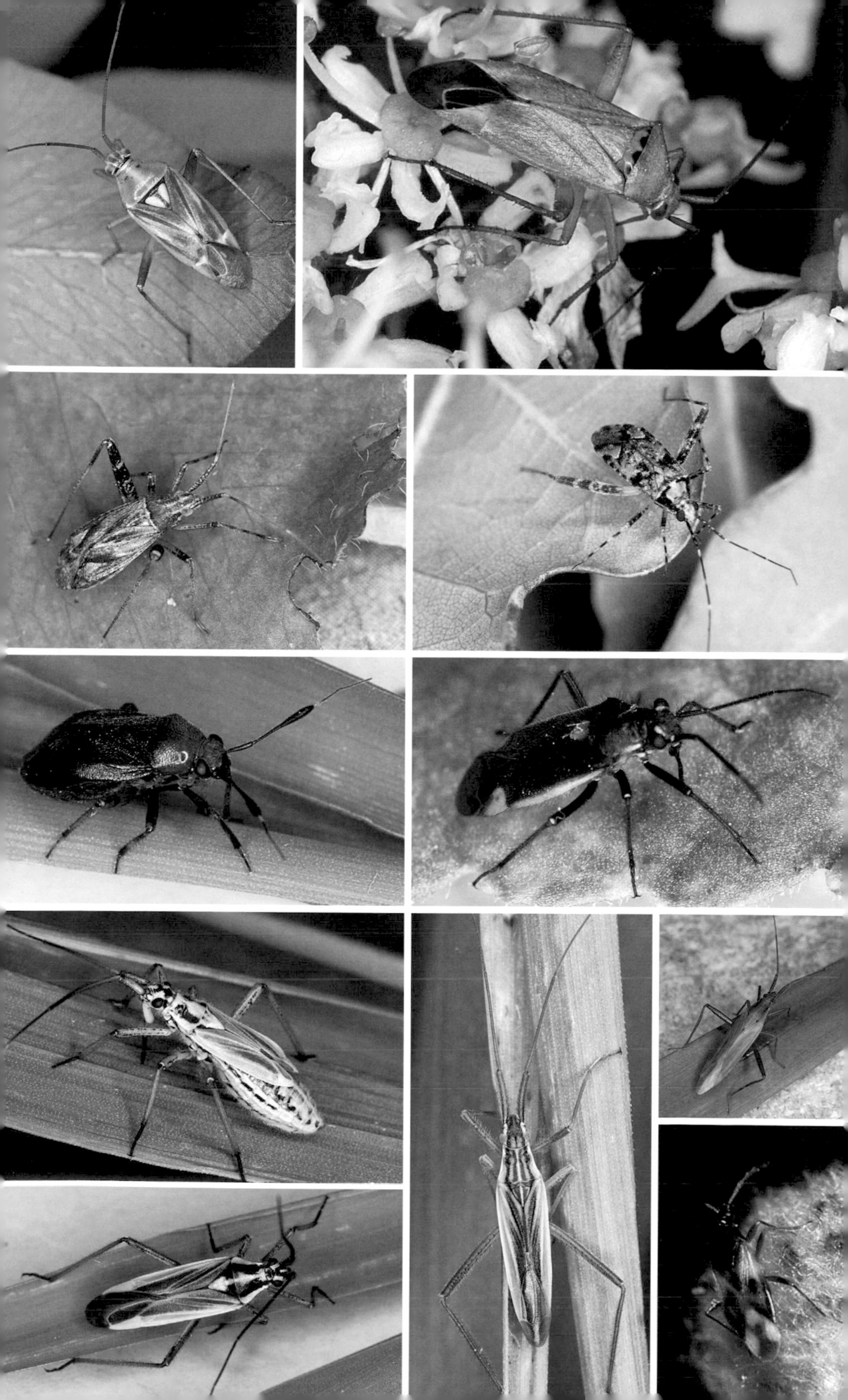

WATER BUGS

Several families of heteropteran bugs have become adapted for life on or in water. Those bugs that live on the surface are furnished with water-repellent hairs that keep them dry and prevent them from sinking into the water. Tufts of water-repellent hairs on the feet produce little dimples where they stretch the water surface without breaking through. All the surface bugs are predators and all have clearly visible antennae. Bugs that live under the water have small antennae that are generally concealed from view. The young stages of these submerged bugs generally breathe by means of gills, and some adults have also evolved a technique for obtaining dissolved oxygen direct from the water. But most adults need to surface periodically to take in air. They come up tail-first and draw fresh supplies into reservoirs under the wings or held in place elsewhere on the body by layers of water-repellent hair. Most of the submerged bugs are predators.

POND SKATER *Gerris lacustris* Family Gerridae 8–10mm
This is the commonest and most widespread of several very similar species that skate over still and slow-moving water. They use their short front legs to capture prey – usually other insects – that fall on the water surface. There are winged and wingless individuals and the winged ones fly well. Adult all year, but dormant in the coldest months, usually in leaf litter and other debris. ***Aquarius najas*** is larger and usually wingless and prefers flowing streams.

WATER MEASURER *Hydrometra stagnorum* Family Hydrometridae 9–12mm
This normally wingless insect walks very slowly over the surface of ponds and streams and usually keeps close to the marginal vegetation. It is much more slender than the pond skaters and has a very long head. It spears mosquito larvae, water fleas and other small prey by jabbing its beak down through the water surface. All year, but dormant in winter.

WATER SCORPION *Nepa cinerea* Family Nepidae *c.* 20mm
This very flat, slow-moving bug lives in shallow, muddy water, where it catches other creatures, including tadpoles and small fish, with its powerful front legs. It breathes by pushing its hollow 'tail' up to the surface. Despite its name and appearance, it is harmless and feigns death when handled. Although fully winged, it rarely flies.

WATER STICK INSECT *Ranatra linearis* 30–35mm
This well-named bug lives in deeper water than the previous species and spends most of its time clinging to the vegetation and waiting for prey to come within reach of its long, grasping front legs. Although it is larger than the Water Scorpion, it usually takes smaller prey, such as mosquito larvae and water fleas. It flies well in warm weather.

COMMON WATER BOATMAN *Corixa punctata* Family Corixidae 12–14mm
This insect and its numerous relatives resemble the backswimmers but swim the right way up. The middle and hind-legs are about the same length. Water boatmen feed mainly by sucking up debris from the bottom of the pond. They fly well and the males stridulate loudly by rubbing their front legs against their heads. They are sometimes called lesser water boatmen.

SAUCER BUG *Ilyocoris cimicoides* Family Naucoridae *c.* 15mm
Identified by its hornlike front legs, this bug lives in mud and dense vegetation and preys on a wide range of animals. It can inflict a painful wound if handled. Although fully winged, it cannot fly because its flight muscles are poorly developed.

COMMON BACKSWIMMER *Notonecta glauca* Family Notonectidae *c.* 15mm
This is the commonest of our 4 backswimmers, so-called because they swim on their backs. They are sometimes called water boatmen. They use their long back legs like oars to row themselves through the water. The forewings of this species are silvery grey and the underside is black, although the air bubble gives it a silvery appearance. It preys on many other animals, including tadpoles and small fish, and will inflict a painful stab if handled. Abundant in still water, it flies well in warm weather. ***N. maculata***, found in stony pools, has mottled reddish-brown forewings.

SUBORDER HOMOPTERA

The forewings of these insects, when present, are either leathery or membranous, but they are always of a uniform texture. They are usually folded back over the body in a tent-like fashion. The antennae are as long as the body in some aphids, but those of most homopteran bugs are short and bristle-like. All the homopterans are vegetarians and tend to be much more sedentary than the heteropteran bugs.

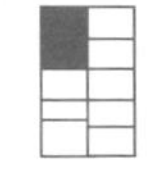

CICADETTA MONTANA Family Cicadidae *c.* 25mm
This is the only British cicada. The pronotum sometimes bears orange streaks. Male cicadas produces a 'song' by rapidly vibrating a tiny membrane on each side of the body, but this species is very quiet and its warbling song is easily missed. Confined to the New Forest in Britain and extremely rare, possibly now extinct. It feeds on numerous trees, shrubs, and herbaceous plants June–August.

HORNED TREEHOPPER *Centrotus cornutus* Family Membracidae *c.* 10mm
As in all treehoppers, the pronotum extends back over the body. In this species – one of only 2 British treehoppers – the pronotum reaches roughly to the tip of the abdomen and also bears a horn on each side. April–August on various herbs and shrubs in woodland rides and clearings. ***Gargara genistae***, found mainly on shrubby legumes, is about 6mm long, with no lateral horns and with the pronotum reaching only about halfway along the abdomen.

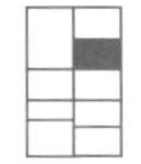

COMMON FROGHOPPER *Philaenus spumarius* Family Cercopidae 5–7mm
This is the commonest of the 10 British froghoppers, named for their jumping abilities and for the vaguely frog-like appearance of most of the species. The pattern varies, but this species differs from most other froghoppers in the strongly curved front edge of the forewing. June–September on vegetation almost everywhere. Froghoppers are also called cuckoo-spit insects because their nymphs surround themselves with protective masses of white froth, once believed made by cuckoos.

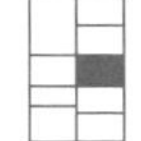

CERCOPIS VULNERATA 9–11mm
This unmistakable froghopper differs from our other species by spending its early life feeding on underground roots, although still surrounded by froth. April–August, mainly in wooded areas.

LEAFHOPPERS: FAMILY CICADELLIDAE

This is a huge family of jumping bugs, superficially like froghoppers but generally smaller and with lots of bristles on their back legs instead of the two stout spines of the froghoppers. The forewings are leathery and usually softer than those of the froghoppers.

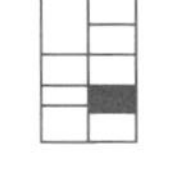

LEDRA AURITA 13–18mm
No other bug has the ear-like lobes on the pronotum. May–September in woodlands, especially oak woods, but hard to find because it is so well camouflaged on lichen-covered twigs.

POTATO LEAFHOPPER *Eupteryx aurata* 4–6mm
Easily recognised by its pattern, this little bug is abundant on stinging nettles and many other plants June–November. Sometimes damages potato crops by puncturing the leaves and destroying the chlorophyll with its toxic saliva.

CICADELLA VIRIDIS 6–9mm
The bicolored pronotum, yellow at the front and green at the rear, distinguishes this from our other green leafhoppers. The forewings are bright green in the female, but usually tinged with deep purple or even black in the male. Common July–October in marshes and damp grassland.

RHODODENDRON LEAFHOPPER *Graphocephala fennahi* 8–10mm
This unmistakable leafhopper was introduced from N America and is now found on rhododendron leaves all over southern Britain; June–October.

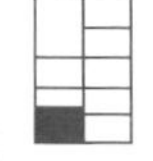

CIXIUS NERVOSUS Family Cixiidae 6–8mm
This is one of several very similar species with strongly-veined membranous wings and three prominent keels on the scutellum. May–October on various deciduous trees.

APHIDS: SUPERFAMILY APHIDOIDEA

The aphids are small homopterans with pear-shaped bodies rarely more than 3–4mm long. All 4 wings, when present, are membranous with few veins. Most aphids possess a pair of horn-like cornicles at the rear that exude waxy defensive secretions. Many have complex life histories, often involving both egg-laying forms and forms that give birth to active young. Two different kinds of food-plant may also be necessary to complete the cycle. A female can give birth when only a few days old, and can produce several babies every day, so what the insects lack in size they make up for in numbers. Wingless individuals predominate while colonies are building up in spring, but winged forms appear later and move to new plants. Winter is usually passed in the egg stage, although some adults can survive mild winters. Many aphids, including the well-known garden greenfly and blackfly, are serious pests. Some are responsible for the formation of galls (*see* p. 38). There are about 550 British species, arranged in 12 families.

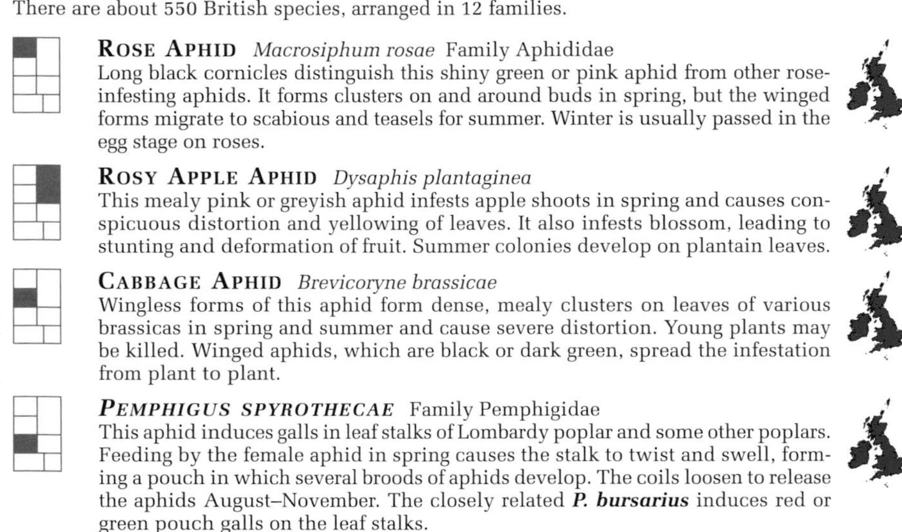

Rose Aphid *Macrosiphum rosae* Family Aphididae
Long black cornicles distinguish this shiny green or pink aphid from other rose-infesting aphids. It forms clusters on and around buds in spring, but the winged forms migrate to scabious and teasels for summer. Winter is usually passed in the egg stage on roses.

Rosy Apple Aphid *Dysaphis plantaginea*
This mealy pink or greyish aphid infests apple shoots in spring and causes conspicuous distortion and yellowing of leaves. It also infests blossom, leading to stunting and deformation of fruit. Summer colonies develop on plantain leaves.

Cabbage Aphid *Brevicoryne brassicae*
Wingless forms of this aphid form dense, mealy clusters on leaves of various brassicas in spring and summer and cause severe distortion. Young plants may be killed. Winged aphids, which are black or dark green, spread the infestation from plant to plant.

Pemphigus spyrothecae Family Pemphigidae
This aphid induces galls in leaf stalks of Lombardy poplar and some other poplars. Feeding by the female aphid in spring causes the stalk to twist and swell, forming a pouch in which several broods of aphids develop. The coils loosen to release the aphids August–November. The closely related ***P. bursarius*** induces red or green pouch galls on the leaf stalks.

Adelges viridis Family Adelgidae
This aphid induces galls on spruce. The galls are formed from the bases of adjacent needles, which swell up and enclose spaces that contain the developing aphids. The gall looks like a small cone and is called a pineapple gall. The cavities open June–July and the aphids fly to larch trees for summer.

JUMPING PLANT LICE or PSYLLIDS: SUPERFAMILY PSYLLOIDEA

These bugs resemble tiny leafhoppers, but they usually have transparent wings and much longer antennae. They leap well with their long back legs. About 80 species, in several families, live in the British Isles.

Apple Psyllid *Psylla mali* Family Psyllidae 3–4mm
Green in spring and early summer, the insect then becomes reddish brown. April–September. Abundant on apple trees, where its feeding punctures damage blossom and young shoots.

WHITEFLIES: SUPERFAMILY ALEYRODOIDEA

These insects, with wings spanning no more than about 3mm, look more like tiny moths than bugs. About 20 species occur in the British Isles, all in the family Aleyrodidae, but they are not easy to distinguish. Nearly all feed gregariously on the undersides of leaves, which become discoloured and distorted if the infestation is heavy.

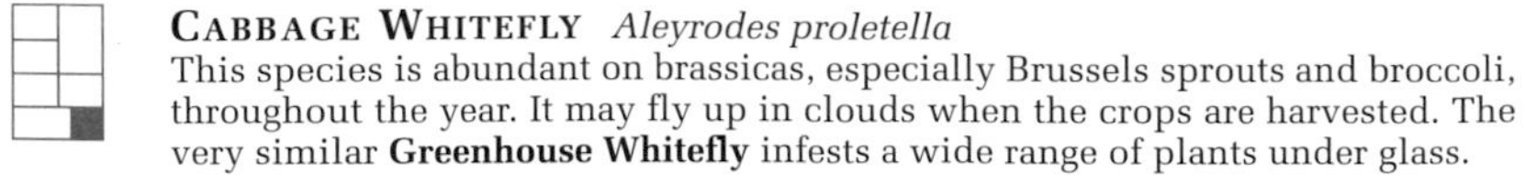

Cabbage Whitefly *Aleyrodes proletella*
This species is abundant on brassicas, especially Brussels sprouts and broccoli, throughout the year. It may fly up in clouds when the crops are harvested. The very similar **Greenhouse Whitefly** infests a wide range of plants under glass.

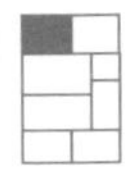

ANT-LION *Euroleon nostras* Family Myrmeleontidae 30–40mm

The size and clubbed antennae immediately identify this insect – the only one of its kind in the British Isles and only discovered on the mainland in recent years, although it exists in the Channel Islands. June–September on heathland, where it plucks other small insects from the vegetation. The insect's name refers to the large-jawed larva, which excavates a conical pit in sandy soil and lies in wait at the bottom for ants and other insects to fall in.

GREEN LACEWINGS: FAMILY CHRYSOPIDAE

This family contains about 18 British species of predominantly green insects, several of which are difficult to identify without detailed examination. The insects are often called golden-eyes because of the brilliant golden or brassy shine from their eyes. In common with most other lacewings, all have conspicuously forked veins at the outer edges of the wings. *Nothochrysa* species are brown, but their venation clearly shows them to belong with the green lacewings. They are much larger than the brown lacewings of the Hemerobiidae (*see* below). Adults and larvae all feed mainly on aphids, and larvae of some species camouflage themselves with the drained skins of their prey.

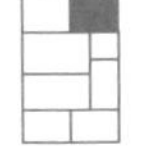

CHRYSOPA PERLA *c.* 15mm

This species can be distinguished from most other green lacewings by its distinctly bluish tinge and extensive black head markings. The 2nd antennal segment is black. May–August; abundant in many well-vegetated habitats, including gardens and hedgerows. ***C. dorsalis*** is very similar but much rarer and usually found on pine trees.

CHRYSOPA PALLENS 15–20mm

This bright green species can be distinguished from our other lacewings by the 7 small black spots on the head, including one between the bases of the antennae. The abdomen is completely green. May–August; abundant in many habitats, including gardens. ***Dichochrysa prasina*** has no more than five black spots on the head and has a dark spot at the base of each wing.

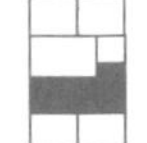

CHRYSOPERLA CARNEA *c.* 15mm

A pale green lacewing with no black veins and no black spot between completely green antennae is likely to be this species, which is probably our commonest lacewing although recent work suggests that *carnea* may actually be two or even three species. All year in a wide variety of habitats, including gardens, although, unlike any of our other green lacewings, the insect hibernates for winter. Prior to hibernation, which often takes place in buildings, the insect becomes flesh-coloured and then cannot be mistaken for any other lacewing.

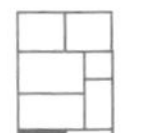

OSMYLUS FULVICEPHALUS Family Osmylidae *c.* 25mm

Our largest lacewing, easily recognised by its size and spotted wings. May–August, rarely far from shady streamsides where the larvae hunt in damp moss and leaf litter.

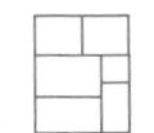

WESMAELIUS SUBNEBULOSUS Family Hemerobiidae 7–8mm

The chequered wing margins and long veins, and the numerous dark spots distinguish the typical form of this common species from most of our other lacewings, although unspotted forms are not uncommon, especially in urban areas. March–October in nettle beds and other dense vegetation, including hedgerows, and also on various deciduous trees. ***W. nervosus*** is indistinguishable without microscopic examination. The brown lacewings of this family can be distinguished from the sponge flies by the numerous cross-veins.

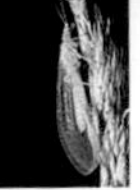
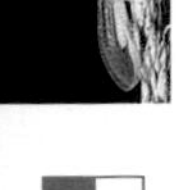

MICROMUS ANGULATUS 6–8mm

This is one of several small and rather similar brown lacewings, most of which can be reliably separated only by examining the genitalia. Some have dark patterns on the forewings, but these are not usually sufficiently constant to be of use in identification. Most species also have conspicuously forked veins at the front edge of each wing. Flies April–October in many well-vegetated habitats, where adults and larvae feed on aphids but it is not common.

MICROMUS VARIEGATUS *c.* 6mm

Three dark smudges near the tip of the hind-wing will usually distinguish this species from other brown lacewings without reference to venation. April–September in dense, low-growing vegetation nearly everywhere, including gardens, but rare in the north.

DREPANEPTERYX PHALAENOIDES *c.* 25mm

This uncommon lacewing cannot be confused with any other species because of its hooked wing-tips, although when at rest it can easily be mistaken for a dead leaf. May–September, mainly in and around light woodland.

ALDERFLIES: ORDER MEGALOPTERA

SIALIS LUTARIA Family Sialidae 10–15mm

This is the commonest of the 3 British species, which can be reliably separated only by examining the genitalia. April–October, breeding in still and slow-moving water with plenty of bottom mud. Our other 2 alderflies prefer running water. All are weak-flying insects with smoky wings and thick, dark veins and they rarely move far from water. Some caddis flies (*see* p. 318) are superficially similar but have hairy wings and never have the series of short veins seen along the front edge of the forewing. Caddis flies also have much longer legs, with prominent spurs.

SNAKEFLIES: ORDER RAPHIDIOPTERA

ATLANTORAPHIDIA MACULICOLLIS Family Raphidiidae 10–15mm

Britain's 4 species of snakefly are all much alike, although they can be separated by examining the venation. The pterostigma in the forewing of this species starts about halfway along the elongated cell behind it and extends well beyond that cell. It contains just 1 cross-vein. May–August, mainly associated with pine trees. ***Phaeostigma notata*** resembles *A. maculicollis* but is a little larger and associated mainly with oak trees. The pterostigma normally has at least 2 cross-veins and does not extend far beyond the cell behind it.

XANTHOSTIGMA XANTHOSTIGMA *c.* 15mm

This species has a pale yellow pterostigma, starting at about the same point as the cell behind it and containing just 1 cross-vein. ***Subilla confinis*** is similar but the pterostigma is brown or black. Both species can be found on oak trees throughout the summer, but are not tied to them. All 4 species feed mainly on aphids and other small insects.

SCORPION FLIES: ORDER MECOPTERA

PANORPA GERMANICA Family Panorpidae 10–15mm

A B

This is probably the commonest of the 3 British scorpion flies. The density of spots varies; some northern specimens have no spots. May–September in shady places nearly everywhere. ***P. communis*** and ***P. cognata*** are very similar and, although *communis* tends to have heavier spotting, the 3 species are impossible to distinguish with certainty without examining the genitalia. All 3 are scavenging insects and, despite the male's scorpion-like tail (A), they are completely harmless. The female's abdomen simply tapers to a point (B).

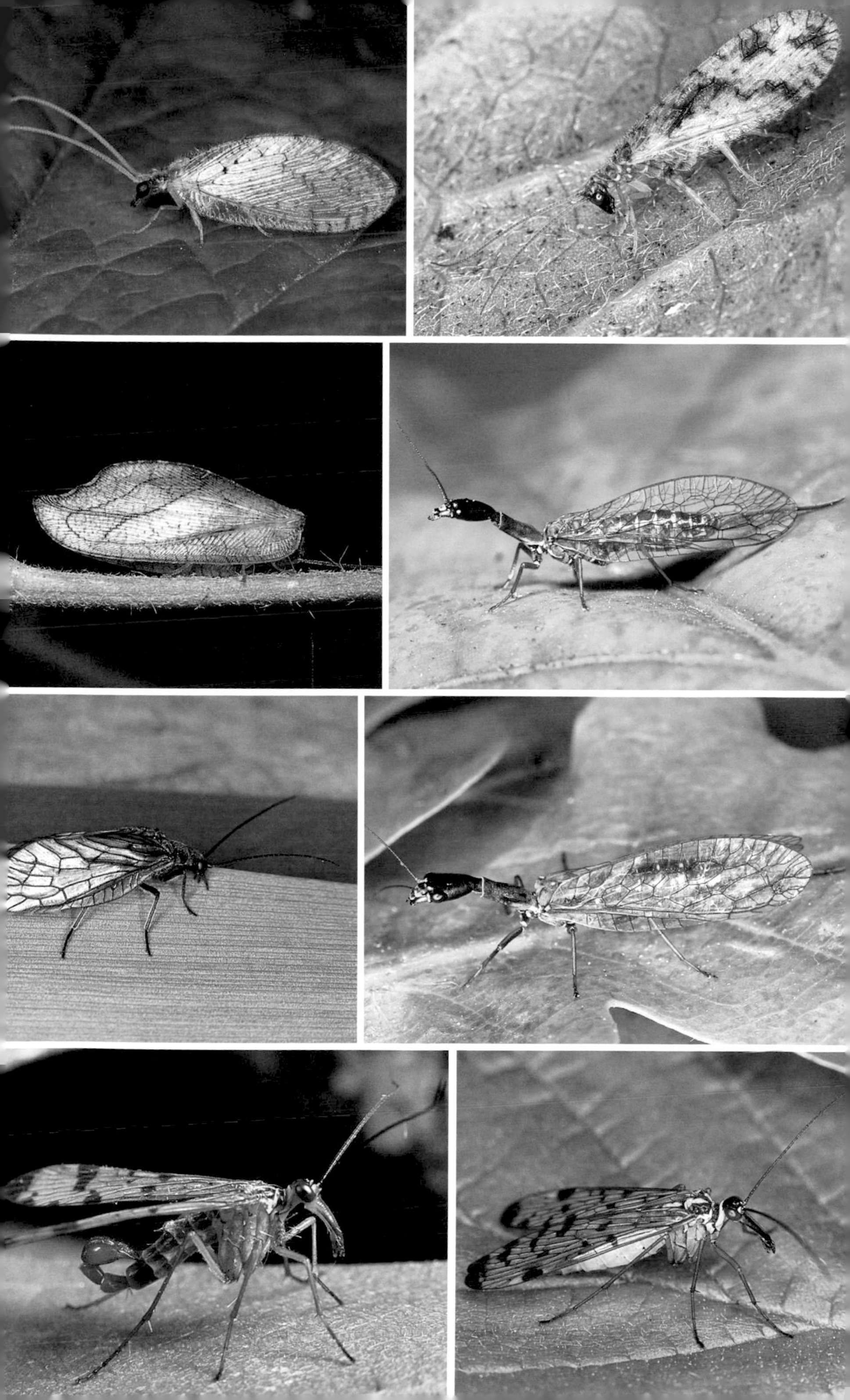

TIGER BEETLES AND GROUND BEETLES: FAMILY CARABIDAE

These fast-running, long-legged beetles are essentially carnivorous, feeding on a variety of other invertebrates. Long sensory bristles spring from various parts of their bodies but these are very slender and visible only at close quarters. Many species have brilliant metallic or iridescent colours. The tiger beetles and some other diurnal species fly well, but many ground beetles are flightless, with no hindwings and often with their elytra fused together. There are about 350 species in the British Isles, differing widely in shape and ranging from about 4–30mm in length.

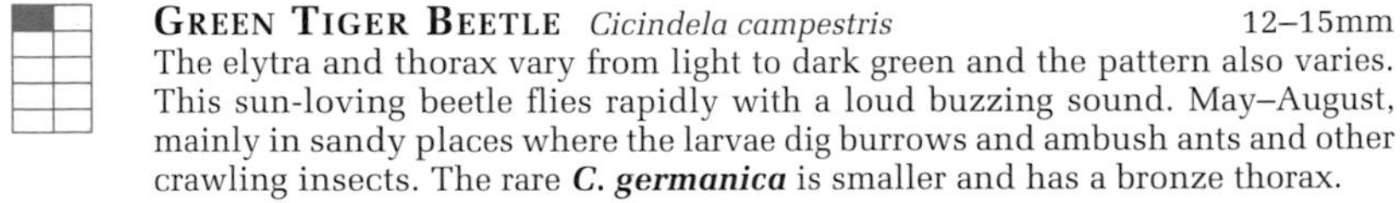

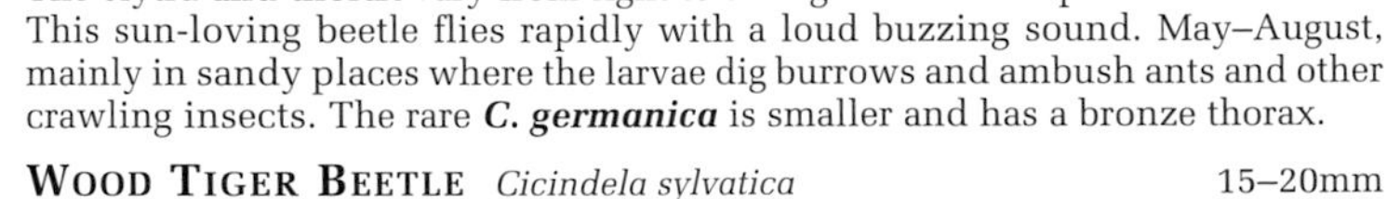

GREEN TIGER BEETLE *Cicindela campestris* 12–15mm
The elytra and thorax vary from light to dark green and the pattern also varies. This sun-loving beetle flies rapidly with a loud buzzing sound. May–August, mainly in sandy places where the larvae dig burrows and ambush ants and other crawling insects. The rare ***C. germanica*** is smaller and has a bronze thorax.

WOOD TIGER BEETLE *Cicindela sylvatica* 15–20mm
The elytra have a strong purplish tinge and 2 isolated cream spots at the rear. The underside is metallic blue or violet. May–September in open pine woods and on heathland, where the larvae behave like those of the Green Tiger Beetle.

CICINDELA HYBRIDA 12–15mm
This beetle is very much like the Wood Tiger Beetle but slightly smaller and more rust-coloured. The cream markings are also more extensive and the underside is metallic green. Heaths and dunes from April–October.

CICINDELA MARITIMA 12–15mm
This species is more slender than *C. hybrida* but the most obvious difference is the sharp backward bend in the pale stripe in the middle of each elytron. April–September, almost entirely confined to bare coastal dunes.

CALOSOMA INQUISITOR 15–22mm
The parallel-sided elytra are usually bronze-coloured with green or red iridescence and a green outer flange, but are sometimes almost black. The flange on the pronotum does not reach the rear margin. The underside is brassy green. May–August in deciduous woodland, especially oak woods, where adults and larvae all feed on caterpillars in the trees. The adults fly well and are largely diurnal.

CALOSOMA SYCOPHANTA 25–30mm
This strong-flying, diurnal beetle is easily recognised by its shiny, more-or-less parallel-sided elytra, which are usually golden green but often exhibit a strong blue or purplish tinge. It lives in trees, mainly in deciduous woodlands, where both adults and larvae are voracious predators of caterpillars. A rare visitor to Britain from the continent, seen mainly in the summer months.

VIOLET GROUND BEETLE *Carabus violaceus* 20–30mm
The almost smooth elytra have a strong violet tinge that often becomes coppery on their flanged edges. The thorax has a similar marginal sheen. All year in many habitats, including gardens, hiding under logs and stones by day and emerging to hunt slugs and other invertebrates at night.

CARABUS NEMORALIS 20–25mm
The bronzy green to black elytra bear numerous narrow ridges and each elytron has 3 rows of small pits. The margins are violet or coppery. All year in many habitats, including gardens, with habits like those of the Violet Ground Beetle.

BADISTER UNIPUSTULATUS 6–9mm
The elytra and heart-shaped pronotum are yellow or brick-coloured and each elytron has 1 large and 2 smaller black spots. Common all year in moss and leaf litter and under dead bark, mainly in damp habitats. ***B. bipustulatus*** is a little smaller and the dark spots are often linked to enclose a pale heart-shaped patch near the rear.

CALATHUS MELANOCEPHALUS 6–9mm
With its brick-coloured pronotum sandwiched between the black head and elytra, this beetle should not be confused with any other. All year in fairly dry grassland and other open habitats.

CYCHRUS CARABOIDES 15–20mm

This beetle has a markedly convex, pear-shaped body, with a very narrow head. It feeds on snails, plunging its narrow front end deep into their shells while eating the flesh. All year, mainly in woodland habitats.

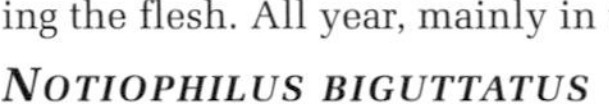

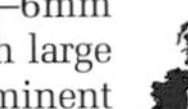

NOTIOPHILUS BIGUTTATUS 5–6mm

This is the commonest of several rather similar diurnal ground beetles with large eyes and a shiny strip running the length of each elytron. There are two prominent pits on each elytron. Most of the leg segments are bronze, but the tibiae are yellow. All year in many habitats, but dormant in winter.

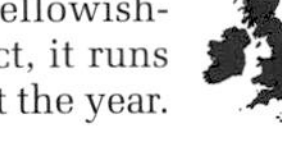

ELAPHRUS RIPARIUS 6–8mm

The shiny, mirror-like eye-spots on the green elytra, together with the yellowish-brown legs, distinguish this beetle from all others. A sun-loving insect, it runs down prey on the sand and mud around ponds and streams throughout the year. It also flies well.

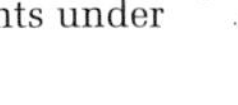

DROMIUS QUADRIMACULATUS 5–7mm

The elytral pattern of this rather flat beetle resembles that of *Badister unipustulatus* (*see* p. 110), but the pale spots are much yellower and the antennae are entirely yellow. All year, mainly in deciduous woodland, where it hunts under loose bark; dormant under bark in winter.

HARPALUS AFFINIS 9–12mm

The elytra range from metallic green or blue, through bronzy red, to almost black, with the outer areas punctate and slightly pubescent. The antennae and legs are reddish yellow. All year in gardens and almost any other open situation. ***H. rubripes*** has darker legs and the elytra are not at all pubescent. ***Chlaenius*** species have similar colours but the elytra are pubescent all over.

PTEROSTICHUS MADIDUS 12–18mm

As in most *Pterostichus* species, the pronotum and elytra are shiny black, but this common flightless species has chestnut femora and the rest of the leg segments are commonly chestnut as well. The rear corners of the pronotum are blunt. All year in most open habitats, including gardens where it augments its animal food by nibbling fruit, especially strawberries.

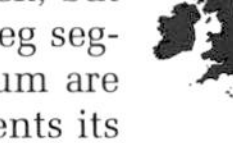

PTEROSTICHUS NIGRITA 8–13mm

This fully-winged beetle is completely black and the rear corners of the pronotum are pointed. All year in damp habitats almost everywhere. There are many similar species, although not all are fully winged and capable of flight.

PTEROSTICHUS CUPREUS 9–14mm

This fully-winged, diurnal beetle ranges from golden green, through coppery red and violet, to almost black, and the first 2 antennal segments are noticeably brown. All year in open grassland, especially in damp areas.

AGONUM SEXPUNCTATUM 7–10mm

With its head and pronotum green and elytra coppery-red, this beetle is easy to recognise. It is named for the row of pits, usually 6, on each elytron, although these are not always easy to see. All year in damp places, especially on the banks of ponds and streams, but dormant in winter.

ODACANTHA MELANURA 6–8mm

The slender blue-green thorax and the more-or-less cylindrical abdominal region give this beetle a bottle-like shape and make it very easy to recognise. It is fully-winged and the dark tips of the elytra reflect blue and green. All year on reeds and other tall waterside vegetation, but dormant in winter.

BOMBARDIER BEETLE *Brachinus crepitans* 6–10mm

The brick-coloured head and thorax and the dark elytra with green or blue iridescence distinguish this beetle from all other species. The beetle gets its name for its extraordinary habit of firing a salvo of hot liquid at its adversaries (*see* p. 46). All year in dry, open country, especially on chalk grassland, but dormant in winter and most often seen in early summer.

ROVE BEETLES: FAMILY STAPHYLINIDAE

These beetles all have very short elytra that leave much of the abdomen exposed. Some have a superficial resemblance to earwigs (*see* p. 84) although their cerci, when present, are never curved and pincer-like. Despite their small size, the elytra generally conceal fully-developed hindwings and most species fly well. Black or black and red are the dominant colours. Adults and larvae are mostly predators or omnivorous scavengers. There are nearly 1,000 British species and they include some of our smallest beetles. Many of them are under 1mm long.

CREOPHILUS MAXILLOSUS 15–22mm

The head and pronotum are shiny black, but the grey and black hairs on the elytra and abdomen will always identify this powerful predator. May–October in many habitats, but the beetle is most common in open areas where it preys on other insects in dung. Rotting vegetation is another regular habitat. The much rarer ***Emus hirtus*** has similar habits but is easily identified by the golden hair on its head and on the rear half of its abdomen.

DEVIL'S COACH-HORSE *Ocypus olens* 20–30mm

This completely black beetle has longer antennae than *Creophilus* and occupies a much wider range of habitats, being common in woods, hedgerows and gardens throughout the year. It is not uncommon in older houses, where its habit of raising its tail and opening its large jaws causes some alarm. It feeds at night on slugs and other debris-inhabiting invertebrates.

PAEDERUS LITTORALIS 7–9mm

Metallic-blue elytra sandwiched between an orange pronotum and largely orange abdomen distinguish this species from most other rove beetles. A flightless predator, it hunts mainly on river banks and in other damp places. Other *Paederus* species are a little smaller and fully winged.

PHILONTHUS MARGINATUS 7–9mm

The orange sides of the pronotum distinguish this rove beetle from other black species. It hunts fly grubs and other insects in dung and decaying vegetation and can be found throughout the year.

STENUS GUTTULA 5–7mm

This is one of several very similar species with an orange or yellow spot on each elytron. All have eyes that completely occupy the sides of the head. They hunt by day around the edges of ponds and streams and can also skim over the water surface: oily secretions released from the rear end lower the surface tension there and the higher surface tension at the front then draws the beetle smartly forward. All year, but rarely active in the winter. ***S. bimaculatus*** differs in having largely orange or yellowish legs.

NICROPHORUS HUMATOR Family Silphidae 15–30mm

This is one of the burying or sexton beetles that bury the bodies of small birds and mammals and then lay their eggs on them. Adults and larvae all feed on the decaying flesh and on the fly grubs and other scavengers that attack it. The antennae are abruptly clubbed and usually orange tipped, and the elytra are clearly truncated, although the beetles generally fly well – usually at night. *N. humator* is the only species with entirely black elytra and orange-tipped antennae. Spring to autumn in many habitats.

NICROPHORUS INVESTIGATOR 15–20mm

The posterior orange band in this species is almost complete, with just a small break where the elytra meet. It occurs in many habitats from spring to autumn. ***N. vespillo*** is very similar but can be recognised by its strongly curved hind tibiae. There are several other species with slightly different patterns.

SILPHA ATRATA 10–15mm

The narrow head of this glossy black beetle is an adaptation to its diet of snails, enabling it to reach deep into the shells to devour the flesh. All year in damp places, although dormant in winter.

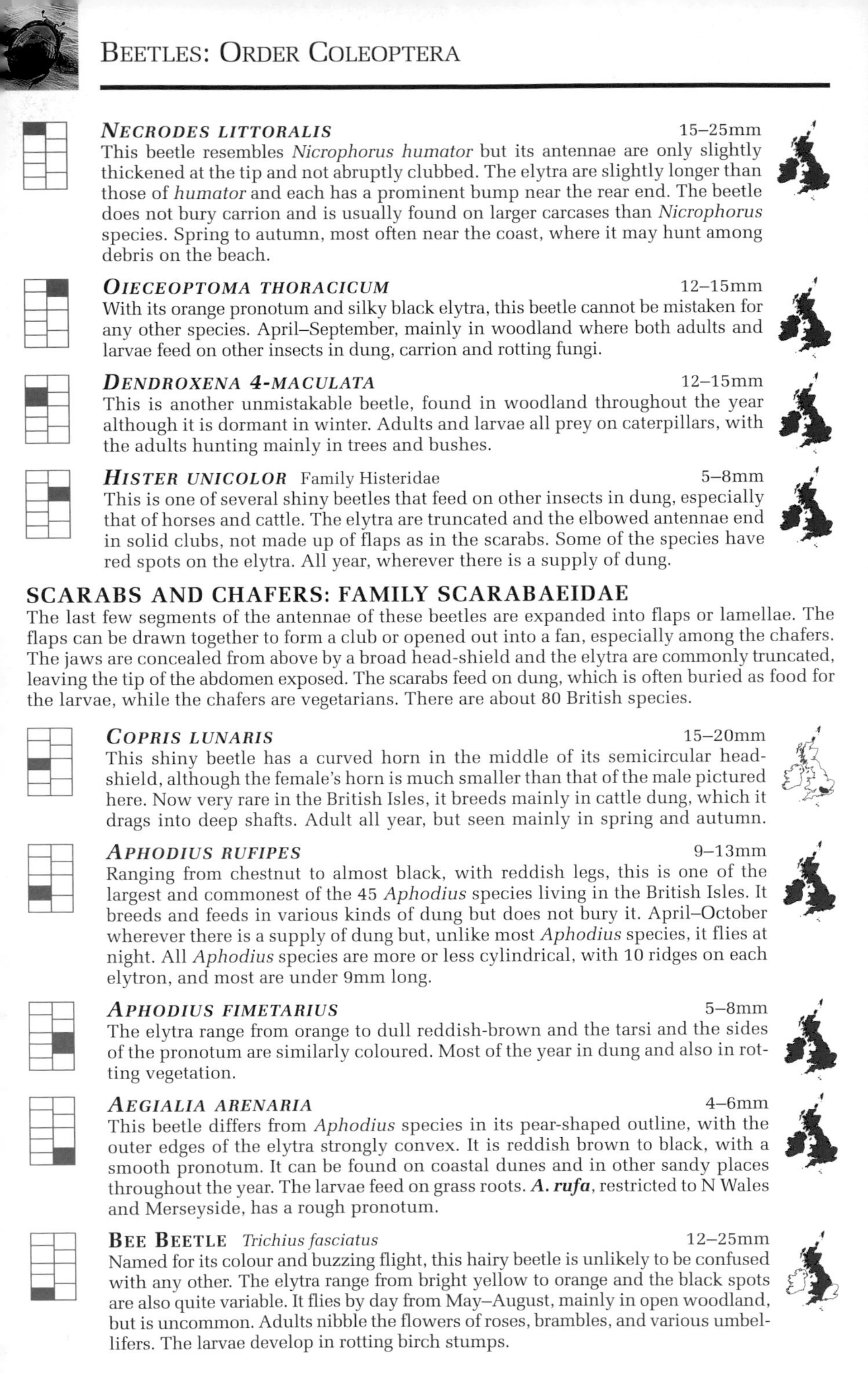

NECRODES LITTORALIS 15–25mm
This beetle resembles *Nicrophorus humator* but its antennae are only slightly thickened at the tip and not abruptly clubbed. The elytra are slightly longer than those of *humator* and each has a prominent bump near the rear end. The beetle does not bury carrion and is usually found on larger carcases than *Nicrophorus* species. Spring to autumn, most often near the coast, where it may hunt among debris on the beach.

OIECEOPTOMA THORACICUM 12–15mm
With its orange pronotum and silky black elytra, this beetle cannot be mistaken for any other species. April–September, mainly in woodland where both adults and larvae feed on other insects in dung, carrion and rotting fungi.

DENDROXENA 4-MACULATA 12–15mm
This is another unmistakable beetle, found in woodland throughout the year although it is dormant in winter. Adults and larvae all prey on caterpillars, with the adults hunting mainly in trees and bushes.

HISTER UNICOLOR Family Histeridae 5–8mm
This is one of several shiny beetles that feed on other insects in dung, especially that of horses and cattle. The elytra are truncated and the elbowed antennae end in solid clubs, not made up of flaps as in the scarabs. Some of the species have red spots on the elytra. All year, wherever there is a supply of dung.

SCARABS AND CHAFERS: FAMILY SCARABAEIDAE

The last few segments of the antennae of these beetles are expanded into flaps or lamellae. The flaps can be drawn together to form a club or opened out into a fan, especially among the chafers. The jaws are concealed from above by a broad head-shield and the elytra are commonly truncated, leaving the tip of the abdomen exposed. The scarabs feed on dung, which is often buried as food for the larvae, while the chafers are vegetarians. There are about 80 British species.

COPRIS LUNARIS 15–20mm
This shiny beetle has a curved horn in the middle of its semicircular head-shield, although the female's horn is much smaller than that of the male pictured here. Now very rare in the British Isles, it breeds mainly in cattle dung, which it drags into deep shafts. Adult all year, but seen mainly in spring and autumn.

APHODIUS RUFIPES 9–13mm
Ranging from chestnut to almost black, with reddish legs, this is one of the largest and commonest of the 45 *Aphodius* species living in the British Isles. It breeds and feeds in various kinds of dung but does not bury it. April–October wherever there is a supply of dung but, unlike most *Aphodius* species, it flies at night. All *Aphodius* species are more or less cylindrical, with 10 ridges on each elytron, and most are under 9mm long.

APHODIUS FIMETARIUS 5–8mm
The elytra range from orange to dull reddish-brown and the tarsi and the sides of the pronotum are similarly coloured. Most of the year in dung and also in rotting vegetation.

AEGIALIA ARENARIA 4–6mm
This beetle differs from *Aphodius* species in its pear-shaped outline, with the outer edges of the elytra strongly convex. It is reddish brown to black, with a smooth pronotum. It can be found on coastal dunes and in other sandy places throughout the year. The larvae feed on grass roots. ***A. rufa***, restricted to N Wales and Merseyside, has a rough pronotum.

BEE BEETLE *Trichius fasciatus* 12–25mm
Named for its colour and buzzing flight, this hairy beetle is unlikely to be confused with any other. The elytra range from bright yellow to orange and the black spots are also quite variable. It flies by day from May–August, mainly in open woodland, but is uncommon. Adults nibble the flowers of roses, brambles, and various umbellifers. The larvae develop in rotting birch stumps.

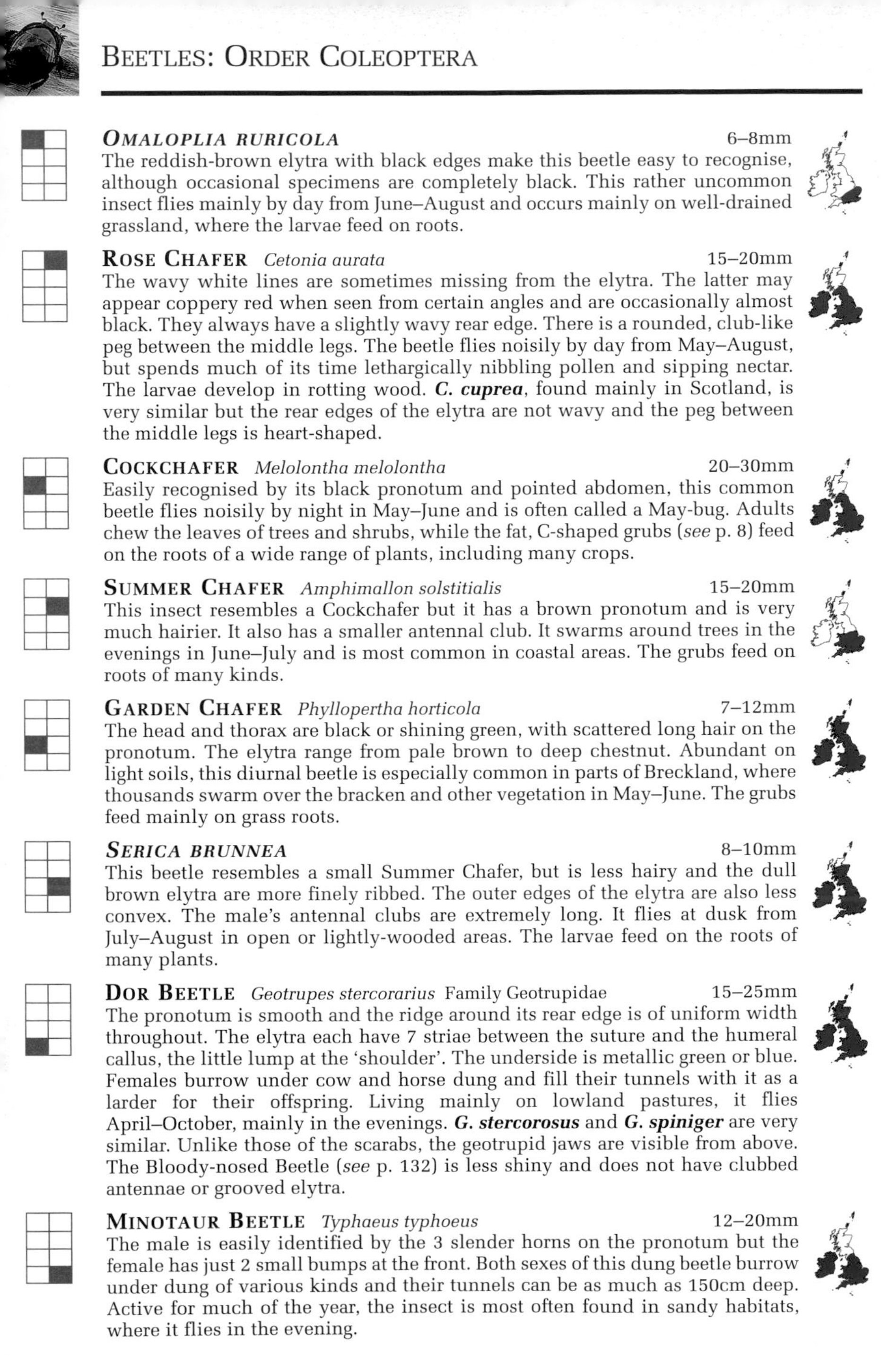

OMALOPLIA RURICOLA 6–8mm
The reddish-brown elytra with black edges make this beetle easy to recognise, although occasional specimens are completely black. This rather uncommon insect flies mainly by day from June–August and occurs mainly on well-drained grassland, where the larvae feed on roots.

ROSE CHAFER *Cetonia aurata* 15–20mm
The wavy white lines are sometimes missing from the elytra. The latter may appear coppery red when seen from certain angles and are occasionally almost black. They always have a slightly wavy rear edge. There is a rounded, club-like peg between the middle legs. The beetle flies noisily by day from May–August, but spends much of its time lethargically nibbling pollen and sipping nectar. The larvae develop in rotting wood. ***C. cuprea***, found mainly in Scotland, is very similar but the rear edges of the elytra are not wavy and the peg between the middle legs is heart-shaped.

COCKCHAFER *Melolontha melolontha* 20–30mm
Easily recognised by its black pronotum and pointed abdomen, this common beetle flies noisily by night in May–June and is often called a May-bug. Adults chew the leaves of trees and shrubs, while the fat, C-shaped grubs (*see* p. 8) feed on the roots of a wide range of plants, including many crops.

SUMMER CHAFER *Amphimallon solstitialis* 15–20mm
This insect resembles a Cockchafer but it has a brown pronotum and is very much hairier. It also has a smaller antennal club. It swarms around trees in the evenings in June–July and is most common in coastal areas. The grubs feed on roots of many kinds.

GARDEN CHAFER *Phyllopertha horticola* 7–12mm
The head and thorax are black or shining green, with scattered long hair on the pronotum. The elytra range from pale brown to deep chestnut. Abundant on light soils, this diurnal beetle is especially common in parts of Breckland, where thousands swarm over the bracken and other vegetation in May–June. The grubs feed mainly on grass roots.

SERICA BRUNNEA 8–10mm
This beetle resembles a small Summer Chafer, but is less hairy and the dull brown elytra are more finely ribbed. The outer edges of the elytra are also less convex. The male's antennal clubs are extremely long. It flies at dusk from July–August in open or lightly-wooded areas. The larvae feed on the roots of many plants.

DOR BEETLE *Geotrupes stercorarius* Family Geotrupidae 15–25mm
The pronotum is smooth and the ridge around its rear edge is of uniform width throughout. The elytra each have 7 striae between the suture and the humeral callus, the little lump at the 'shoulder'. The underside is metallic green or blue. Females burrow under cow and horse dung and fill their tunnels with it as a larder for their offspring. Living mainly on lowland pastures, it flies April–October, mainly in the evenings. ***G. stercorosus*** and ***G. spiniger*** are very similar. Unlike those of the scarabs, the geotrupid jaws are visible from above. The Bloody-nosed Beetle (*see* p. 132) is less shiny and does not have clubbed antennae or grooved elytra.

MINOTAUR BEETLE *Typhaeus typhoeus* 12–20mm
The male is easily identified by the 3 slender horns on the pronotum but the female has just 2 small bumps at the front. Both sexes of this dung beetle burrow under dung of various kinds and their tunnels can be as much as 150cm deep. Active for much of the year, the insect is most often found in sandy habitats, where it flies in the evening.

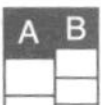

STAG BEETLE *Lucanus cervus* Family Lucanidae 30–70mm (including the antlers)

One of Britain's bulkiest insects, the male (A) is unmistakable. It is said to use its enormously enlarged jaws for grappling with other males when disputing 'ownership' of females, although such behaviour is rarely observed in the wild. The female (B) has a similar body but lacks the big jaws. The middle tibiae of both sexes bear 3 small spines. The beetle breeds in dead tree stumps and other rotting wood, including large fence posts, but has become rare in recent decades. Adults feed on ripe fruit and oozing sap and fly June–August, usually around dusk.

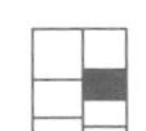

LESSER STAG BEETLE *Dorcus parallelopipedus* 20–30mm

This species resembles the female Stag Beetle but the elytra are blacker and the middle tibiae have only 1 spine. The pronotum is more-or-less parallel-sided and not much wider than the head. It breeds in rotting timber, mainly in deciduous woodland. The nocturnal adults feed largely on sap and can be found throughout the year, although they fly mainly in summer.

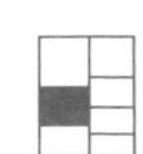

SINODENDRON CYLINDRICUM 10–20mm

This distinctly cylindrical beetle has 5 forward-pointing projections on the front of the pronotum. The male also has a curved horn on its head. It breeds in dead wood of various kinds. Adults can be found throughout the year but are most likely to be seen feeding on sap on summer evenings.

CLICK BEETLES: FAMILY ELATERIDAE

These slender and often rather downy beetles have the ability to spring into the air and right themselves if they fall on to their backs. They do not always land the right way up, but they go on leaping until they do. The action is accompanied by a loud click, from which the beetles get their name. The head is largely concealed by the pronotum, which usually bears sharp spines at the rear corners, and the antennae are toothed or comb-like. Adult click beetles feed mainly on pollen and nectar, while the short-legged larvae include herbivores, carnivores and omnivores. Some larvae live in decaying wood and many others, commonly known as wireworms, live in the soil and feed on roots, often causing serious damage to crops. There are about 65 British species.

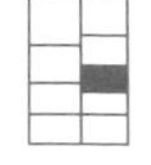

CTENICERA PECTINICORNIS 15–18mm

The very shiny elytra are usually deep green, but sometimes coppery. As in many other click beetles, the female antennae are much less comb-like than those of the male pictured here. May–July, mainly in damp grassland where the larvae feed on grass roots.

CTENICERA CUPREA 12–15mm

The pronotum is always coppery or violet, but the elytra exhibit 2 colour forms, entirely metallic or brown with a coppery or violet patch at the rear. May–July on rough grassland, especially in upland areas, where the larvae eat roots and various soil-dwelling animals.

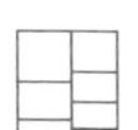

AMPEDUS BALTEATUS 7–9mm

The black pronotum and bicoloured elytra readily identify this species, found in and around coniferous woods and heathland. Adults frequent flowers, especially umbellifers, May–July. The larvae live in decaying stumps, usually those of coniferous trees.

AMPEDUS SANGUINOLENTUS *c.* 10mm

The black mark on the elytra is sometimes missing. Adults frequent flowers, especially hawthorn blossom, May–June and occur mainly in and around woods and heathland. The larvae live as predators in rotting timber. Specimens lacking the black mark could be confused with ***A. cinnabarinus***, although the latter is usually larger and the scarlet elytra are clothed with reddish-brown hairs. ***A. sanguineus***, associated mainly with coniferous woodland, has black hairs on the elytra.

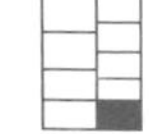

AGRYPNUS MURINA 12–18mm

This fairly broad and flat click beetle has dark brown or black elytra patterned with grey and light brown scales. The basal segment of each antenna is red and the rest is black. Reddish tarsi contrast with the black tibiae. April–August, in rough grassland, where the larvae feed on roots. Mainly coastal in north.

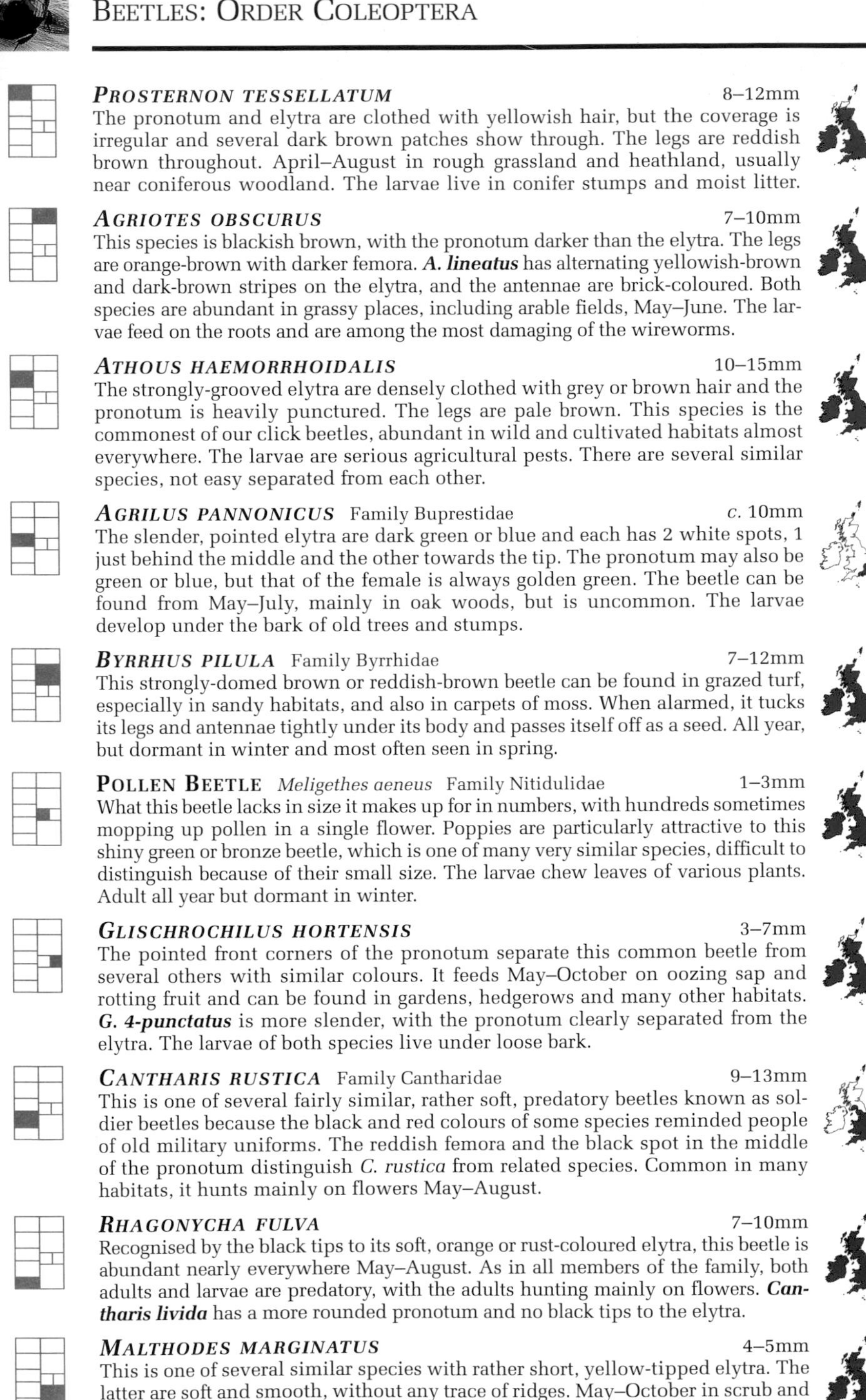

PROSTERNON TESSELLATUM 8–12mm
The pronotum and elytra are clothed with yellowish hair, but the coverage is irregular and several dark brown patches show through. The legs are reddish brown throughout. April–August in rough grassland and heathland, usually near coniferous woodland. The larvae live in conifer stumps and moist litter.

AGRIOTES OBSCURUS 7–10mm
This species is blackish brown, with the pronotum darker than the elytra. The legs are orange-brown with darker femora. ***A. lineatus*** has alternating yellowish-brown and dark-brown stripes on the elytra, and the antennae are brick-coloured. Both species are abundant in grassy places, including arable fields, May–June. The larvae feed on the roots and are among the most damaging of the wireworms.

ATHOUS HAEMORRHOIDALIS 10–15mm
The strongly-grooved elytra are densely clothed with grey or brown hair and the pronotum is heavily punctured. The legs are pale brown. This species is the commonest of our click beetles, abundant in wild and cultivated habitats almost everywhere. The larvae are serious agricultural pests. There are several similar species, not easy separated from each other.

AGRILUS PANNONICUS Family Buprestidae *c.* 10mm
The slender, pointed elytra are dark green or blue and each has 2 white spots, 1 just behind the middle and the other towards the tip. The pronotum may also be green or blue, but that of the female is always golden green. The beetle can be found from May–July, mainly in oak woods, but is uncommon. The larvae develop under the bark of old trees and stumps.

BYRRHUS PILULA Family Byrrhidae 7–12mm
This strongly-domed brown or reddish-brown beetle can be found in grazed turf, especially in sandy habitats, and also in carpets of moss. When alarmed, it tucks its legs and antennae tightly under its body and passes itself off as a seed. All year, but dormant in winter and most often seen in spring.

POLLEN BEETLE *Meligethes aeneus* Family Nitidulidae 1–3mm
What this beetle lacks in size it makes up for in numbers, with hundreds sometimes mopping up pollen in a single flower. Poppies are particularly attractive to this shiny green or bronze beetle, which is one of many very similar species, difficult to distinguish because of their small size. The larvae chew leaves of various plants. Adult all year but dormant in winter.

GLISCHROCHILUS HORTENSIS 3–7mm
The pointed front corners of the pronotum separate this common beetle from several others with similar colours. It feeds May–October on oozing sap and rotting fruit and can be found in gardens, hedgerows and many other habitats. ***G. 4-punctatus*** is more slender, with the pronotum clearly separated from the elytra. The larvae of both species live under loose bark.

CANTHARIS RUSTICA Family Cantharidae 9–13mm
This is one of several fairly similar, rather soft, predatory beetles known as soldier beetles because the black and red colours of some species reminded people of old military uniforms. The reddish femora and the black spot in the middle of the pronotum distinguish *C. rustica* from related species. Common in many habitats, it hunts mainly on flowers May–August.

RHAGONYCHA FULVA 7–10mm
Recognised by the black tips to its soft, orange or rust-coloured elytra, this beetle is abundant nearly everywhere May–August. As in all members of the family, both adults and larvae are predatory, with the adults hunting mainly on flowers. ***Cantharis livida*** has a more rounded pronotum and no black tips to the elytra.

MALTHODES MARGINATUS 4–5mm
This is one of several similar species with rather short, yellow-tipped elytra. The latter are soft and smooth, without any trace of ridges. May–October in scrub and grassland, especially in the vicinity of woodland. The individual species cannot be distinguished with certainty without looking at the genitalia.

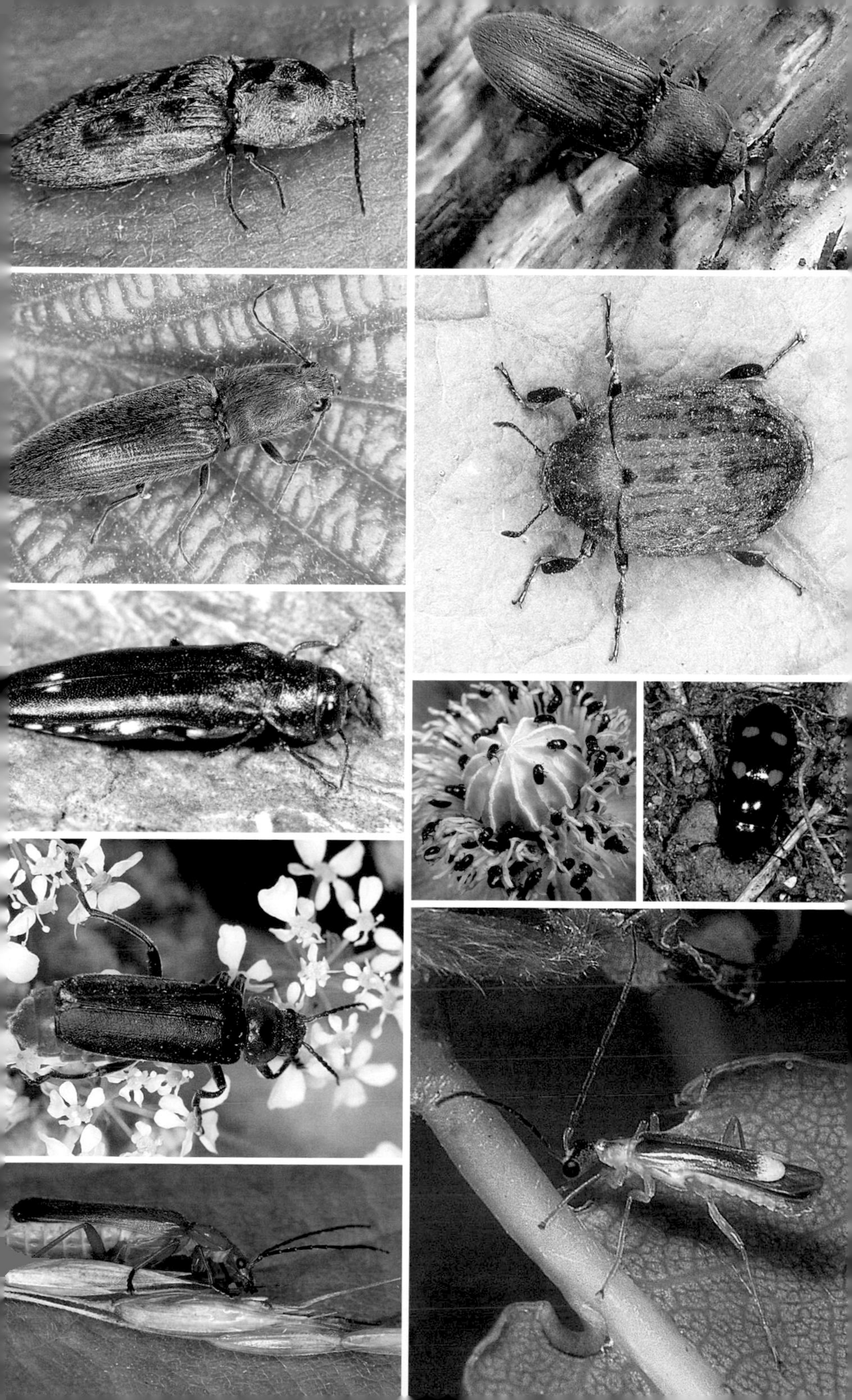

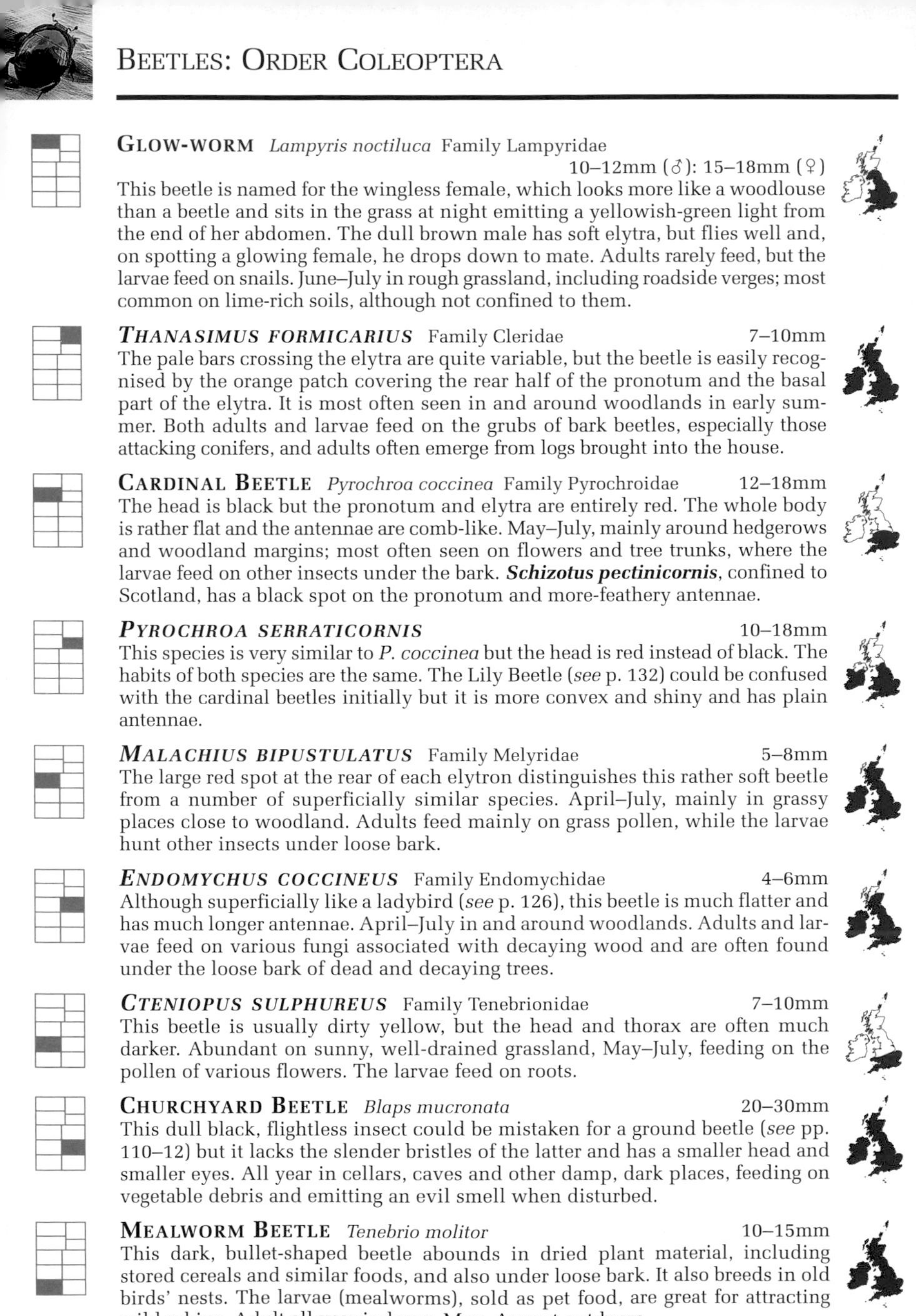

Glow-worm *Lampyris noctiluca* Family Lampyridae 10–12mm (♂): 15–18mm (♀)
This beetle is named for the wingless female, which looks more like a woodlouse than a beetle and sits in the grass at night emitting a yellowish-green light from the end of her abdomen. The dull brown male has soft elytra, but flies well and, on spotting a glowing female, he drops down to mate. Adults rarely feed, but the larvae feed on snails. June–July in rough grassland, including roadside verges; most common on lime-rich soils, although not confined to them.

Thanasimus formicarius Family Cleridae 7–10mm
The pale bars crossing the elytra are quite variable, but the beetle is easily recognised by the orange patch covering the rear half of the pronotum and the basal part of the elytra. It is most often seen in and around woodlands in early summer. Both adults and larvae feed on the grubs of bark beetles, especially those attacking conifers, and adults often emerge from logs brought into the house.

Cardinal Beetle *Pyrochroa coccinea* Family Pyrochroidae 12–18mm
The head is black but the pronotum and elytra are entirely red. The whole body is rather flat and the antennae are comb-like. May–July, mainly around hedgerows and woodland margins; most often seen on flowers and tree trunks, where the larvae feed on other insects under the bark. ***Schizotus pectinicornis***, confined to Scotland, has a black spot on the pronotum and more-feathery antennae.

Pyrochroa serraticornis 10–18mm
This species is very similar to *P. coccinea* but the head is red instead of black. The habits of both species are the same. The Lily Beetle (*see* p. 132) could be confused with the cardinal beetles initially but it is more convex and shiny and has plain antennae.

Malachius bipustulatus Family Melyridae 5–8mm
The large red spot at the rear of each elytron distinguishes this rather soft beetle from a number of superficially similar species. April–July, mainly in grassy places close to woodland. Adults feed mainly on grass pollen, while the larvae hunt other insects under loose bark.

Endomychus coccineus Family Endomychidae 4–6mm
Although superficially like a ladybird (*see* p. 126), this beetle is much flatter and has much longer antennae. April–July in and around woodlands. Adults and larvae feed on various fungi associated with decaying wood and are often found under the loose bark of dead and decaying trees.

Cteniopus sulphureus Family Tenebrionidae 7–10mm
This beetle is usually dirty yellow, but the head and thorax are often much darker. Abundant on sunny, well-drained grassland, May–July, feeding on the pollen of various flowers. The larvae feed on roots.

Churchyard Beetle *Blaps mucronata* 20–30mm
This dull black, flightless insect could be mistaken for a ground beetle (*see* pp. 110–12) but it lacks the slender bristles of the latter and has a smaller head and smaller eyes. All year in cellars, caves and other damp, dark places, feeding on vegetable debris and emitting an evil smell when disturbed.

Mealworm Beetle *Tenebrio molitor* 10–15mm
This dark, bullet-shaped beetle abounds in dried plant material, including stored cereals and similar foods, and also under loose bark. It also breeds in old birds' nests. The larvae (mealworms), sold as pet food, are great for attracting wild robins. Adult all year indoors, May–August outdoors.

Oedemera nobilis Family Oedemeridae 8–10mm
The narrow, gaping elytra distinguish this from other bright green beetles. Only the male has the swollen hind-legs. April–August in open habitats, feeding on the pollen of many flowers. The larvae live in the stems of various herbs. ***O. lurida*** is duller and the male lacks swollen hind-legs.

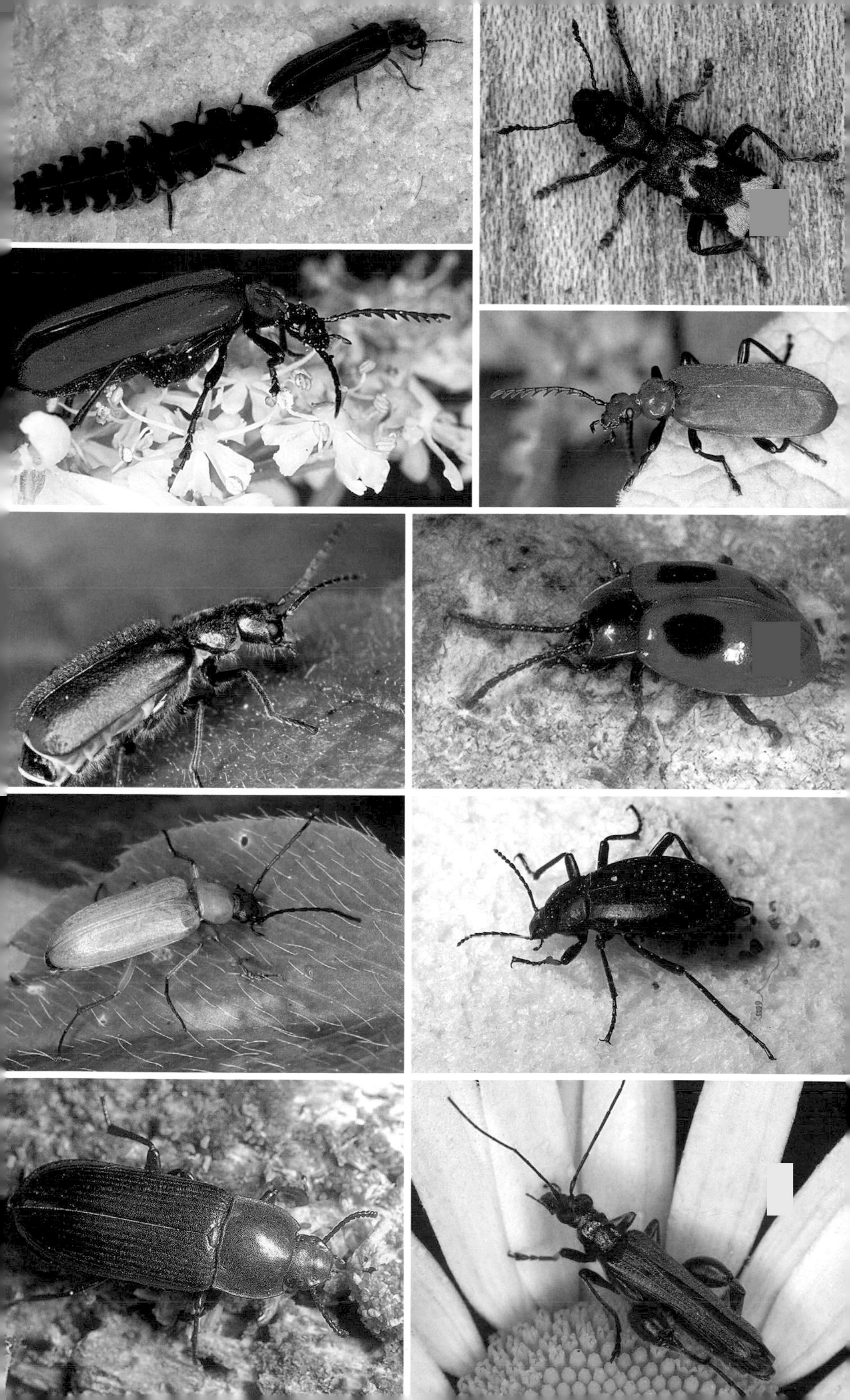

LADYBIRDS: FAMILY COCCINELLIDAE

The ladybirds are among the best-known and best-liked of the beetles because of their bright colours and the aphid-eating habits of most species, although not all are predatory. Most species are strongly domed and more or less hemispherical, with the head almost completely concealed by the pronotum. The antennae are short and slightly clubbed. Their bright colours advertise their bitter taste, and when handled most species exude a pungent fluid with a long-lasting odour (*see* p. 47). All the British ladybirds pass the winter as dormant adults, often sleeping in large aggregations. Eggs are usually laid in the spring and new adults begin to appear in early summer, so active ladybirds can be found from early spring until late in the autumn. Some leaf beetles (*see* pp. 132–4) may be confused with ladybirds, but most leaf beetles have longer antennae and their foot structure is quite different: a lens will reveal that in ladybirds the 2nd tarsal segment is enlarged and heart-shaped, while in leaf beetles it is the 3rd segment that is like this. The family contains 42 British species, although only 24 of these are true ladybirds.

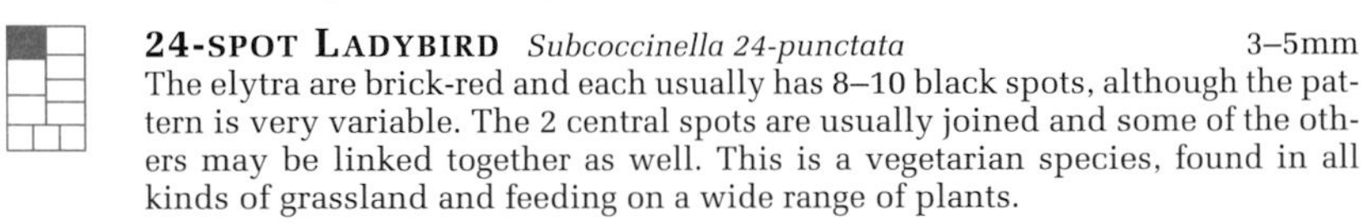

24-SPOT LADYBIRD *Subcoccinella 24-punctata* 3–5mm
The elytra are brick-red and each usually has 8–10 black spots, although the pattern is very variable. The 2 central spots are usually joined and some of the others may be linked together as well. This is a vegetarian species, found in all kinds of grassland and feeding on a wide range of plants.

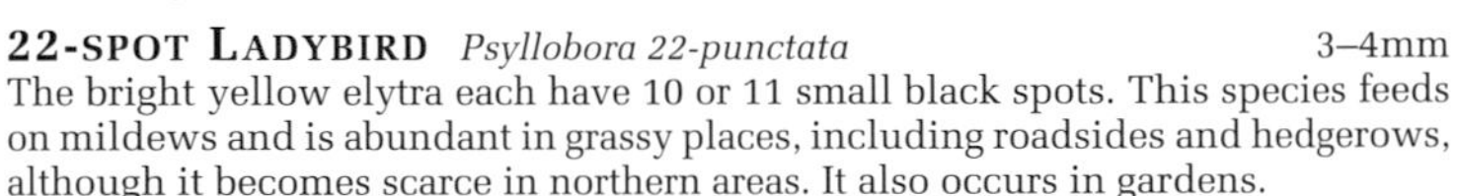

22-SPOT LADYBIRD *Psyllobora 22-punctata* 3–4mm
The bright yellow elytra each have 10 or 11 small black spots. This species feeds on mildews and is abundant in grassy places, including roadsides and hedgerows, although it becomes scarce in northern areas. It also occurs in gardens.

7-SPOT LADYBIRD *Coccinella 7-punctata* 5–8mm
The adult (A) is less variable than most other ladybirds and the 7 black spots rarely show any sign of fusion, although their size may vary a good deal. It is one of our commonest ladybirds, abundant in all kinds of habitats. The steely blue larva (B), resembling that of many other ladybirds, feeds on a wide variety of aphids.

2-SPOT LADYBIRD *Adalia 2-punctata* 3–6mm
The typical form has just 1 black spot on each elytron, but the species is *extremely* variable and often has black elytra bearing 2, 4, or 6 red spots. The legs are black. Abundant nearly everywhere.

10-SPOT LADYBIRD *Adalia 10-punctata* 3–6mm
The typical form has a fairly pale pronotum and 10 black spots on red elytra, but the species is extremely variable and is very often black with just 2 red spots. Other individuals have the black spots linked up to form a network. The legs are always reddish brown. Woods and hedges are the main habitats of this common species.

PINE LADYBIRD *Exochomus 4-pustulatus* 3–5mm
Also called the 4-spotted Ladybird, this species usually has 4 red spots, the front 2 of which are comma-shaped. There is a distinct rim around the edges of the elytra. It is very common on pines and other conifers in the S, but rare in the N.

EYED LADYBIRD *Anatis ocellata* 8–10mm
The size and the pale rings around the black spots will normally identify our largest ladybird, although the rings are sometimes absent. There are most commonly 18 spots, although the number varies. Common on and under pine trees.

CREAM-SPOT LADYBIRD *Calvia 14-guttata* 4–6mm
There are always 7 cream spots on each brick-red elytron, those just before the middle aligned in a fairly clear arc. Common in deciduous trees and hedgerows. The much rarer **Orange Ladybird** (*Halyza 16-guttata*) is usually paler and never has 3 spots in an arc.

16-SPOT LADYBIRD *Micraspis 16-punctata* c. 2.5mm
The ground colour of this mildew-feeder is dull yellow or cream, often heavily dusted with grey, and there is a broad black line along the suture. There are eight black spots on each elytron, with the outer three on each side usually joined in a line. Common in grassland in the south but rare in the north. The *Propylea 14-punctata* is larger and has a crown-like pattern on the pronotum..

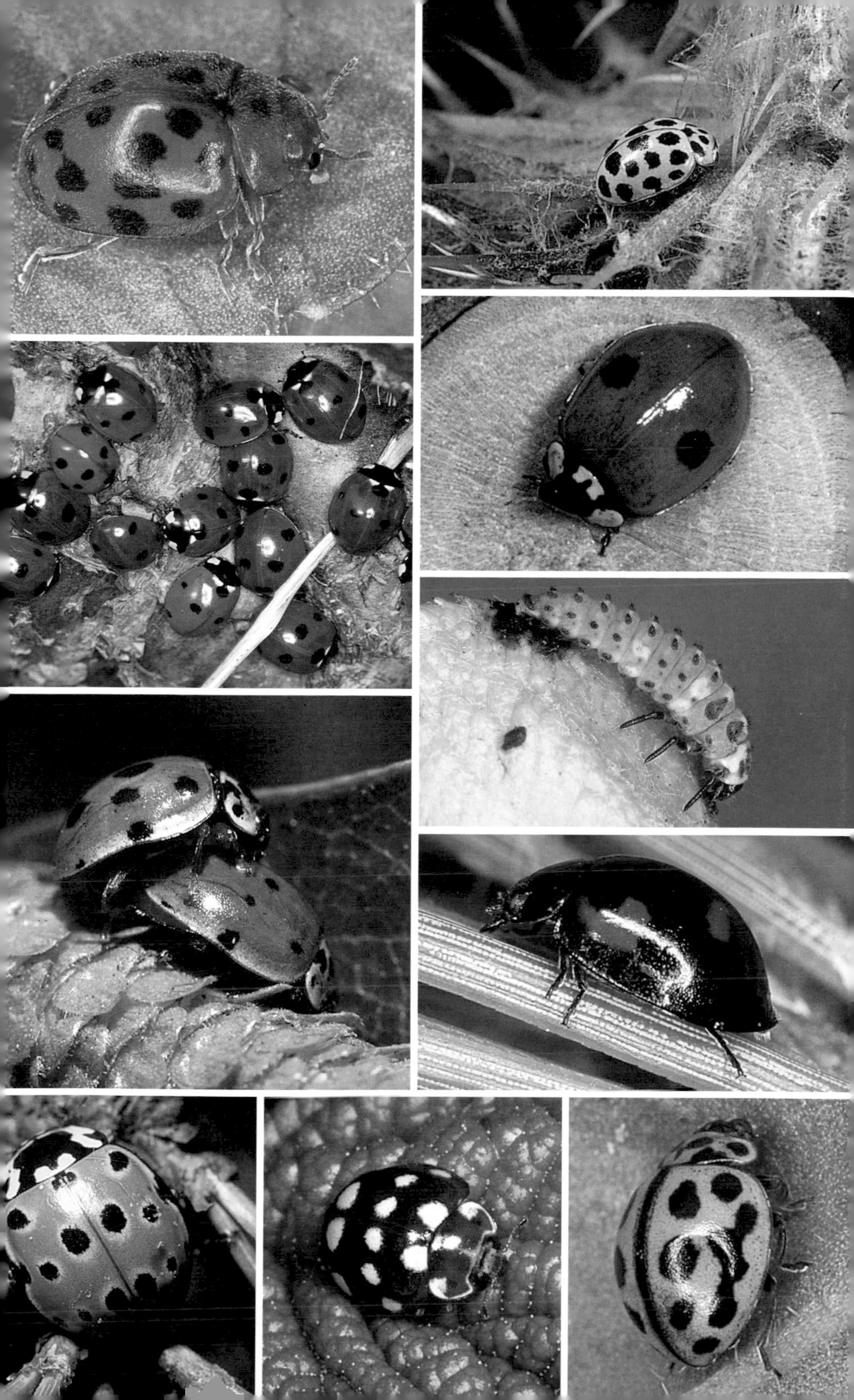

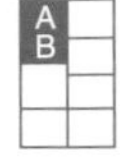

VARIED CARPET BEETLE *Anthrenus verbasci* Family Dermestidae 1.5–3.5mm
The elytra of this common little beetle are clothed with grey, yellow and black scales, and the legs are black. The adults (A) feed mainly on pollen and congregate on flowers, but the larvae (B), known as Woolly Bears, feed on dried animal matter, including woollen carpets and insect collections. The insects often breed in old birds' nests. The **Museum Beetle** (*A. museorum*) has rust-coloured legs.

FURNITURE BEETLE *Anobium punctatum* Family Anobiidae 3–5mm
With its head tucked right under the humped pronotum, this beetle is easily mistaken for a seed. Its larva is the infamous woodworm that does so much damage to furniture and building timbers, but the insect is equally abundant in the wild, where it plays a major role in breaking down dead wood. Adults usually emerge from the timber May–July, leaving exit holes about 2mm across.

DEATH-WATCH BEETLE *Xestobium rufovillosum* 5–9mm
Hairs on the elytra are responsible for the mottled appearance of this destructive beetle, which, like the Furniture Beetle, hides its head under the pronotum. It breeds in old timber, especially oak, and its larvae cause severe damage in old buildings. Adults emerge through holes 3–4mm across April–June and attract mates by tapping their heads on the surrounding timber.

SPANISH FLY *Lytta vesicatoria* Family Meloidae 10–20mm
The iridescent green elytra, each with 2 prominent ridges, may have a golden or bluish sheen, and the abdomen often projects well beyond them. June–July, feeding on ash and privet leaves, but uncommon. It gives out a strong mouse-like smell, and also emits a blistering liquid when alarmed. The larvae live in the nests of various solitary bees, feeding on the stored pollen. The Musk Beetle (*see* below) may look similar but has much longer antennae.

OIL BEETLE *Meloe proscarabaeus* 10–30mm
This is one of several flightless beetles with short, gaping elytra. They emit a pungent oily liquid when alarmed. Males are smaller than females and, in this species, the male antennae are swollen in the middle. May–July, chewing leaves in grassy places. The larvae live in the nests of solitary bees. ***M. violaceus*** is noticeably bluer, while ***M. variegatus*** is metallic green with a coppery sheen.

LONGHORN BEETLES: FAMILY CERAMBYCIDAE

Most of these beetles have fairly slender bodies and very long antennae, those of the males often being longer than the body. Most longhorns fly well, often at dusk, and many stridulate loudly by rubbing one part of the thorax against another. The larvae nearly all feed in timber and some are serious forest pests. They may take several years to mature and adult size depends on the quality of the larval food. The adults feed mainly on pollen, although some hardly feed at all. There are about 70 British species, mostly associated with wooded habitats.

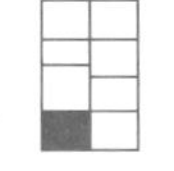

TANNER BEETLE *Prionus coriarius* 20–50mm
This is our largest longhorn and unlikely to be mistaken for any other species. July–August, mainly in wooded areas, although not common. The larvae live in the roots of various trees.

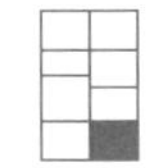

MUSK BEETLE *Aromia moschata* 10–25mm
Named for its musk-like scent, this beetle ranges from bright green to blackish bronze, sometimes with a coppery or bluish sheen. The pronotum bears a large spine on each side. June–August in the vicinity of willows. The Spanish Fly (*see* above) has shorter antennae and no thoracic spines.

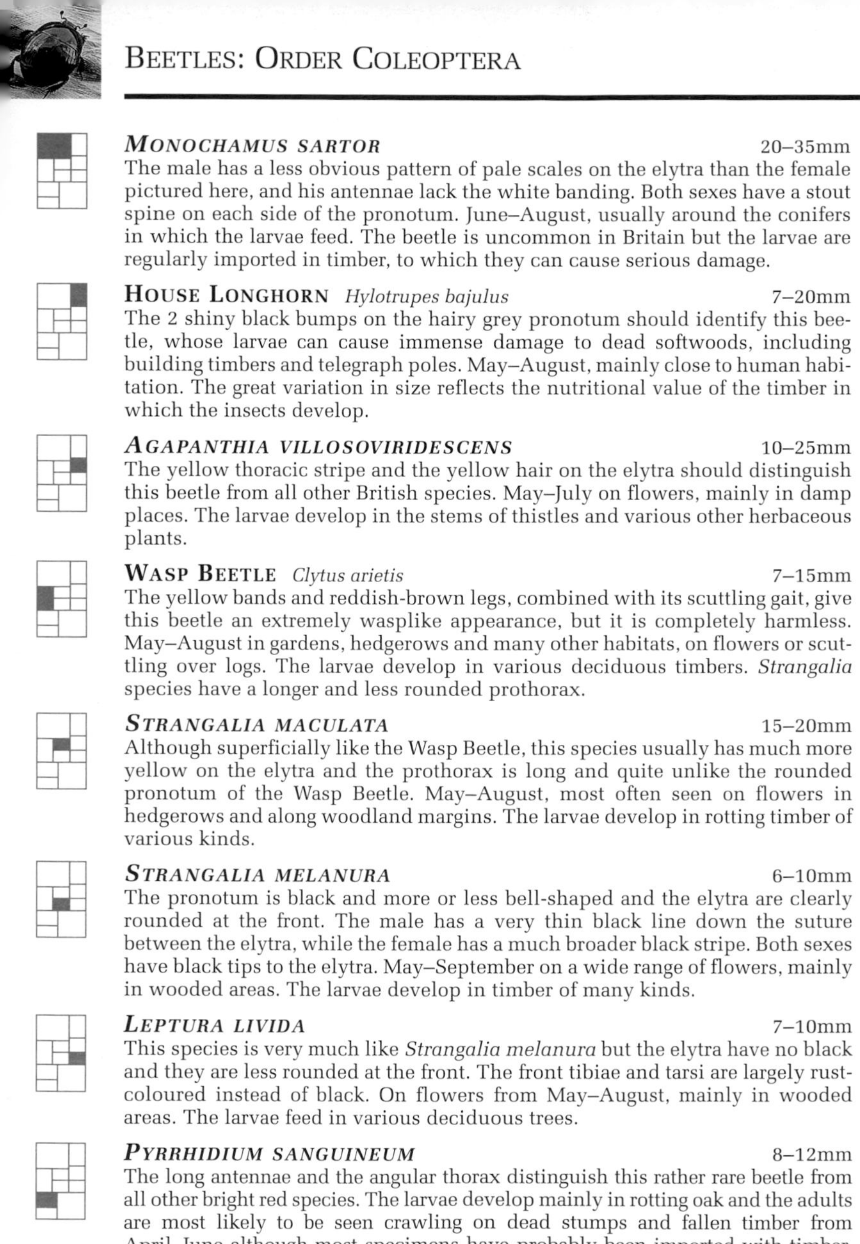

MONOCHAMUS SARTOR 20–35mm
The male has a less obvious pattern of pale scales on the elytra than the female pictured here, and his antennae lack the white banding. Both sexes have a stout spine on each side of the pronotum. June–August, usually around the conifers in which the larvae feed. The beetle is uncommon in Britain but the larvae are regularly imported in timber, to which they can cause serious damage.

HOUSE LONGHORN *Hylotrupes bajulus* 7–20mm
The 2 shiny black bumps on the hairy grey pronotum should identify this beetle, whose larvae can cause immense damage to dead softwoods, including building timbers and telegraph poles. May–August, mainly close to human habitation. The great variation in size reflects the nutritional value of the timber in which the insects develop.

AGAPANTHIA VILLOSOVIRIDESCENS 10–25mm
The yellow thoracic stripe and the yellow hair on the elytra should distinguish this beetle from all other British species. May–July on flowers, mainly in damp places. The larvae develop in the stems of thistles and various other herbaceous plants.

WASP BEETLE *Clytus arietis* 7–15mm
The yellow bands and reddish-brown legs, combined with its scuttling gait, give this beetle an extremely wasplike appearance, but it is completely harmless. May–August in gardens, hedgerows and many other habitats, on flowers or scuttling over logs. The larvae develop in various deciduous timbers. *Strangalia* species have a longer and less rounded prothorax.

STRANGALIA MACULATA 15–20mm
Although superficially like the Wasp Beetle, this species usually has much more yellow on the elytra and the prothorax is long and quite unlike the rounded pronotum of the Wasp Beetle. May–August, most often seen on flowers in hedgerows and along woodland margins. The larvae develop in rotting timber of various kinds.

STRANGALIA MELANURA 6–10mm
The pronotum is black and more or less bell-shaped and the elytra are clearly rounded at the front. The male has a very thin black line down the suture between the elytra, while the female has a much broader black stripe. Both sexes have black tips to the elytra. May–September on a wide range of flowers, mainly in wooded areas. The larvae develop in timber of many kinds.

LEPTURA LIVIDA 7–10mm
This species is very much like *Strangalia melanura* but the elytra have no black and they are less rounded at the front. The front tibiae and tarsi are largely rust-coloured instead of black. On flowers from May–August, mainly in wooded areas. The larvae feed in various deciduous trees.

PYRRHIDIUM SANGUINEUM 8–12mm

The long antennae and the angular thorax distinguish this rather rare beetle from all other bright red species. The larvae develop mainly in rotting oak and the adults are most likely to be seen crawling on dead stumps and fallen timber from April–June although most specimens have probably been imported with timber.

RHAGIUM BIFASCIATUM 12–22mm

Each elytron has 3 shiny raised lines and there are usually 2 oblique yellowish patches with a black or slightly rusty area between them. There is a sharp spine on each side of the pronotum and, as in all *Rhagium* species, the antennae are quite short for longhorn beetles. May–July, mainly in wooded areas where the larvae live in decaying coniferous timber.

RHAGIUM MORDAX 12–22mm

This species is much hairier than *R. bifasciatum* and the pale markings, varying from cream to deep orange, are less distinct. The thoracic spines are shorter and blunter. May–July, mainly on flowers in and around woodlands. The larvae develop under the bark of dead stumps and fallen trees.

LEAF BEETLES: FAMILY CHRYSOMELIDAE

Most of the beetles in this huge family are smooth and shiny, with rounded outlines. Many are brightly coloured. The tarsi are 5-segmented, but the 4th segment is very small and more or less hidden in the enlarged 3rd segment. This feature should distinguish the leaf beetles from the ladybirds (*see* p.126) and all other superficially similar beetles. Adults and larvae nearly all eat leaves and many species are serious agricultural and horticultural pests. The larvae are soft and slug-like and move very slowly on their short legs. There are over 250 British species.

DONACIA VULGARIS 6–8mm

The elytra of this beetle are metallic green or brassy and, when viewed from certain angles, there is a purplish or red stripe on each side of the suture. May–June on waterside plants, especially bur-reed. The larvae live in bur-reed stems. Several similar species live on other aquatic plants.

OULEMA MELANOPA 4–5mm

The head and elytra range from bright blue to deep navy or almost black. Abundant in grassy places April–September; the adults often bask on sunny walls and fences in the spring. Adults and their slimy black larvae feed on grasses and may cause damage to cereal crops.

ASPARAGUS BEETLE *Crioceris asparagi* 5–7mm

The usual impression is that of a black beetle with 3 yellow spots on each elytron, but the amount of black varies and some individuals have more yellow than black. There is always an orange border. All year, but dormant in winter. The beetle can ruin asparagus crops in some areas.

LILY BEETLE *Lilioceris lilii* 6–8mm

Adults and larvae chew lily leaves and flower buds April–August, and are serious pests in many gardens. The larvae cover themselves with their slimy black excrement and are often mistaken for slugs or even bird droppings. The cardinal beetles (*see* p. 124) are flatter and have toothed antennae.

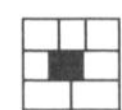

CLYTRA 4-PUNCTATA 7–12mm

With its stubby antennae and the 2 comma-shaped black spots near the rear, this beetle is not likely to be confused with any other British species. May–August, usually in wooded areas where the larvae live as scavengers in the nests of wood ants.

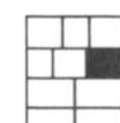

CRYPTOCEPHALUS HYPOCHAERIDIS *c.* 5mm

As the generic name suggests, the head of this very shiny beetle and its relatives is completely hidden under the pronotum. May–August in grassy habitats, where the beetles commonly sit on the flowers of hawkweeds and other yellow composites. The larvae feed on various plants.

BLOODY-NOSED BEETLE *Timarcha tenebricosa* 10–20mm

Britain's largest leaf beetle, this species gets its name for its habit of releasing a drop of red blood from its mouth when disturbed (*see* p. 47). A flightless species, it walks very slowly and feeds on bedstraws in open habitats. The long antennae and smooth elytra readily distinguish it from dung beetles.

CHRYSOMELA POPULI 10–12mm

The pronotum ranges from bronzy green to almost black and each elytron normally has a small black spot at the rear. May–August on poplars and sometimes on sallows.

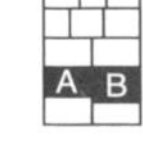

GASTROPHYSA VIRIDULA 4–8mm

This green beetle often has a strong golden or bluish sheen. The male (A) exhibits the typical leaf-beetle shape, but mated females (B) become grossly distended, with the elytra perched saddle-like on top of the shiny black abdomen. May–August on docks and related plants, usually in waterside habitats.

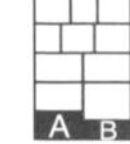

COLORADO BEETLE *Leptinotarsa decemlineata* *c.* 10mm

This unmistakable beetle (A) and its fleshy pink grubs (B) cause severe damage to potato crops on the continent. They also feed on tomatoes and nightshades. All year, but dormant in winter. The beetle is rarely seen in Britain and if it is found it should be reported to the authorities.

CHRYSOLINA POLITA 6–9mm
The pronotum is a beautiful metallic green, while the elytra range from bright red to deep chestnut. Abundant on mints and other labiates, especially in damp habitats. All year, but dormant in winter.

CHRYSOLINA MENTHASTRI *c.* 10mm
A brilliant-green beetle feeding on mints and other labiates is likely to be this species. May–September, mainly in waterside habitats, although it sometimes appears in gardens. ***C. varians***, found on St John's-worts, is similar, although it often has a blue or violet sheen. ***C. hyperici*** also occurs on St John's-worts, but has black legs instead of green.

CHRYSOLINA BANKSI 6–12mm
This shiny bronze beetle, the largest British *Chrysolina* species, is common on roadsides and waste places. It feeds on black horehound, May–August.

LOCHMAEA CAPREAE 4–6mm
This fairly shiny, hairless beetle has the pronotum sharply separated from the elytra. It abounds on sallows and other willows in damp habitats April–September and is sometimes sufficiently numerous for the adults and larvae to defoliate the trees. Closely related species live on hawthorn and heather. ***Galerucella lineola*** lives on willows and looks similar but its elytra are downy.

CASSIDA RUBIGINOSA 6–8mm
This species is very much like the Green Tortoise Beetle but has a small rusty patch around the scutellum and a thin rusty line along the suture. The rear angles of the pronotum are also more pointed. It is common on thistles from May–August.

GREEN TORTOISE BEETLE *Cassida viridis* 7–10mm
The tortoise beetles are named for their shape, with the pronotum and elytra extending beyond the body like a shell. At rest, the margins are pulled tightly down on to the leaves and the beetles are very well camouflaged. The pronotum and elytra are entirely green in *C. viridis*, which feeds on various labiates June–September. The spiky larvae camouflage themselves with their own excrement.

WEEVILS: SUPERFAMILY CURCULIONOIDEA

The weevils are mostly quite small beetles in which the front of the head is drawn out to form a beak or rostrum with the jaws at its tip. The antennae are elbowed and normally attached part way along the rostrum. The elytra are often clothed with scales. Many species are flightless and their elytra are commonly fused together. The majority of weevils are vegetarians, their larvae mostly without legs and generally living inside the stems, fruits, or seeds of their food-plants. There are nearly 600 British species, arranged in several families.

VINE WEEVIL *Otiorhynchus sulcatus* Family Curculionidae 8–12mm
The scape of the antenna is longer than the width of the pronotum. The elytra are parallel-sided with irregular patches of yellowish scales and each bears a row of shiny raised tubercles. The femora are toothed below. This weevil is a troublesome garden pest: the adults chew the leaves and shoots of many plants, but the larvae do the most damage by destroying roots, mainly those of pot plants and greenhouse crops. There are several similar species.

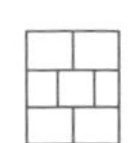

PINE WEEVIL *Hylobius abietis* 10–13mm
This weevil resembles the Vine Weevil but has a much longer and narrower rostrum and the scape of the antenna is much shorter than the width of the pronotum. The pale, hairy patches range from cream to chestnut. Common in coniferous woods and plantations, where the adults interfere with growth by chewing the bark of young shoots. The larvae live in dead stumps.

PEA WEEVIL *Sitona lineatus* 4–5mm
The conspicuous light and dark stripes on the elytra distinguish this from several otherwise similar species. Abundant in grassy places, the adults nibble semicircular holes in the edges of the leaves of peas, clovers and other legumes. The larvae live in the root nodules of these plants. All year, often swarming on sunny walls and fences before overwintering in the turf.

PHYLLOBIUS POMACEUS 7–10mm

The scales clothing this weevil may be golden green or bluish green, but they are often lost with age and the insect may then look largely black. Each front femur bears a strong tooth. April–August; abundant on stinging nettles, where the larvae feed on the roots. Many similar species live in trees.

NUT WEEVIL *Curculio nucum* 6–10mm

The rostrum of this golden-brown weevil is extremely long, nearly as long as the body in the female, and its distal half is strongly curved. The female gnaws into unripe hazelnuts and lays an egg in each. The male rostrum is shorter and the antennae are attached nearer to the tip than in the female. In both sexes the scales form a ridge around the rear half of the suture. May–August wherever hazel grows, but most common in woodland.

ACORN WEEVIL *Curculio glandium* 4–8mm

This weevil is very like the Nut Weevil but its larvae grow up in acorns. The scales do not form a ridge around the rear half of the suture. April–July, in and around oak woods.

CIONUS HORTULANUS 3–5mm

The 2 black spots on the elytra are more or less equal in size. June–September on figwort and dark mullein, where the larvae feed and pupate under the protection of gelatinous grey domes. There are several similar species but the 2 black spots are unequal.

APODERUS CORYLI Family Attelabidae 5–8mm

The bell-shaped black head and narrow neck make this weevil unmistakable. There is no obvious rostrum. May–July on hazel, especially in woodland. The larvae live in leaves that are rolled up by the egg-laying females.

RHYNCHITES AEQUATUS 2.5–5mm

The brick-coloured elytra are quite hairy and there is a dark line along the suture, especially near the front. The rostrum is as long as or longer than the head and pronotum together. April–August, mainly on hawthorn but also on other rosaceous trees and shrubs.

ATTELABUS NITENS 4–6mm

This weevil resembles *Apoderus coryli* in colour, but the head is rectangular and has a short rostrum and the legs are entirely black. May–July on oak trees, especially young ones, where the larvae live in tightly rolled leaves.

PLATYRHINUS RESINOSUS Family Anthribidae 8–13mm

The broad white rostrum and the clubbed antennae distinguish this weevil from most other British species. There is a blunt tooth on each side of the pronotum. When disturbed, the beetle rolls over, pulls in its legs, and looks remarkably like a bird dropping. It breeds in fungi on dead trees, mainly in woodland.

PLATYSTOMOS ALBINUS 7–10mm

With its white rostrum and white rear, this weevil is superficially like *Platyrhinus resinosus* but its antennae are longer especially in the male, and there are no teeth on the pronotum. Mainly in woodland, where the larvae feed in dead timber.

ELM BARK BEETLE *Scolytus scolytus* Family Scolytidae 3–6mm

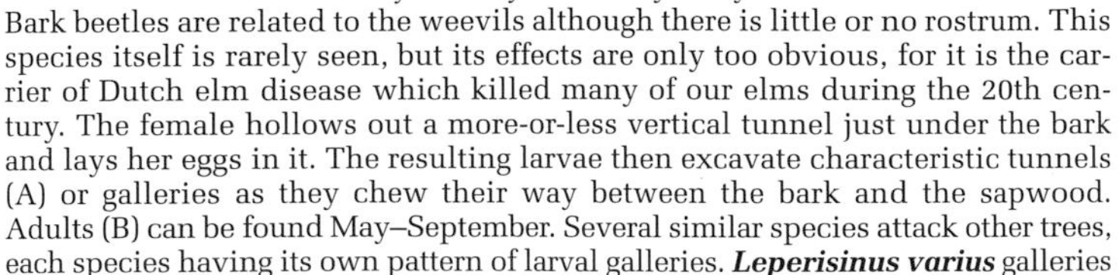

Bark beetles are related to the weevils although there is little or no rostrum. This species itself is rarely seen, but its effects are only too obvious, for it is the carrier of Dutch elm disease which killed many of our elms during the 20th century. The female hollows out a more-or-less vertical tunnel just under the bark and lays her eggs in it. The resulting larvae then excavate characteristic tunnels (A) or galleries as they chew their way between the bark and the sapwood. Adults (B) can be found May–September. Several similar species attack other trees, each species having its own pattern of larval galleries. ***Leperisinus varius*** galleries in ash resemble those of the Elm Bark Beetle but the main tunnels are horizontal.

WATER BEETLES

Beetles of several families have taken to life on or in the water. Some are fierce predators, while others are herbivores or scavengers. Most have smooth outlines and swim well with the aid of broad, hairy hind-legs. Submerged species carry air supplies with them, either under the elytra or trapped among hairs on the underside of the body. The spiracles open into these reservoirs and allow the air to enter the body. When the oxygen has been used up, the beetles return to the surface for fresh supplies, and this is when they are most often seen. A few small species get their oxygen direct from the water and can survive without surfacing. Most water beetles are active throughout the year, although they may become dormant in the coldest weather. Many species fly well.

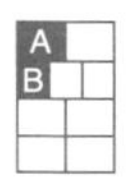

GREAT DIVING BEETLE *Dytiscus marginalis* Family Dytiscidae 25–35mm

This is the commonest of 6 very similar species and can be distinguished from most of them by its deep yellow underside, the complete yellow border to the pronotum, and the lack of yellow around the eyes. The male has very smooth, shiny green elytra but the female (A) is duller and her elytra are ribbed. Inhabiting still waters with plenty of vegetation, it attacks almost anything that moves, including frogs and fish. The larvae (B), known as water tigers, are equally fierce.

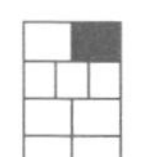

ACILIUS SULCATUS 15–20mm

The male's elytra are smooth and orange with variable and often indistinct black patterns. The female's elytra are strongly ribbed, with orange hair between the ribs. A predator in still and slow-moving water. ***A. canaliculatus,*** of boggy habitats, has the hind femora yellow instead of partly black.

PLATAMBUS MACULATUS 7–9mm

This beetle can usually be recognised by its bright black and yellow pattern, although the amounts of black and yellow vary to some extent. The hind-legs are fringed and well-adapted for swimming. Predatory, mainly in running water with a sandy or stony bottom.

NEBRIOPORUS ELEGANS 4–5mm

This beetle often resembles *Platambus maculatus* in colour although its head is yellow instead of black. It is also smaller and very hairy and the elytra are slightly toothed at the tip. The slender legs are much less adapted for swimming and the beetle is most likely to be found crawling on the bottom of lakes and streams with sandy or gravelly beds.

WHIRLIGIG BEETLE *Gyrinus natator* Family Gyrinidae 5–7mm

This well-named shiny beetle zooms round and round on the surface of still and slow-moving water with the aid of its short, paddle-like middle and hind-legs. Each eye is divided into 2 parts, one looking across the water surface and the other down into the water. The beetle is active by day and feeds mainly on mosquito larvae. All year but dormant in winter. The **Hairy Whirligig** (*Orectochilus villosus*) behaves similarly but is nocturnal.

SCREECH BEETLE *Hygrobia herrmanni* Family Hygrobiidae 8–10mm

Named for its ability to squeak by rubbing the tip of its abdomen against its elytra, this beetle can be recognised by the rusty borders to the otherwise black elytra. It is a predator in muddy ponds and streams.

GREAT SILVER BEETLE *Hydophilus piceus* Family Hydrophilidae 35–50mm

This is one of our largest beetles. It is very shiny and usually tinged with green above, but it gets its name for the silvery appearance of its underside which is due to the layer of air trapped there. Although it sometimes attacks water snails and other small invertebrates, it feeds largely on plant debris on the bottom of weedy ponds. The Great Diving Beetle and other dytiscids hang upside-down from the surface to renew their air supplies, but the Great Silver Beetle and its allies surface head-first.

OMOPHRON LIMBATUM Family Carabidae 4–7mm

Although this is a ground beetle (*see* pp. 110–12) rather than a water beetle, it is never found far from the waterside, where it burrows in damp sand and silt and emerges to feed at night. Long, slender legs, long antennae, and rather circular elytra distinguish it from true water beetles with similar coloration.

SKIPPER BUTTERFLIES: FAMILY HESPERIIDAE

Named for their swift, bouncing flight, which is very different from that of other butterflies, the skippers are small insects with relatively stout, moth-like bodies. Their antennae are widely separated at the base and strongly curved or hooked at the tip. Of the 8 British species, 5 have orange or golden uppersides and are called golden skippers. These all bask with their forewings partly raised and the males have conspicuous patches of scent-emitting scales.

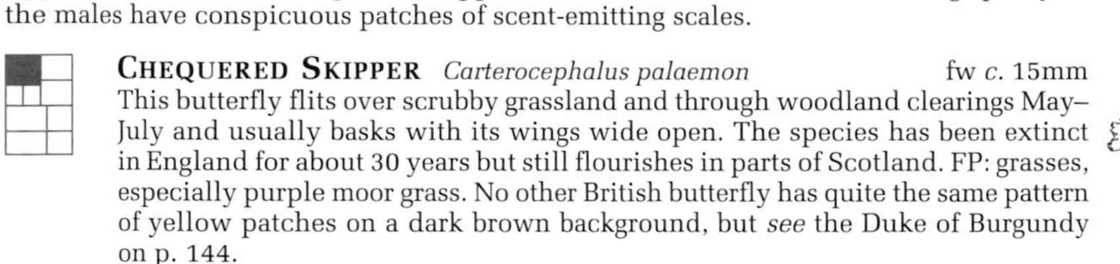

CHEQUERED SKIPPER *Carterocephalus palaemon* fw *c.* 15mm

This butterfly flits over scrubby grassland and through woodland clearings May–July and usually basks with its wings wide open. The species has been extinct in England for about 30 years but still flourishes in parts of Scotland. FP: grasses, especially purple moor grass. No other British butterfly has quite the same pattern of yellow patches on a dark brown background, but *see* the Duke of Burgundy on p. 144.

SILVER-SPOTTED SKIPPER *Hesperia comma* fw *c.* 15mm

This rare butterfly is named for the silvery spots on its olive-green underside. The female is much browner on the upperside than the male pictured here, lacks the dark streak of scent scales and also has more pale spots. Flies June–September on chalk and limestone slopes with short grass and plenty of bare soil. FP: sheep's fescue grass.

LARGE SKIPPER *Ochlodes venatus* fw 12–18mm

This common butterfly resembles the silver-spotted skipper but the spots on the underside are dull yellow. The sexes are quite different, with only the male (A, female, B) having the black scent scales. The insect flies in all sorts of grassy places June–September. FP: various coarse grasses. Caterpillar p. 278.

SMALL SKIPPER *Thymelicus sylvestris* fw 12–16mm

This very common skipper is golden orange above, but only the male has the conspicuous black scent scales. The underside is also golden, with a dusting of greyish-green scales, and the underside of the antennal club is orange. It flies in grassy places June–August. FP: various tall grasses, especially Yorkshire fog. Caterpillar p. 278.

ESSEX SKIPPER *Thymelicus lineola* fw 12–15mm

Very similar to the Small Skipper, but the underside is greyer and the antennal tip is black underneath. Despite its name, the butterfly is widely distributed in southern England and quite common in many grassy places; flies June–August. FP: cocksfoot and other tall grasses.

LULWORTH SKIPPER *Thymelicus acteon* fw 10–13mm

Differs from the 2 previous species in its slightly browner coloration and the small arc of yellow dots near the tip of the forewing, although this is often very faint in the male. Confined to rough grassland on the south coast; flies June–September. FP: tor grass.

GRIZZLED SKIPPER *Pyrgus malvae* fw 10–15mm

This skipper has an unmistakable pattern, similar on both surfaces although somewhat duller on the underside. It flies in all kinds of grassy places May–September and basks with its wings wide open. Although unlike any other British butterfly, it is often mistaken for a moth. FP: wild strawberry, tormentil and creeping cinquefoil. Caterpillar p. 278.

DINGY SKIPPER *Erynnis tages* fw 12–15mm

This well-named species is also moth-like, and the only skipper in Ireland. The dull brown upperside is edged with white dots and may be relieved by patches of grey. The light brown underside is unmarked apart from a few white dots. It flies in grassy places May–August and basks with wings wide open. FP: bird's-foot trefoil. Caterpillar p. 278.

WHITE AND YELLOW BUTTERFLIES: FAMILY PIERIDAE

These insects are basically white or yellow and, apart from the Brimstone, are all marked with black. The sexes usually differ in the amount and arrangement of the black markings and may also differ in colour. The British Isles have 6 resident species, and a further 4 are summer visitors, although only the Clouded Yellow can be regarded as a regular visitor.

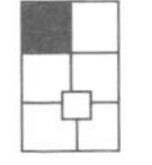

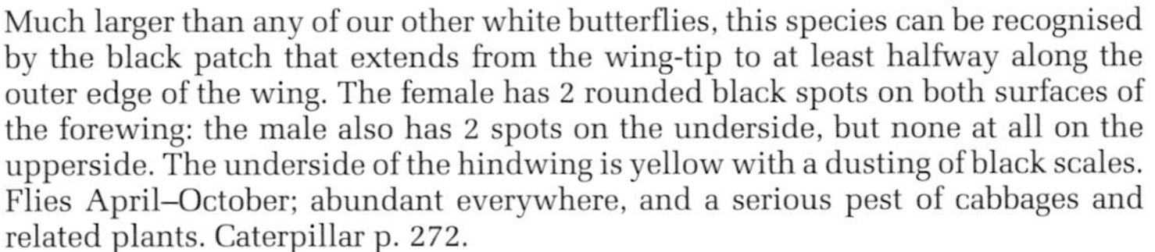

LARGE WHITE *Pieris brassicae* fw 25–35mm

Much larger than any of our other white butterflies, this species can be recognised by the black patch that extends from the wing-tip to at least halfway along the outer edge of the wing. The female has 2 rounded black spots on both surfaces of the forewing: the male also has 2 spots on the underside, but none at all on the upperside. The underside of the hindwing is yellow with a dusting of black scales. Flies April–October; abundant everywhere, and a serious pest of cabbages and related plants. Caterpillar p. 272.

SMALL WHITE *Pieris rapae* fw 15–27mm

This species resembles the Large White, but the wing-tip patch is less dense, especially in the spring brood, and does not reach anywhere near the centre of the outer margin. The female has 2 prominent black spots on the upperside of the forewing, but the male has only 1 faint spot or none at all. Flies in all kinds of open habitats April–October; a serious pest of cabbages and related plants. Caterpillar p. 272.

GREEN-VEINED WHITE *Pieris napi* fw 18–25mm

Named for the lines of scales following the veins under the hindwings, and easy to recognise with its wings closed. The upperside resembles that of the Small White although the veins are more prominent, especially in summer insects. Flies April–October in gardens and many other open habitats, although in the north and west of the British Isles it is found mainly in areas of damp grassland. FP: a variety of wild crucifers, including garlic mustard and watercress: rarely attacks cabbages or other cultivated brassicas.

WOOD WHITE *Leptidea sinapis* fw 18–25mm

This very delicate butterfly has narrower wings than the other whites. The male has a sooty smudge at the wing-tip, but the female has no more than a few dark streaks and the upperside may be completely white in the summer. The underside is smudged with grey and/or yellow in both sexes and is quite green in some Irish specimens. Flies May–June, and sometimes again in August in the south. It flutters weakly over low-growing vegetation in woodland rides and clearings in Britain, but in Ireland it can be found in open grassland. FP: meadow vetchling and other related plants. Caterpillar p. 272.

ORANGE-TIP *Anthocharis cardamines* fw 18–25mm

The orange-tipped male (A) is unmistakable, but the female (B) could be mistaken for one of the other whites unless the mottled underside is visible. The wing-tips are more rounded than those of other whites. With the forewings pulled right down, the mottled underside affords the insect excellent camouflage. Flies April–June in many habitats, including gardens, but favours damp grassland and woodland margins. FP: cuckoo-flower, garlic mustard, garden honesty and many other cruciferous plants. Caterpillar p. 272.

BATH WHITE *Pontia daplidice* fw 20–25mm

This rare visitor could be mistaken for a female Orange-tip, but has more-pointed wings and more black on the upperside: look especially for the white spots inside the black wing-tips. It flies in open habitats throughout the summer. FP: various wild cruciferous plants and mignonette. The Marbled White (*see* p. 154) has much more black, and no green pattern under the hindwing.

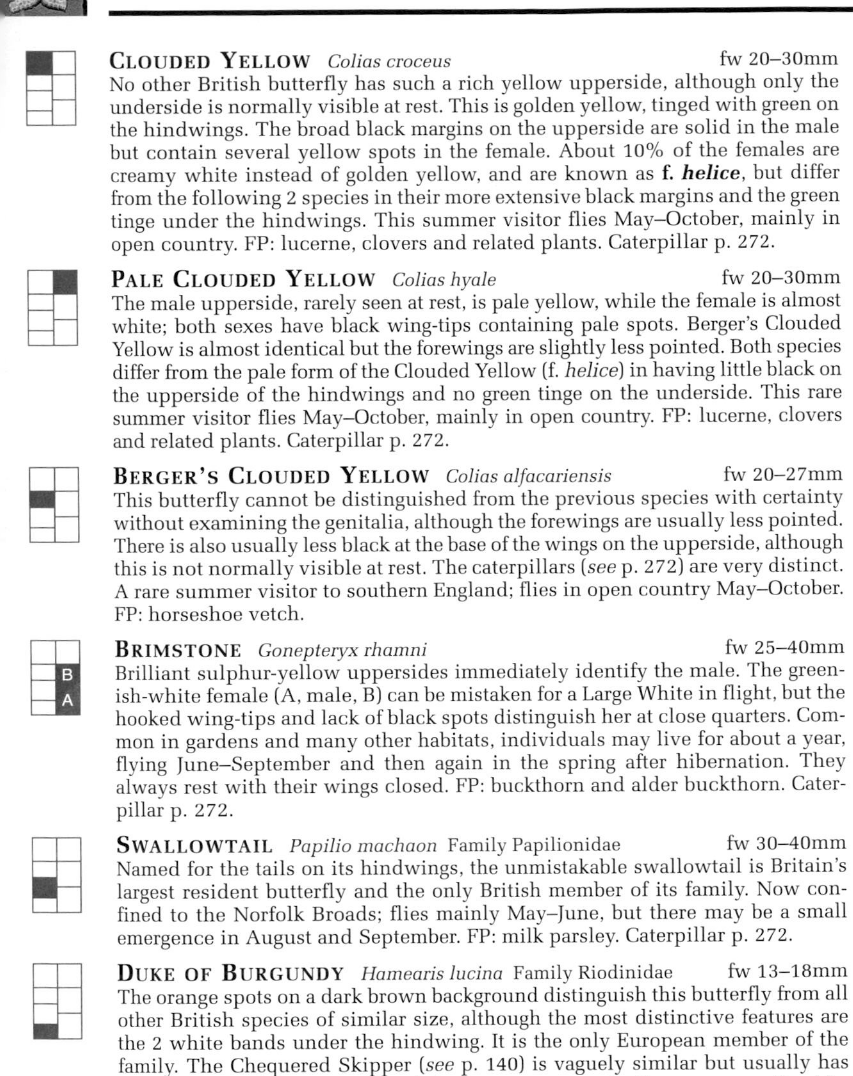

CLOUDED YELLOW *Colias croceus* fw 20–30mm
No other British butterfly has such a rich yellow upperside, although only the underside is normally visible at rest. This is golden yellow, tinged with green on the hindwings. The broad black margins on the upperside are solid in the male but contain several yellow spots in the female. About 10% of the females are creamy white instead of golden yellow, and are known as **f. *helice***, but differ from the following 2 species in their more extensive black margins and the green tinge under the hindwings. This summer visitor flies May–October, mainly in open country. FP: lucerne, clovers and related plants. Caterpillar p. 272.

PALE CLOUDED YELLOW *Colias hyale* fw 20–30mm
The male upperside, rarely seen at rest, is pale yellow, while the female is almost white; both sexes have black wing-tips containing pale spots. Berger's Clouded Yellow is almost identical but the forewings are slightly less pointed. Both species differ from the pale form of the Clouded Yellow (f. *helice*) in having little black on the upperside of the hindwings and no green tinge on the underside. This rare summer visitor flies May–October, mainly in open country. FP: lucerne, clovers and related plants. Caterpillar p. 272.

BERGER'S CLOUDED YELLOW *Colias alfacariensis* fw 20–27mm
This butterfly cannot be distinguished from the previous species with certainty without examining the genitalia, although the forewings are usually less pointed. There is also usually less black at the base of the wings on the upperside, although this is not normally visible at rest. The caterpillars (*see* p. 272) are very distinct. A rare summer visitor to southern England; flies in open country May–October. FP: horseshoe vetch.

BRIMSTONE *Gonepteryx rhamni* fw 25–40mm
Brilliant sulphur-yellow uppersides immediately identify the male. The greenish-white female (A, male, B) can be mistaken for a Large White in flight, but the hooked wing-tips and lack of black spots distinguish her at close quarters. Common in gardens and many other habitats, individuals may live for about a year, flying June–September and then again in the spring after hibernation. They always rest with their wings closed. FP: buckthorn and alder buckthorn. Caterpillar p. 272.

SWALLOWTAIL *Papilio machaon* Family Papilionidae fw 30–40mm
Named for the tails on its hindwings, the unmistakable swallowtail is Britain's largest resident butterfly and the only British member of its family. Now confined to the Norfolk Broads; flies mainly May–June, but there may be a small emergence in August and September. FP: milk parsley. Caterpillar p. 272.

DUKE OF BURGUNDY *Hamearis lucina* Family Riodinidae fw 13–18mm
The orange spots on a dark brown background distinguish this butterfly from all other British species of similar size, although the most distinctive features are the 2 white bands under the hindwing. It is the only European member of the family. The Chequered Skipper (*see* p. 140) is vaguely similar but usually has paler spots on the upperside and does not now occur in the same area. Scrubby grassland and sunny woodland rides and clearings are the normal habitats of this increasingly rare species; flies mainly May–June. FP: cowslips and primroses. Caterpillar p. 272.

HAIRSTREAKS, COPPERS AND BLUES: FAMILY LYCAENIDAE

These small butterflies are mostly brightly coloured and many exhibit a metallic sheen on the upperside. The hairstreaks, named for the fine streaks on the underside, have short 'tails' on their hindwings. The sexes are often very different: female blues, for example, are usually brown and nothing like the shiny blue males. Hairstreaks usually frequent trees and shrubs, but most coppers and blues fly close to the ground. Britain has 14 resident species, but only 7 of these reach Ireland. A further 3 species are rare summer visitors.

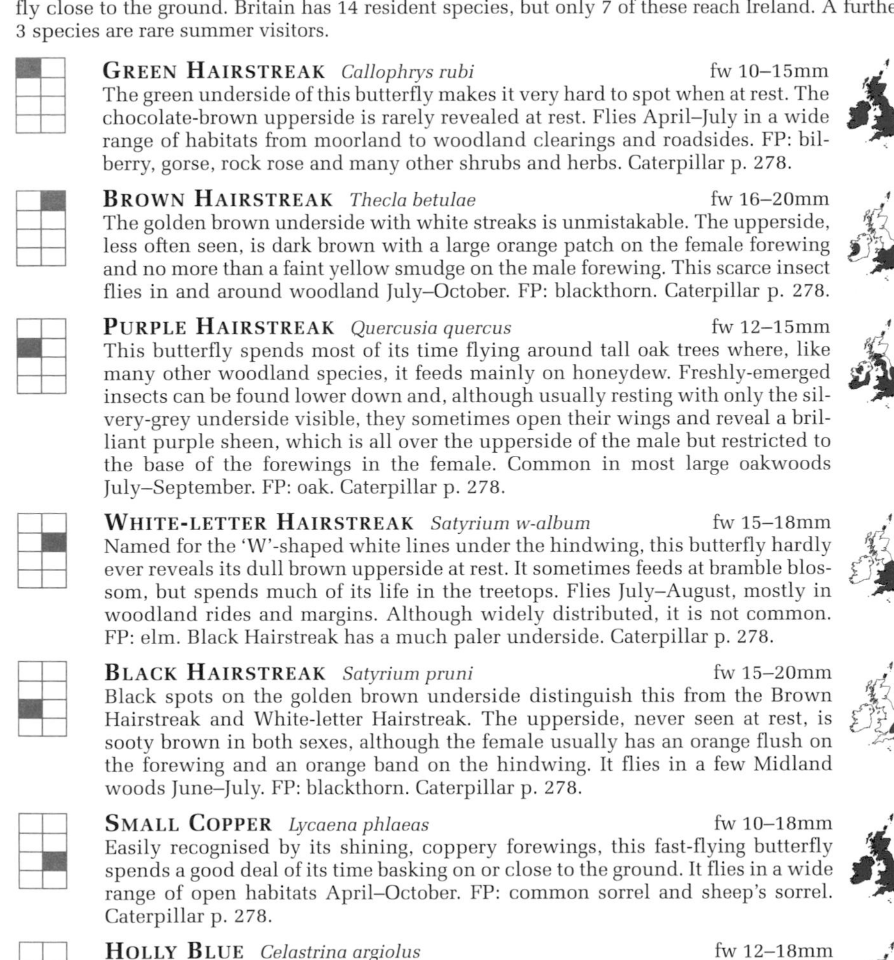

GREEN HAIRSTREAK *Callophrys rubi* fw 10–15mm

The green underside of this butterfly makes it very hard to spot when at rest. The chocolate-brown upperside is rarely revealed at rest. Flies April–July in a wide range of habitats from moorland to woodland clearings and roadsides. FP: bilberry, gorse, rock rose and many other shrubs and herbs. Caterpillar p. 278.

BROWN HAIRSTREAK *Thecla betulae* fw 16–20mm

The golden brown underside with white streaks is unmistakable. The upperside, less often seen, is dark brown with a large orange patch on the female forewing and no more than a faint yellow smudge on the male forewing. This scarce insect flies in and around woodland July–October. FP: blackthorn. Caterpillar p. 278.

PURPLE HAIRSTREAK *Quercusia quercus* fw 12–15mm

This butterfly spends most of its time flying around tall oak trees where, like many other woodland species, it feeds mainly on honeydew. Freshly-emerged insects can be found lower down and, although usually resting with only the silvery-grey underside visible, they sometimes open their wings and reveal a brilliant purple sheen, which is all over the upperside of the male but restricted to the base of the forewings in the female. Common in most large oakwoods July–September. FP: oak. Caterpillar p. 278.

WHITE-LETTER HAIRSTREAK *Satyrium w-album* fw 15–18mm

Named for the 'W'-shaped white lines under the hindwing, this butterfly hardly ever reveals its dull brown upperside at rest. It sometimes feeds at bramble blossom, but spends much of its life in the treetops. Flies July–August, mostly in woodland rides and margins. Although widely distributed, it is not common. FP: elm. Black Hairstreak has a much paler underside. Caterpillar p. 278.

BLACK HAIRSTREAK *Satyrium pruni* fw 15–20mm

Black spots on the golden brown underside distinguish this from the Brown Hairstreak and White-letter Hairstreak. The upperside, never seen at rest, is sooty brown in both sexes, although the female usually has an orange flush on the forewing and an orange band on the hindwing. It flies in a few Midland woods June–July. FP: blackthorn. Caterpillar p. 278.

SMALL COPPER *Lycaena phlaeas* fw 10–18mm

Easily recognised by its shining, coppery forewings, this fast-flying butterfly spends a good deal of its time basking on or close to the ground. It flies in a wide range of open habitats April–October. FP: common sorrel and sheep's sorrel. Caterpillar p. 278.

HOLLY BLUE *Celastrina argiolus* fw 12–18mm

A B

The only blue butterfly likely to be seen in trees in Britain. It differs from most other blues in its silvery blue underside with small black dots and no orange (B). The upperside is shiny blue, with narrow black margins in the male (A) and much broader ones in the female. Flies March–October in scrubby and wooded habitats, including parks and gardens. FP: holly, dogwood, gorse and many other shrubs in spring, but usually ivy in autumn. Caterpillar p. 276.

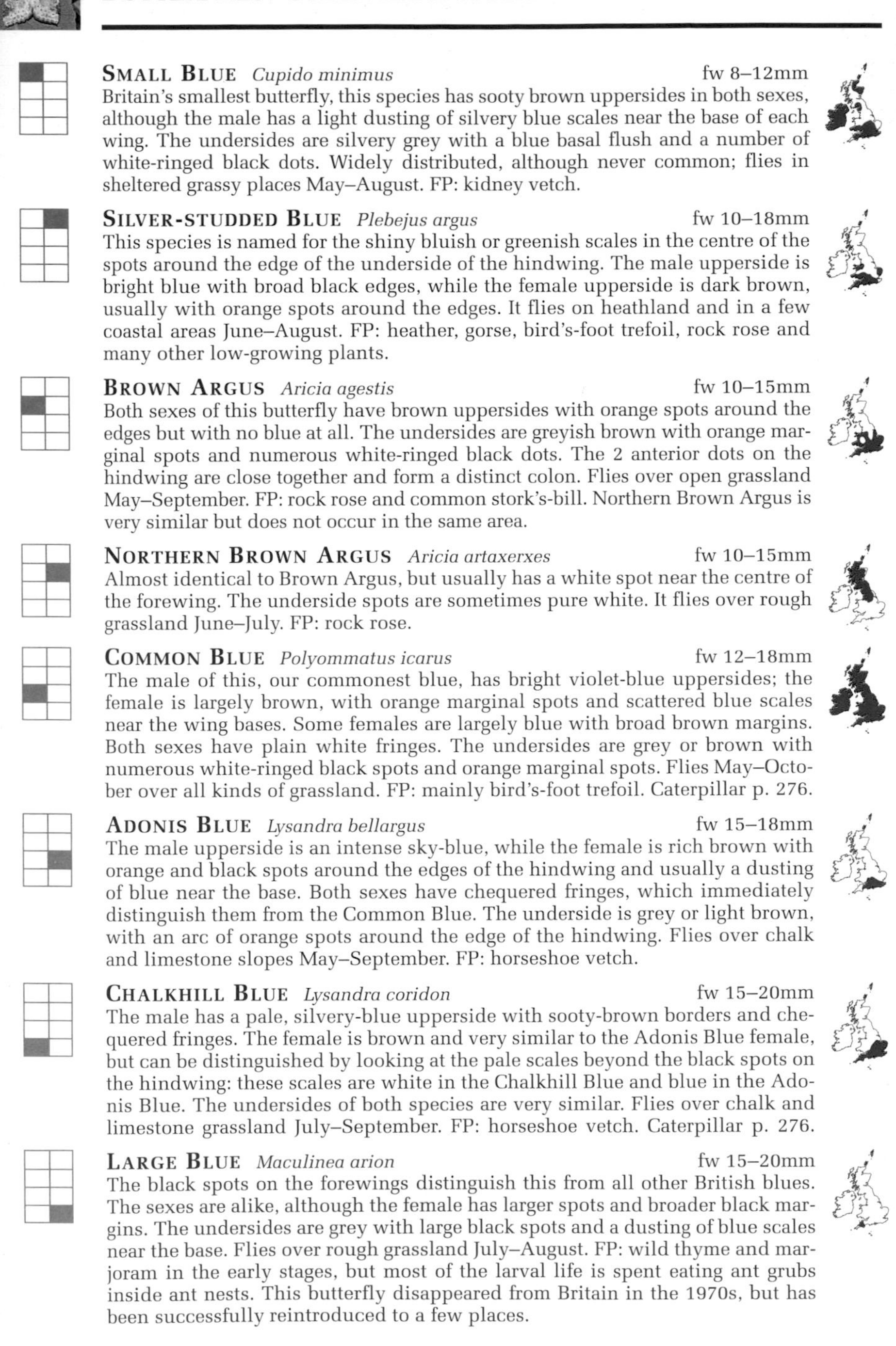

SMALL BLUE *Cupido minimus* fw 8–12mm
Britain's smallest butterfly, this species has sooty brown uppersides in both sexes, although the male has a light dusting of silvery blue scales near the base of each wing. The undersides are silvery grey with a blue basal flush and a number of white-ringed black dots. Widely distributed, although never common; flies in sheltered grassy places May–August. FP: kidney vetch.

SILVER-STUDDED BLUE *Plebejus argus* fw 10–18mm
This species is named for the shiny bluish or greenish scales in the centre of the spots around the edge of the underside of the hindwing. The male upperside is bright blue with broad black edges, while the female upperside is dark brown, usually with orange spots around the edges. It flies on heathland and in a few coastal areas June–August. FP: heather, gorse, bird's-foot trefoil, rock rose and many other low-growing plants.

BROWN ARGUS *Aricia agestis* fw 10–15mm
Both sexes of this butterfly have brown uppersides with orange spots around the edges but with no blue at all. The undersides are greyish brown with orange marginal spots and numerous white-ringed black dots. The 2 anterior dots on the hindwing are close together and form a distinct colon. Flies over open grassland May–September. FP: rock rose and common stork's-bill. Northern Brown Argus is very similar but does not occur in the same area.

NORTHERN BROWN ARGUS *Aricia artaxerxes* fw 10–15mm
Almost identical to Brown Argus, but usually has a white spot near the centre of the forewing. The underside spots are sometimes pure white. It flies over rough grassland June–July. FP: rock rose.

COMMON BLUE *Polyommatus icarus* fw 12–18mm
The male of this, our commonest blue, has bright violet-blue uppersides; the female is largely brown, with orange marginal spots and scattered blue scales near the wing bases. Some females are largely blue with broad brown margins. Both sexes have plain white fringes. The undersides are grey or brown with numerous white-ringed black spots and orange marginal spots. Flies May–October over all kinds of grassland. FP: mainly bird's-foot trefoil. Caterpillar p. 276.

ADONIS BLUE *Lysandra bellargus* fw 15–18mm
The male upperside is an intense sky-blue, while the female is rich brown with orange and black spots around the edges of the hindwing and usually a dusting of blue near the base. Both sexes have chequered fringes, which immediately distinguish them from the Common Blue. The underside is grey or light brown, with an arc of orange spots around the edge of the hindwing. Flies over chalk and limestone slopes May–September. FP: horseshoe vetch.

CHALKHILL BLUE *Lysandra coridon* fw 15–20mm
The male has a pale, silvery-blue upperside with sooty-brown borders and chequered fringes. The female is brown and very similar to the Adonis Blue female, but can be distinguished by looking at the pale scales beyond the black spots on the hindwing: these scales are white in the Chalkhill Blue and blue in the Adonis Blue. The undersides of both species are very similar. Flies over chalk and limestone grassland July–September. FP: horseshoe vetch. Caterpillar p. 276.

LARGE BLUE *Maculinea arion* fw 15–20mm
The black spots on the forewings distinguish this from all other British blues. The sexes are alike, although the female has larger spots and broader black margins. The undersides are grey with large black spots and a dusting of blue scales near the base. Flies over rough grassland July–August. FP: wild thyme and marjoram in the early stages, but most of the larval life is spent eating ant grubs inside ant nests. This butterfly disappeared from Britain in the 1970s, but has been successfully reintroduced to a few places.

VANESSIDS AND FRITILLARIES: FAMILY NYMPHALIDAE

This large family contains 2 fairly distinct groups, the multicoloured vanessids pictured on this page and the largely orange and black fritillaries. The front legs are very small and often bear tufts of hairlike scales, so the insects are sometimes called brush-footed butterflies. Many of their caterpillars bear branched spines. Fifteen species regularly breed in the British Isles, and 3 more are irregular visitors from the continent.

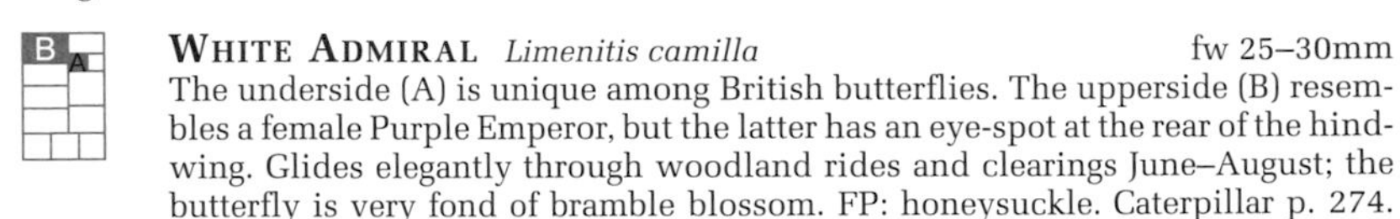
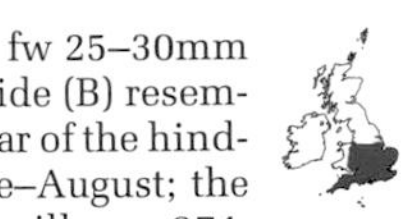

WHITE ADMIRAL *Limenitis camilla* fw 25–30mm
The underside (A) is unique among British butterflies. The upperside (B) resembles a female Purple Emperor, but the latter has an eye-spot at the rear of the hindwing. Glides elegantly through woodland rides and clearings June–August; the butterfly is very fond of bramble blossom. FP: honeysuckle. Caterpillar p. 274.

PURPLE EMPEROR *Apatura iris* fw 30–40mm
The upperside is brown with white splashes, but the male gleams deep purple when seen from certain angles (A). Flies in well-wooded areas July–August; usually keeps to the treetops, but may come down to drink from muddy pools or even carrion. FP: sallow. Caterpillar p. 274.

RED ADMIRAL *Vanessa atalanta* fw *c.* 30mm
This unmistakable butterfly flies March–November in a wide range of habitats, including parks and gardens. Adults hibernate and were once thought to be unable to survive the British winter, but they now seem to do so in increasing numbers. Most of our red admirals are the progeny of spring arrivals from the continent. FP: stinging nettle. Caterpillar p. 274.

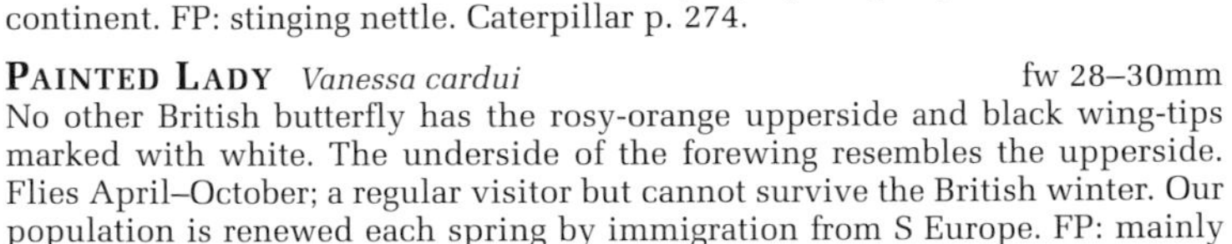

PAINTED LADY *Vanessa cardui* fw 28–30mm
No other British butterfly has the rosy-orange upperside and black wing-tips marked with white. The underside of the forewing resembles the upperside. Flies April–October; a regular visitor but cannot survive the British winter. Our population is renewed each spring by immigration from S Europe. FP: mainly thistles. Caterpillar p. 274.

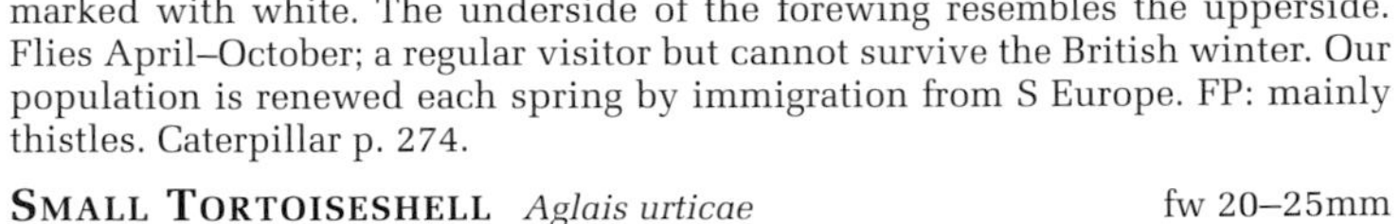

SMALL TORTOISESHELL *Aglais urticae* fw 20–25mm
The rich orange ground-colour and blue marginal spots readily identify this very common butterfly. It flies almost everywhere February–October, and hibernates as an adult in hollow trees and buildings. FP: stinging nettle. Caterpillar p. 274.

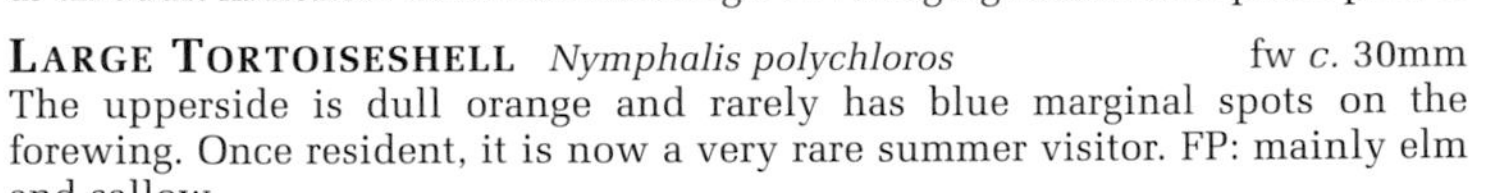
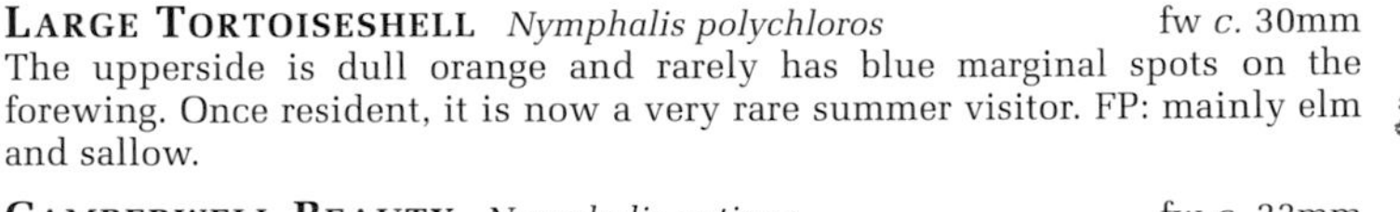

LARGE TORTOISESHELL *Nymphalis polychloros* fw *c.* 30mm
The upperside is dull orange and rarely has blue marginal spots on the forewing. Once resident, it is now a very rare summer visitor. FP: mainly elm and sallow.

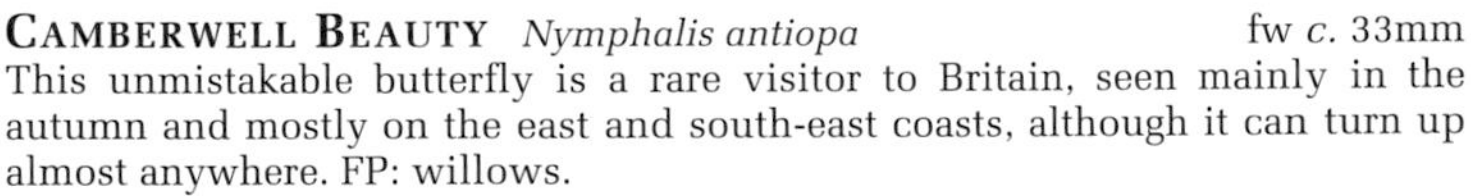

CAMBERWELL BEAUTY *Nymphalis antiopa* fw *c.* 33mm
This unmistakable butterfly is a rare visitor to Britain, seen mainly in the autumn and mostly on the east and south-east coasts, although it can turn up almost anywhere. FP: willows.

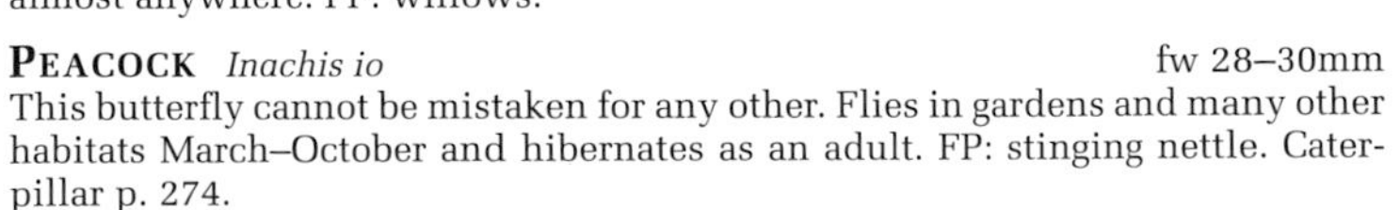
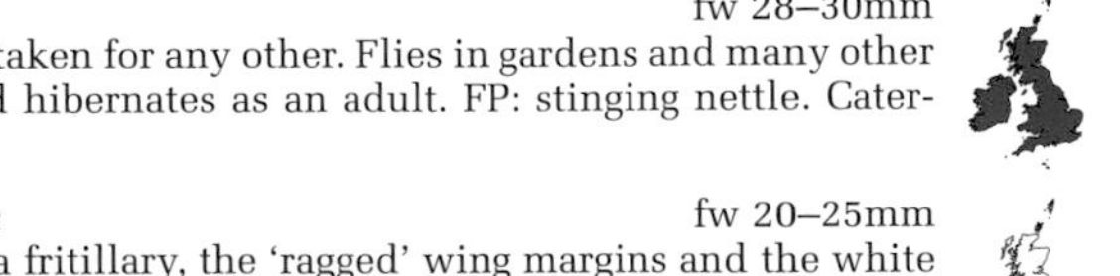

PEACOCK *Inachis io* fw 28–30mm
This butterfly cannot be mistaken for any other. Flies in gardens and many other habitats March–October and hibernates as an adult. FP: stinging nettle. Caterpillar p. 274.

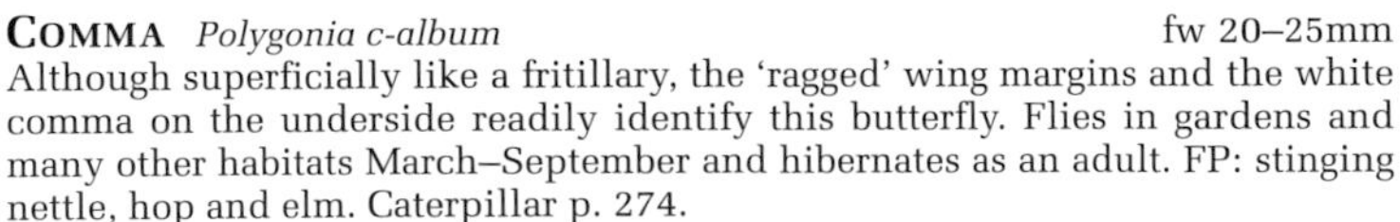

COMMA *Polygonia c-album* fw 20–25mm
Although superficially like a fritillary, the 'ragged' wing margins and the white comma on the underside readily identify this butterfly. Flies in gardens and many other habitats March–September and hibernates as an adult. FP: stinging nettle, hop and elm. Caterpillar p. 274.

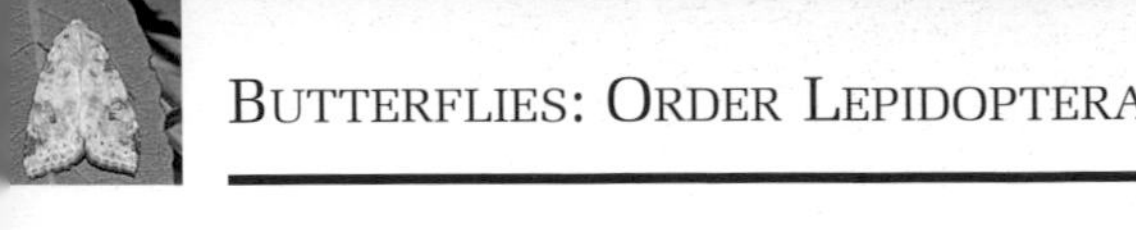

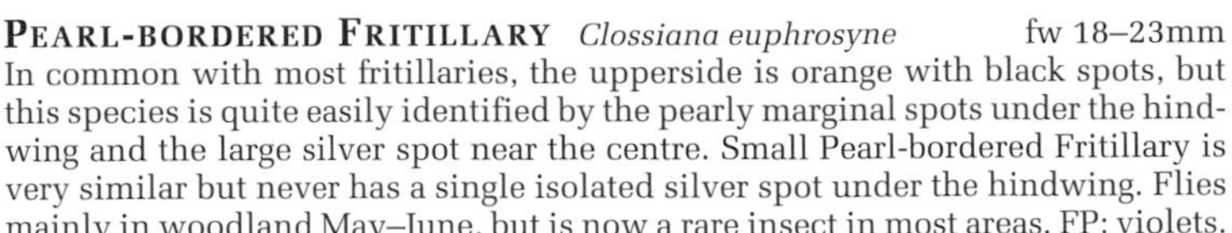

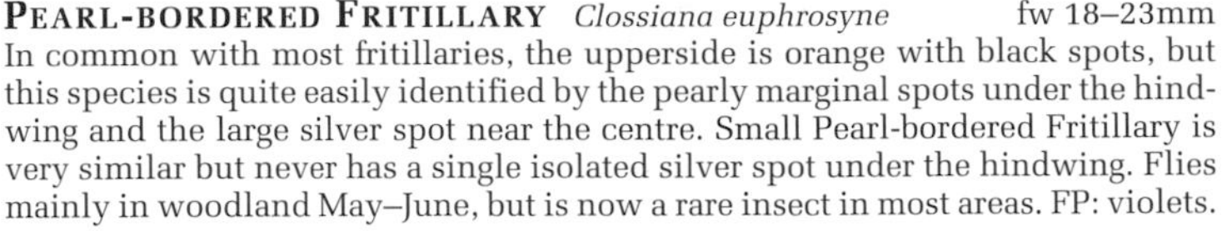

PEARL-BORDERED FRITILLARY *Clossiana euphrosyne* fw 18–23mm
In common with most fritillaries, the upperside is orange with black spots, but this species is quite easily identified by the pearly marginal spots under the hindwing and the large silver spot near the centre. Small Pearl-bordered Fritillary is very similar but never has a single isolated silver spot under the hindwing. Flies mainly in woodland May–June, but is now a rare insect in most areas. FP: violets.

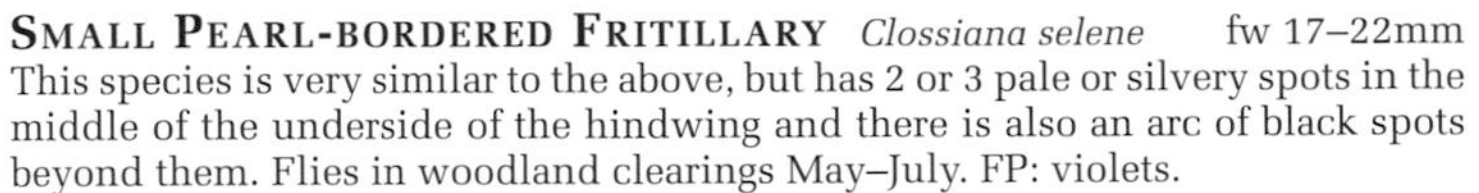

SMALL PEARL-BORDERED FRITILLARY *Clossiana selene* fw 17–22mm
This species is very similar to the above, but has 2 or 3 pale or silvery spots in the middle of the underside of the hindwing and there is also an arc of black spots beyond them. Flies in woodland clearings May–July. FP: violets.

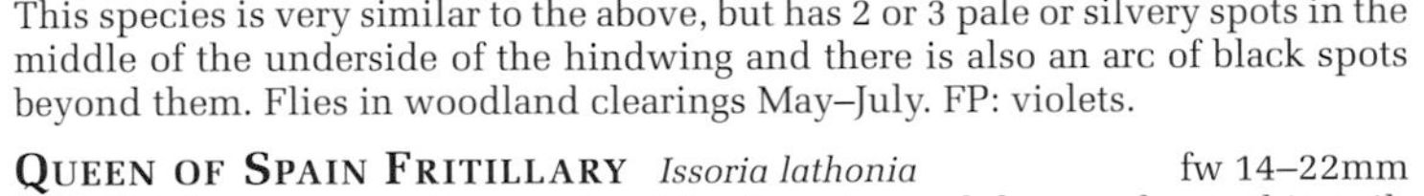

QUEEN OF SPAIN FRITILLARY *Issoria lathonia* fw 14–22mm
The slightly concave outer margin of the forewing and the very large, shiny silver spots under the hindwing distinguish this occasional visitor from our resident fritillaries. Flies in open flowery habitats April–September. FP: violets.

HIGH BROWN FRITILLARY *Argynnis adippe* fw 25–30mm
The best diagnostic feature is an arc of red spots with silver centres under the hindwing. There are also several larger silvery spots on a green and brown background. This rather rare insect flies in open woods and scrubby habitats June–July. FP: violets.

DARK GREEN FRITILLARY *Argynnis aglaja* fw 25–30mm
This species is very similar to the High Brown Fritillary, but the underside of the hindwing has no red spots and the basal half containing the silver spots is largely green. Flies over rough grassland and other open country June–August. FP: violets. Caterpillar p. 274.

SILVER-WASHED FRITILLARY *Argynnis paphia* fw *c.* 35mm
This species is named for the silver streaks among the green scales under the hindwing. The male (the lower insect in the left-hand picture) is brighter than the female and easily recognised by the ridges of black scent scales on its forewings. Flies in and around woodlands July–August. FP: violets. Caterpillar p. 274.

HEATH FRITILLARY *Mellicta athalia* fw 15–25mm
Unlike our other fritillaries, this species has a network of brown lines and no distinct spots on the upperside. A broad band of white cells crosses the underside of the hindwing. Restricted to a few woodland and heathland habitats; flies June–July. FP: mainly cow-wheat and plantain.

GLANVILLE FRITILLARY *Melitaea cinxia* fw 15–25mm
The orange or sandy-brown upperside has a reticulate pattern like that of the Heath Fritillary but the hindwing has a conspicuous arc of orange spots with dark centres. Confined to coastal cliffs on the Isle of Wight; flies May–June. FP: ribwort plantain.

MARSH FRITILLARY *Eurodryas aurinia* fw 15–25mm
The mixture of orange and yellow on the upperside distinguishes this from all our other fritillaries. There is a clear arc of black dots on both surfaces of the hindwing. It flies over open grassland, both wet and dry, May–July. FP: devil's-bit scabious.

MONARCH *Danaus plexippus* Family Danaidae fw 35–50mm
This beautiful butterfly, also known as the Milkweed, is a rare visitor to Britain from the Canary Islands or even from North America. It is most likely to be seen in the south-west and late summer is the most likely time for it to reach us. It cannot breed in Britain.

BROWNS: FAMILY SATYRIDAE

These insects are mainly brown but their most characteristic features are the eye-spots on the wings (*see* p. 43). The front legs are reduced to hairy 'brushes'. The larvae feed on grasses or sedges. There are 11 species in the British Isles.

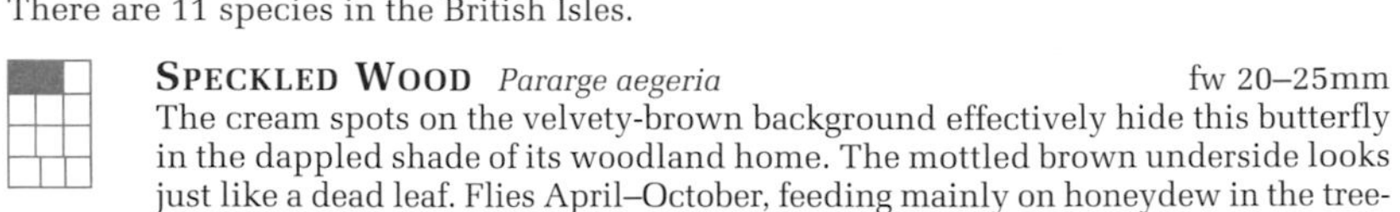

SPECKLED WOOD *Pararge aegeria* fw 20–25mm

The cream spots on the velvety-brown background effectively hide this butterfly in the dappled shade of its woodland home. The mottled brown underside looks just like a dead leaf. Flies April–October, feeding mainly on honeydew in the treetops. FP: various grasses. Caterpillar p. 276.

WALL BROWN *Lasiommata megera* fw 17–25mm

The prominent eye-spots readily distinguish this butterfly from the similarly-coloured fritillaries. The hindwing is silvery grey below with zig-zag lines and an arc of large eye-spots. Flies May–October in grassy places and likes to bask on bare ground. FP: various coarse grasses.

MOUNTAIN RINGLET *Erebia epiphron* fw 16–22mm

The upperside is dark brown with a variably-developed orange band containing small black spots. The underside is similar, although the orange bands are less obvious. Flies weakly over mountain slopes June–July, usually keeping close to the ground. FP: mat grass.

SCOTCH ARGUS *Erebia aethiops* fw 20–25mm

Resembles a large Mountain Ringlet, but the eye-spots have white pupils and those under the hindwing are set in a silvery-grey band. Flies over upland bogs and moors July–August. FP: mainly blue moor grass and purple moor grass.

MARBLED WHITE *Melanargia galathea* fw 20–30mm

No other British butterfly has this pattern. Eye-spots are present on both surfaces, although not always obvious on the upperside. Flies over rough grassland June–September. FP: red fescue and other grasses. Caterpillar p. 276.

GRAYLING *Hipparchia semele* fw 20–30mm

This butterfly rests on the ground or low vegetation, beautifully camouflaged by its mottled grey undersides. The eye-spot is usually concealed. The orange and brown uppersides are never revealed at rest. Flies over heaths and rough grassland July–September. FP: various grasses. Caterpillar p. 276.

MEADOW BROWN *Maniola jurtina* fw 20–27mm

The male is darker than the female shown here and may lack the orange patch. The insect is abundant in all kinds of grassy places June–September. FP: various grasses. Caterpillar p. 276.

GATEKEEPER *Pyronia tithonus* fw 17–25mm

Although often confused with the Meadow Brown, this butterfly's upperside is much more orange and the eye-spot usually has 2 pupils. The female forewing lacks the central patch of dark scent scales seen here in the male. It frequents hedgerows and other scrubby places July–August. FP: various grasses.

SMALL HEATH *Coenonympha pamphilus* fw 12–18mm

The plain orange upperside is rarely seen at rest, and with the forewing pulled down to conceal the eye-spot, the butterfly is very hard to see. Flies over heaths and grasslands May–October. FP: mainly fine-leaved grasses.

LARGE HEATH *Coenonympha tullia* fw 15–22mm

Less brightly coloured than the previous species, this is a variable butterfly and specimens from the more northerly parts of the range lack the prominent eye-spots under the hindwing. It flies in boggy habitats June–August. FP: mainly cotton grass and other sedges. Caterpillar p. 276.

RINGLET *Aphantopus hyperantus* fw 20–25mm

The yellow-ringed eye-spots on the underside are characteristic. The blackish-brown upperside bears several indistinct eye-spots. Flies July–August, in and around woods and hedgerows, often feeding at bramble blossom. FP: various grasses. Caterpillar p. 276.

SWIFT MOTHS: FAMILY HEPIALIDAE

These are fast-flying moths in which all 4 wings are fairly long and narrow. The antennae are very short and there is no proboscis, so the adults cannot feed. The insects fly mainly at dusk. Females tend to be larger than males. Eggs are scattered in flight and the caterpillars feed on the roots of a wide range of herbaceous plants, including ferns. There are 5 species in the British Isles, although only 4 of them occur in Ireland.

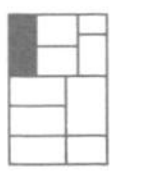

GHOST SWIFT *Hepialus humuli* fw 20–35mm

The upperside of the male is pure white, contrasting strongly with the brown underside and creating an eerie ghostlike effect when the moth flies. The white flashes are thought to attract the females, which are butter-coloured with irregular orange lines. June–August in many habitats, including gardens where the caterpillar damages both crops and weeds.

ORANGE SWIFT *Hepialus sylvina* fw 12–25mm

Only the male (A) forewing is really orange, crossed by a dark-edged white 'V'. The female (B) is duller and her pattern is less regular. June–September; common in many well-vegetated habitats, including gardens. FP: bracken is one of many acceptable food-plants.

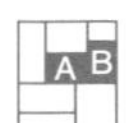

GOLD SWIFT *Hepialus hecta* fw 12–15mm

The male's forewing (A) is golden orange, crossed by 2 diagonal rows of cream or silvery spots. The inner row is often short or broken but, unlike in the Orange Swift and Common Swift, the rows are more or less parallel. There may also be a row of silvery dots close to the outer edge. The female (B) is dull brown, with paler diagonal stripes instead of silvery spots. June–July, mainly on heathland or in open woodland. FP: mainly bracken but also grasses.

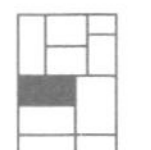

COMMON SWIFT *Hepialus lupulinus* fw 10–20mm

This is a very variable moth, with forewings ranging from dirty white to mid-brown. 2 rather irregular white lines usually form a 'V' and there may be a central white dot, but these markings are often very faint or even absent, especially in females. Abundant in grassy places and gardens May–July and regularly attracted to lights. FP: many wild and cultivated plants. Caterpillar p. 280.

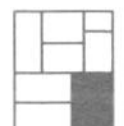

MAP-WINGED SWIFT *Hepialus fusconebulosa* fw 15–25mm

This species is named for the irregular white pattern, sometimes thought to resemble a map. Unlike those of other swift moths, the forewings have conspicuously chequered fringes, which are present even if other markings are faint. The sexes are alike, apart from size. May–August on moors, heaths, and rough grassland. FP: mainly bracken.

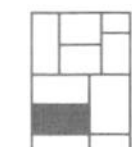

LEOPARD MOTH *Zeuzera pyrina* Family Cossidae fw 20–35mm

The 3 large black spots on each side of the thorax and the chequered wings immediately identify this common moth. June–August in many wooded areas. The caterpillar feeds in the trunks and branches of a wide range of deciduous trees, including apples and other cultivated fruit trees. The Puss Moth (*see* p. 210) has a similar coloration but a very different wing pattern.

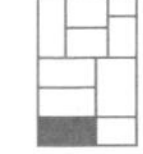

GOAT MOTH *Cossus cossus* fw 30–45mm

The wings of this heavily-built moth have the colour and pattern of cracked bark and provide excellent camouflage for the resting insect. June–July in wooded areas, including gardens and orchards, although it is not common. The caterpillar feeds in the trunks of many deciduous trees and has a strong goat-like smell.

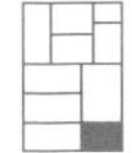

FESTOON *Apoda limacodes* Family Limacodidae fw 10–13mm

The orange-brown forewing is crossed by 2 darker lines, converging strongly towards the front edge and usually enclosing a dark area. The wing-tips are pressed against the substrate when at rest. June–July, flying by day as well as night. FP: mainly oak. The Lackey (*see* p. 162) is superficially similar, but is larger and the more or less parallel cross-lines run from the front to the rear margin.

CLEARWING MOTHS: FAMILY SESIIDAE

These day-flying moths get their name because most of the scales fall from their wings during their first flight, usually leaving just a dark border and a few dark spots. Both at rest and in flight, many clearwings bear a remarkable similarity to various wasps and sawflies, often emitting a characteristic buzz as they fly (*see* p. 51). The caterpillars all live in roots and stems, usually of trees and shrubs, and may take 2 or more years to mature. There are 14 resident species in the British Isles.

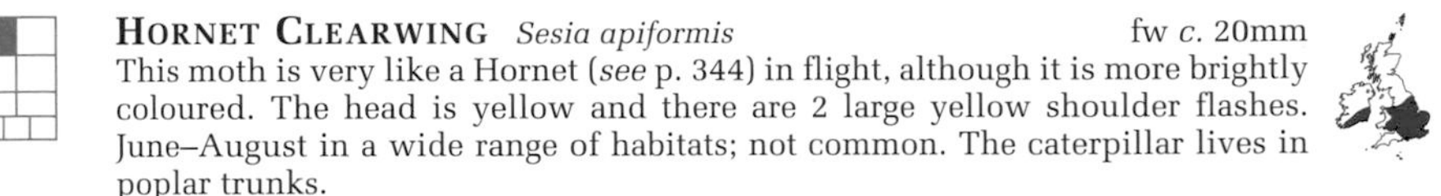

HORNET CLEARWING *Sesia apiformis* fw *c.* 20mm

This moth is very like a Hornet (*see* p. 344) in flight, although it is more brightly coloured. The head is yellow and there are 2 large yellow shoulder flashes. June–August in a wide range of habitats; not common. The caterpillar lives in poplar trunks.

LUNAR HORNET CLEARWING *Sesia bembeciformis* fw 15–20mm

This moth is very like the Hornet Clearwing but there are no yellow shoulder flashes. There is also a bright yellow collar. July–August, mainly in damp woodlands; much commoner than the previous species. The caterpillar lives in the trunks of willows and, less often, poplars.

CURRANT CLEARWING *Synanthedon tipuliformis* fw 8–10mm

This small clearwing has yellowish or orange tips to the forewings and 3 or 4 narrow yellow bands on the abdomen. There is also a thin yellow collar and the thorax usually has 2 slender yellow stripes. June–July, often basking on leaves with its wings open. The caterpillar lives in the twigs of currants and gooseberries.

SALLOW CLEARWING *Synanthedon flaviventris* fw 8–9mm

This species resembles the Currant Clearwing but lacks the yellowish wing-tips and yellow collar and never has yellow stripes on the thorax. June–July, mainly in damp woodlands. The caterpillar lives in the trunks and branches of sallows.

ORANGE-TAILED CLEARWING *Synanthedon andrenaeformis* fw *c.* 10mm

The tuft of orange-tipped scales at the end of the abdomen separates this from most other clearwings, although the orange is less extensive in the female than the male. The abdomen also bears thin yellow bands and the antennae are completely black. May–July on chalk and limestone, but generally rare. The caterpillar lives in the twigs and branches of the wayfaring tree and, less often, guelder rose.

YELLOW-LEGGED CLEARWING *Synanthedon vespiformis* fw 10–12mm

The largely yellow legs and orange or red bar crossing the outer part of the forewing distinguish this from our other clearwings. There are 4 yellow abdominal bands and the female also has a conspicuous yellow tail-tuft. May–August in wooded areas. The caterpillar feeds in various deciduous trees but most often in oak.

RED-TIPPED CLEARWING *Synanthedon formicaeformis* fw *c.* 10mm

The red-tipped forewings together with the single red abdominal band readily identify this species. May–August in damp woodland and riversides, but uncommon. The caterpillar lives in various willows.

THRIFT CLEARWING *Synansphecia muscaeformis* fw 6–8mm

This tiny, rather grey clearwing has 3 or 4 very narrow pale bands on the abdomen. It resembles the Currant Clearwing but has no orange on the wing-tips. June–August on and around cliffs and rocky coasts. FP: thrift.

WELSH CLEARWING *Synanthedon scoliaeformis* fw 12–15mm

This very rare moth resembles the Orange-tailed Clearwing in having an orange-brown tuft at the rear, but it is a larger insect and the dark patch in the middle of the forewing has a very pointed inner margin. The antennae have pale tips, especially in the female. June–July in open birchwoods and on moorland. FP: downy birch.

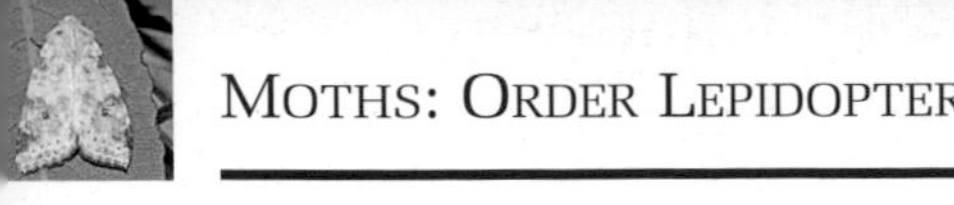

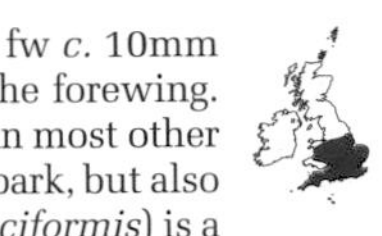

RED-BELTED CLEARWING *Synanthedon myopaeformis* fw *c.* 10mm
The abdomen has a single red band and there is no trace of red on the forewing. June–August, often in orchards; more likely to be seen on the wing than most other clearwings. The caterpillar feeds in apple trees, usually just under the bark, but also attacks other rosaceous trees. The **Large Red-belted Clearwing** (*S. culiciformis*) is a little larger and has a dusting of reddish scales at the base of the forewing.

SIX-BELTED CLEARWING *Bembecia ichneumoniformis* fw 9–12mm
Named for the 6 pale abdominal bands, it has a fairly extensive dusting of orange scales on the outer part of the forewing. June–August in rough grassland, especially on chalk and limestone, but always rare. FP: bird's-foot trefoil and kidney vetch.

FIERY CLEARWING *Pyropteron chrysidiformis* fw 9–12mm
The only clearwing with fiery red scales coating much of the forewing. There is also an orange tuft at the end of the abdomen. June–July in sunny coastal areas, but now endangered. The caterpillar lives in the roots of docks, especially curled dock, and sorrel.

BURNET AND FORESTER MOTHS: FAMILY ZYGAENIDAE

These brightly-coloured, day-flying moths are protected by distasteful poisons. They are rather sluggish (*see* p. 46) and spend a lot of time resting on flowers. When they do take to the air, their narrow wings whirr rapidly, but flight is slow and they seem to drift from flower to flower. Burnets pupate in papery cocoons, often freely exposed on grass stems. There are 10 British species.

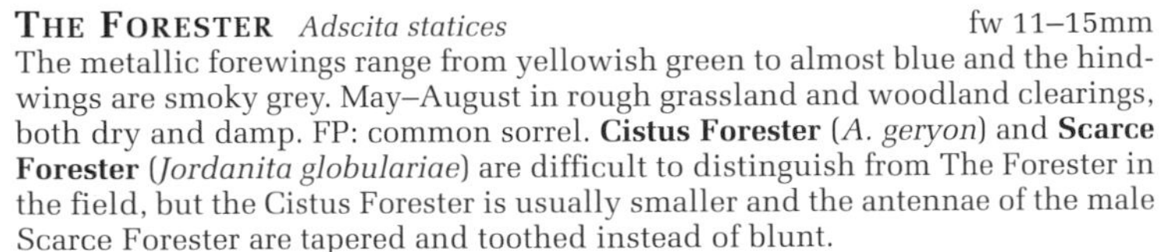

THE FORESTER *Adscita statices* fw 11–15mm
The metallic forewings range from yellowish green to almost blue and the hindwings are smoky grey. May–August in rough grassland and woodland clearings, both dry and damp. FP: common sorrel. **Cistus Forester** (*A. geryon*) and **Scarce Forester** (*Jordanita globulariae*) are difficult to distinguish from The Forester in the field, but the Cistus Forester is usually smaller and the antennae of the male Scarce Forester are tapered and toothed instead of blunt.

SIX-SPOT BURNET *Zygaena filipendulae* fw 15–20mm
The forewings are black with a strong blue or green sheen when fresh, and each has 6 red spots, although the basal pair are separated by only a very thin black line. As in all our burnets, the hindwings are red with a black border. The antennae are clubbed in all burnets. June–August in grassy places. FP: mainly bird's-foot trefoil. Caterpillar p. 280.

FIVE-SPOT BURNET *Zygaena trifolii* fw 14–20mm

Five-spot Burnet
Narrow-bordered Five-spot Burnet

The forewings have 5 red spots, although the 2 central ones are sometimes joined. May–August in all kinds of grassy places, including moorland. FP: bird's-foot trefoil and greater bird's-foot trefoil. The **Narrow-bordered Five-spot Burnet** (*Z. lonicerae*) is very similar but the hindwing border is narrower and, except in Ireland, it is rare for the 2 middle spots to touch. Insects seen before late June are likely to be Five-spot Burnets.

TRANSPARENT BURNET *Zygaena purpuralis* fw *c.* 15mm
The wings are thinly-scaled and the forewing has 3 long streaks instead of spots, although these are sometimes very faint. Scottish specimens tend to be blacker than Irish. June–July on grassy slopes. FP: wild thyme.

SCOTCH BURNET *Zygaena exulans* fw 10–15mm
Also called the Mountain Burnet, this moth has thinly-scaled wings with 5 distinct red spots on each forewing, although these spots are often quite small. June–July on high mountain slopes, above about 700m. FP: mainly crowberry, but also cowberry and bilberry.

SLENDER SCOTCH BURNET *Zygaena loti* fw *c.* 15mm
This species is thinly scaled and 5-spotted like the Scotch Burnet, but the outermost spot on the forewing is very large and often triangular. The hindwing border is narrower and more distinct. June–July, but now very rare and confined to a few coastal localities. FP: bird's-foot trefoil.

LAPPETS AND EGGARS: FAMILY LASIOCAMPIDAE

Most of these moths are heavily-built and rather hairy. The males, which are noticeably smaller than the females, have large, feathery antennae. The insects have no tongue, so cannot feed. The larvae are clothed with irritating hairs, many of which are later incorporated into the cocoons. There are 10 resident British species.

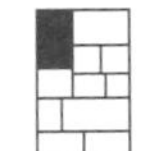

DECEMBER MOTH *Poecilocampa populi* fw 15–25mm

The thinly-scaled, sooty black wings, crossed by 2 wavy cream lines, are unlike those of any other moth flying in the winter. October–January; most frequent in woodland, but not uncommon in parks and gardens. FP: a wide range of deciduous trees. Caterpillar p. 280.

PALE EGGAR *Trichiura crataegi* fw 14–18mm

The gently undulating outer edge of the central band on the forewing should separate this moth from similar species. Females are somewhat browner than the male pictured here, and northern specimens may be darker. August–September, mainly in open woods and scrub. FP: mainly birch, but also many other trees and shrubs.

SMALL EGGAR *Eriogaster lanestris* fw 15–20mm

Although superficially like the December Moth, this species is much browner and can always be distinguished by the white spot near the middle of the forewing. The female has a large tuft of grey hair at the tip of the abdomen. February–March in lightly wooded areas, scrub and hedgerows. FP: mainly blackthorn and hawthorn. Caterpillar p. 280.

THE LACKEY *Malacosoma neustria* fw 12–20mm

The wings range from straw-coloured to brick-red and the forewings are crossed by 2 more-or-less parallel lines, which are dark when the wings are pale and vice versa. The outer fringe bears 2 conspicuous white patches. June–August; abundant in many habitats, including towns and gardens. FP: many deciduous trees and shrubs, including orchard trees. Caterpillar p. 280.

OAK EGGAR *Lasiocampa quercus* fw 25–40mm

Rich brown wings, each crossed by a yellow band, readily identify the male (A). The female (B) is yellowish brown, but the pale bands are still visible. Both sexes have a bright white spot on the forewing. May–August, on heaths and moors and in open woodland. The male flies by day. FP: a wide range of shrubs including bramble, blackthorn and heather. Caterpillar p. 280.

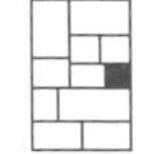

GRASS EGGAR *Lasiocampa trifolii* fw 20–30mm

The wings of this rare moth are normally brick-red with a white spot on the forewing and a curved, pale cross-line beyond it. There may be another line near the base. The hindwing is unmarked. August–September, mainly on sand dunes. FP: many herbaceous plants.

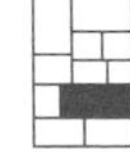

FOX MOTH *Macrothylacia rubi* fw 20–30mm

The forewings are crossed by 2 thin, more or less parallel lines. The hindwings are unmarked. Males are reddish brown and females greyish brown. May–June on heaths and grassland and in open woods, the male flying by day. FP: bramble and many other plants.

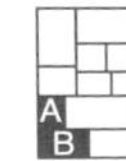

THE DRINKER *Euthrix potatoria* fw 20–35mm

Named for the larval habit of sipping dew, this moth can be recognised by the 2 white spots and the diagonal cross-line running to the wing-tip. Males (A) are much darker than females (B). July–August, common in grassy places. FP: various coarse grasses. Caterpillar p. 280.

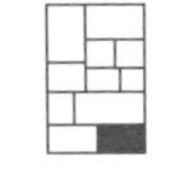

THE LAPPET *Gastropacha quercifolia* fw 25–45mm

The serrated wing margins and the unusual resting position, with the hindwings flat and the forewings held roof-like, readily identify this moth. It is usually purplish brown, but pale brown individuals occasionally occur. June–August in wooded areas. FP: mainly rosaceous trees, including apples.

HOOKTIP MOTHS: FAMILY DREPANIDAE

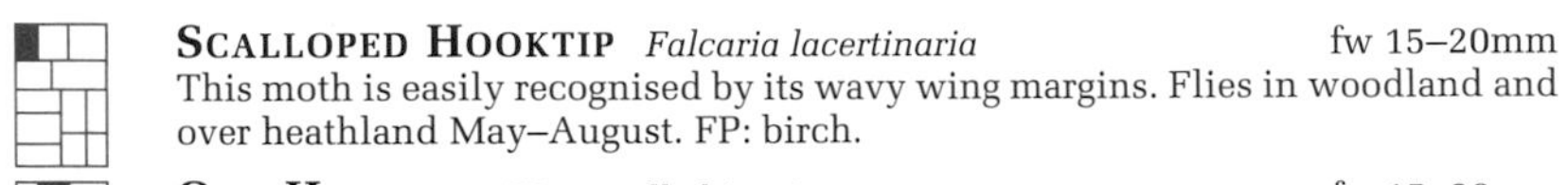

Named for the slightly hooked wing-tips of most of the species, these rather flimsy moths are represented in the British Isles by just 6 resident species. Females are usually noticeably larger than males. Their caterpillars (*see* p. 280) rest with both front and rear ends raised and the rear end tapers to a point without functional claspers.

SCALLOPED HOOKTIP *Falcaria lacertinaria* fw 15–20mm
This moth is easily recognised by its wavy wing margins. Flies in woodland and over heathland May–August. FP: birch.

OAK HOOKTIP *Watsonalla binaria* fw 15–20mm
The 2 small black spots on each forewing easily separate this moth from the Barred Hooktip. May–August in parks and woodlands. FP: oak. Caterpillar p. 280.

BARRED HOOKTIP *Watsonalla cultraria* fw 15–20mm
This is the only one of our hooktips in which the forewing is crossed by a broad band, usually considerably darker than the rest of the wing. May–August, mainly in beech woods. FP: beech. Caterpillar p. 280.

PEBBLE HOOKTIP *Drepana falcataria* fw 15–20mm
The ground colour ranges from white to rusty brown, but the brown stripe and dark 'pebble' near the centre of the forewing distinguish it from the Scarce Hooktip. Northern specimens tend to be paler than southern ones. May–August in woodlands and heathlands. FP: birch and alder.

SCARCE HOOKTIP *Sabra harpagula* fw *c.* 20mm
The dark smudge on the outer margin of the strongly hooked forewing readily identifies this very local insect, which flies June–July in a small area of the Wye Valley. FP: small-leaved lime.

CHINESE CHARACTER *Cilix glaucata* fw *c.* 10mm
This moth lacks hooked wing-tips and is named for the delicate silvery 'scribbles' on its forewing. At rest, it is remarkably like a bird dropping and it is rarely taken by birds. May–August in shrubby places, including gardens. FP: hawthorn, blackthorn and other rosaceous shrubs.

BUFF ARCHES *Habrosyne pyritoides* Family Thyatiridae fw *c.* 20mm
Readily identified by the white-banded wings and the prominent collar, it flies in woods, gardens and other scrubby places June–August. FP: mainly bramble.

PEACH BLOSSOM *Thyatira batis* fw *c.* 18mm
This moth is named for the delicate pink or buff flower-like spots on its forewings. The pattern provides surprisingly good camouflage when the moth is at rest on a twig. May–September in wooded areas. FP: bramble.

FIGURE-OF-EIGHTY *Tethea ocularis* fw 15–20mm
The ground colour ranges from light to dark brown, but the white figure 80 in the centre of each forewing readily identifies this common moth. May–July in wooded areas, including parks and gardens. FP: poplars. Figure-of-eight (*see* p. 214) has a similar but bolder pattern and flies in autumn.

FROSTED GREEN *Polyploca ridens* fw *c.* 16mm
The forewings are basically white with greenish-black marbling and chequered fringes. In many specimens the central region is completely dark but the zigzag white line usually remains visible towards the outer margin. April–May in and around mature woodlands. FP: oak.

POPLAR LUTESTRING *Tethea or* fw 16–20mm
The markings vary in intensity but the wavy cross-lines, the 'lute strings', in the outer part of the forewing are usually clear and distinguish the Poplar from other lutestrings. May–August, mainly in and around damp woodlands. FP: aspen.

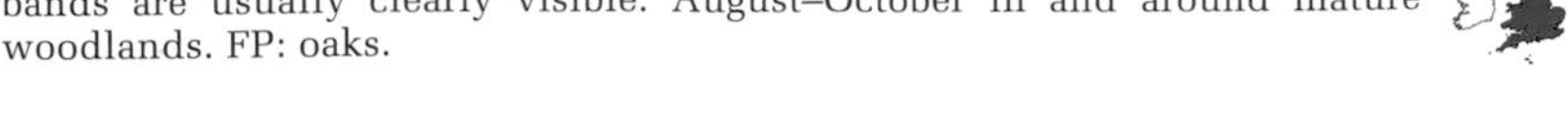

OAK LUTESTRING *Cymatophorima diluta* fw *c.* 16mm
The forewings range from pale grey to brownish but the two dark-edged cross-bands are usually clearly visible. August–October in and around mature woodlands. FP: oaks.

GEOMETER MOTHS: FAMILY GEOMETRIDAE

These moths are mostly quite flimsy and slender-bodied. Most rest with their wings spread and pressed against the surface, often with the hindwings clearly visible. Some females are wingless. The larvae have little or no hair and only 2 pairs of prolegs at the rear. They are often called loopers from the way in which they stretch forward to grip with their front legs and then arch the body as the rear legs are brought forward (*see* p. 282). Many of these caterpillars are incredibly twig-like. There are over 300 British species in the family.

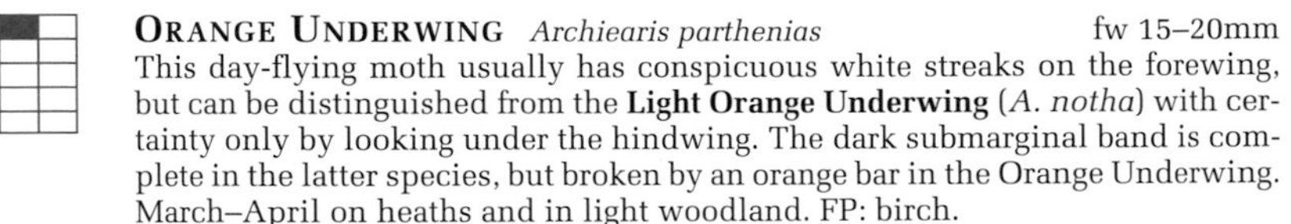

ORANGE UNDERWING *Archiearis parthenias* fw 15–20mm
This day-flying moth usually has conspicuous white streaks on the forewing, but can be distinguished from the **Light Orange Underwing** (*A. notha*) with certainty only by looking under the hindwing. The dark submarginal band is complete in the latter species, but broken by an orange bar in the Orange Underwing. March–April on heaths and in light woodland. FP: birch.

MARCH MOTH *Alsophila aescularia* fw 16–20mm
Only the male is winged, and rests in an unusual position for geometer moths, with the forewings partly overlapping. No other spring-flying species has the prominent white zigzag on the forewing. February–April in most wooded areas. FP: many deciduous trees and shrubs.

GRASS EMERALD *Pseudoterpna pruinata* fw 15–20mm
The dark cross-lines distinguish this from our other emeralds. The ground colour is also much bluer, although, in common with all emeralds, it tends to fade with age and after death. June–September in many open habitats. FP: mainly gorse and broom.

LARGE EMERALD *Geometra papilionaria* fw 25–30mm
Its size readily distinguishes this moth from our other emeralds. It usually rests with its wings partly raised and is often mistaken for a butterfly. June–August, mainly in wooded country and heathland. FP: mainly birch and hazel.

BLOTCHED EMERALD *Comibaena bajularia* fw *c.* 15mm
The wing pattern of this rather local species is unmistakable. June–July in and around mature woodlands. FP: oaks.

COMMON EMERALD *Hemithea aestivaria* fw 14–17mm
The chequered brown margins and the rather pointed hindwings readily distinguish this moth from our other emeralds apart from the Sussex Emerald. The male antennae are hairlike. June–July almost everywhere. FP: hawthorn, blackthorn and many other deciduous trees.

SUSSEX EMERALD *Thalera fimbrialis* fw *c.* 15mm
This is very like the Common Emerald, but the fringes are chequered with red and the margin of the hindwing has 2 points with a slight concavity between them. The male antennae are feathered. July–August. FP: wild carrot. A very rare resident and occasional immigrant.

SMALL GRASS EMERALD *Chlorissa viridata* fw *c.* 12mm
The front edge of the forewing is very straight, with a distinct golden sheen, and the outer white band is almost straight. The male antennae are hairlike. June–July on damp heaths and grassland. FP: various low-growing shrubs.

SMALL EMERALD *Hemistola chrysoprasaria* fw 17–20mm
This moth is very like the Small Grass Emerald, but larger and with a curved front margin to the forewing. The outer cross-line is also gently curved and the male antennae are feathered at the base. June–September in and around woods and hedgerows. FP: clematis.

LITTLE EMERALD *Jodis lactearia* fw *c.* 12mm
The palest of our emeralds, often fading to almost white, this species also differs from most in having 2 cross-lines on the hindwing. May–June in wooded areas. FP: birch and many other deciduous trees.

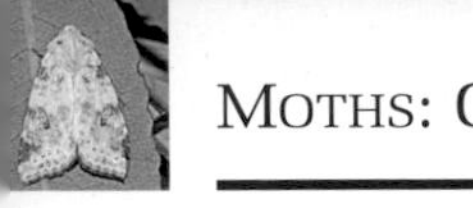

THE MOCHA *Cyclophora annularia* fw 11–14mm
This moth is unlikely to be confused with any other. None of its superficially similar relatives has an outer dark band on the hindwing. May–August in and around woods and hedgerows. FP: field maple.

BIRCH MOCHA *Cyclophora albipunctata* fw 12–14mm
The ground colour of the wings is usually some shade of grey and many specimens have a darker, often reddish band running across each wing and enclosing the central rings. May–August, mainly in open woodland and on heathland. FP: birch.

FALSE MOCHA *Cyclophora porata* fw 12–14mm
The ground colour of the wings is pale brown, often heavily stippled with black and sometimes with a pink flush on the forewings. The dark band may be faint or even missing. May–September, mainly in woodland. FP: oak.

MAIDEN'S BLUSH *Cyclophora punctaria* fw 12–15mm
The ground colour of the wings is off-white or pale yellow, usually heavily speckled with grey. There is a single reddish-brown cross-line on each wing, but the pink blush on the forewing is often very faint, especially in spring specimens. Dark blotches may appear at the rear corner of the forewing. May–September in and around oakwoods. FP: oak.

CLAY TRIPLE-LINES *Cyclophora linearia* fw *c.* 15mm
The ground colour is yellowish brown in spring insects, but 2nd brood moths are duller. There are usually 3 cross-lines but the inner and outer lines are often missing. Some summer insects have a small dark central ring on the hindwing. May–October in woods and parkland. FP: beech.

BLOOD-VEIN *Timandra griseata* fw 15–20mm
This well-named and common moth is easily recognised by its pink fringes and the pink or purplish line, the 'vein', crossing all 4 wings. May–October in most open habitats. FP: docks and related plants.

SMALL BLOOD-VEIN *Scopula imitaria* fw 12–13mm
The ground colour is sandy brown, with a small black central dot on each wing. A dark 'vein' crosses each wing, but it runs from the front edge of the forewing rather than from the wing-tip. July–September in many habitats. FP: various shrubs, including privet.

TAWNY WAVE *Scopula rubiginata* fw *c.* 10mm
The ground colour of this rare moth ranges from bright pink to a rather dull reddish brown and each wing is usually crossed by 3 narrow dark bands. The area between the 2 outer bands is often paler than the rest of the wing. June–September, mainly on heathland. FP: unknown.

MULLEIN WAVE *Scopula marginepunctata* fw 12–15mm
The creamy white wings have a variable dusting of dark scales and their appearance ranges from dirty white to almost grey, but the wavy lines are usually visible in the outer part of each wing, as are the marginal dots. There is a conspicuous central black dot on each wing and there are usually 1 or 2 on the front edge of the forewing. The **Dotted Border Wave** (*Idaea sylvestraria*) has smaller and more rounded marginal dots. June–September, mainly in coastal habitats. FP: a wide range of herbaceous plants, including marjoram and plantains.

LACE BORDER *Scopula ornata* fw *c.* 12mm
This moth, with its intricate border, cannot be mistaken for any other. May–September in grassy places, especially on chalk and limestone. FP: thyme, marjoram and other related plants.

SMOKY WAVE *Scopula ternata* fw 12–15mm
Named for its dull grey appearance, it is heavily dusted with dark scales but lacks the black spots of many of its relatives. The 3 wavy cross-lines are often very hard to see. The Common Wave (*see* p. 202) has a similar colour but less-pointed wings. June–July, mainly on heaths and moors. FP: mainly heather and bilberry.

CREAM WAVE *Scopula floslactata* fw *c.* 15mm
The distinctly curved front edge of the forewing, the black dots on the outer margins of all the wings, and the small but distinct black dot in the centre of the hindwing combine to distinguish this from other similar moths. May–July in woods and hedgerows. FP: mainly bedstraws.

LESSER CREAM WAVE *Scopula immutata* fw *c.* 12mm
Size is the best distinction from the previous species, but the cross-lines are often stronger in the outer part of each wing and a small black central dot is often present on all 4 wings. June–August, mainly in damp places. FP: meadowsweet.

PURPLE-BORDERED GOLD *Idaea muricata* fw 8–10mm
The extent of the gold varies and some specimens are almost entirely purple, with just a small gold spot on each wing. The Gold Fringe (*see* p. 268) and some other micro-moths have similar colours but are usually much darker. June–July in damp habitats. FP: various herbs.

LEAST CARPET *Idaea rusticata* fw *c.* 10mm
Although the dark markings vary in intensity, the smoky brown patch at the base of the forewing and the central black spot on the hindwing separate this moth from the other small carpets. May–October in scrubby areas, including hedgerows and gardens. FP: mainly traveller's joy.

ROSY WAVE *Scopula emutaria* fw *c.* 12mm
The wings commonly have a delicate pink tinge and a dusting of grey scales. The forewing has a soft grey central cross-line that usually dies away before reaching the front edge. Both wings have a line of small dark dots in the outer part; the hindwing is clearly pointed. June–July on heaths, bogs and salt-marshes. FP: unknown.

SMALL FAN-FOOTED WAVE *Idaea biselata* fw *c.* 10mm
This moth is very like the Dotted Border Wave, but often whiter and the outer cross-lines form a grey lace-like border, usually with a white line running through it. June–August in wooded areas and hedgerows. FP: various low-growing plants.

DWARF CREAM WAVE *Idaea fuscovenosa* fw *c.* 10mm
The dark patches at the base of the forewing and along its front edge help to identify this little moth. There is a small black dot in the centre of each wing. June–July in rough grassy areas and scrub. FP: various low-growing plants.

SMALL DUSTY WAVE *Idaea seriata* fw *c.* 10mm
The wings are usually heavily dusted with grey. Each wing has a black dot near the centre, with a cross-line of dots or dashes just outside it and another on the outer margin. June–September, especially in gardens and on waste ground. FP: various low-growing plants.

SINGLE-DOTTED WAVE *Idaea dimidiata* fw *c.* 10mm
The dotted margins and the dark smudge at the rear corner of each forewing are characteristic of this species. June–August in wooded and scrubby areas, including gardens. FP: probably many plants, including the flowers of various umbellifers.

SATIN WAVE *Idaea subsericeata* fw 10–12mm
Named for its somewhat silky white wings, this moth has no obvious markings other than the faint grey cross-lines. The Small White Wave (*see* p. 192) is similar but has more obvious brown cross-lines. June–September in scrubby areas. FP: probably a variety of low-growing plants.

TREBLE BROWN-SPOT *Idaea trigeminata* fw *c.* 10mm
The dark flash at the base of the forewing and the dark outer band, more or less broken into three separate blotches, characterise this rather local moth. May–September, mainly in open woodland. FP: ? ivy.

SMALL SCALLOP *Idaea emarginata* fw *c.* 12mm
The moth is unlikely to be confused with any other because of its scalloped outline. The female is often darker than the male and may have a dark brown linear smudge near the centre of each wing. June–August, mainly in damp grassland. FP: mainly bedstraws.

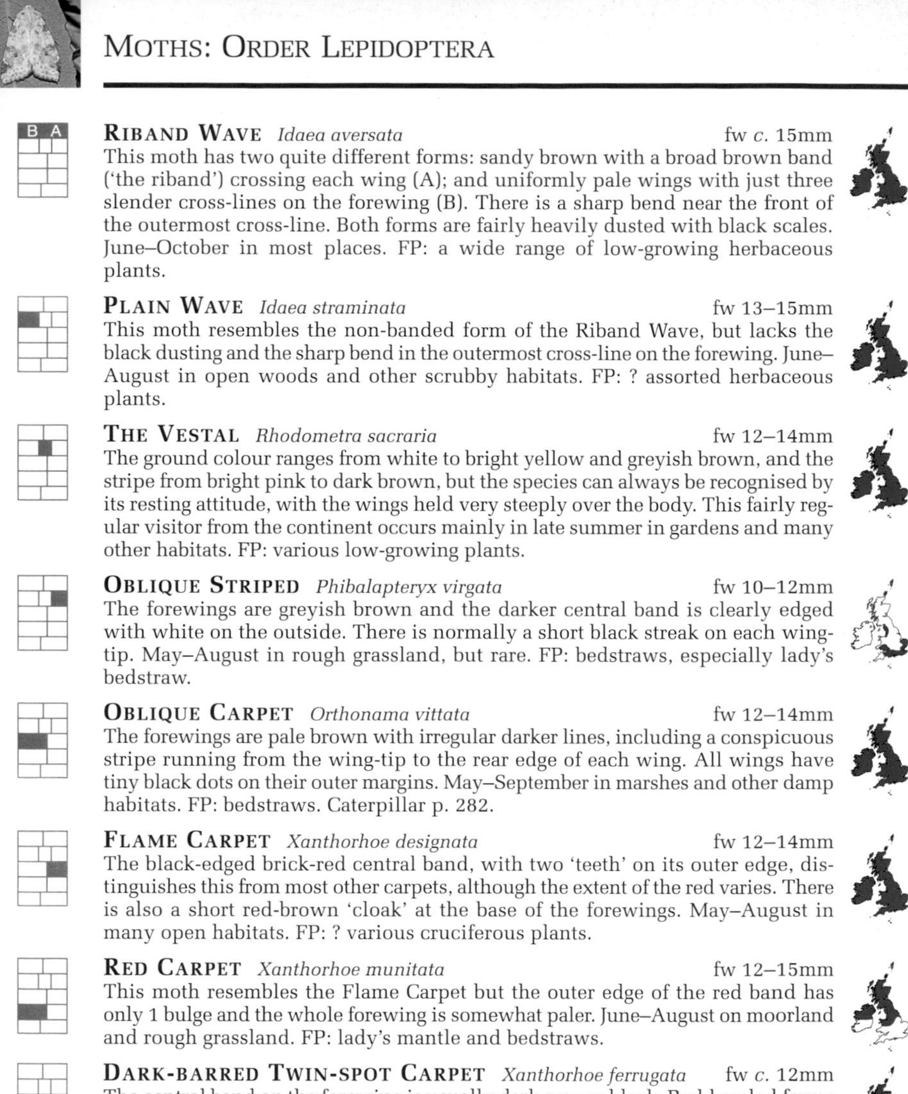

RIBAND WAVE *Idaea aversata* fw *c.* 15mm
This moth has two quite different forms: sandy brown with a broad brown band ('the riband') crossing each wing (A); and uniformly pale wings with just three slender cross-lines on the forewing (B). There is a sharp bend near the front of the outermost cross-line. Both forms are fairly heavily dusted with black scales. June–October in most places. FP: a wide range of low-growing herbaceous plants.

PLAIN WAVE *Idaea straminata* fw 13–15mm
This moth resembles the non-banded form of the Riband Wave, but lacks the black dusting and the sharp bend in the outermost cross-line on the forewing. June–August in open woods and other scrubby habitats. FP: ? assorted herbaceous plants.

THE VESTAL *Rhodometra sacraria* fw 12–14mm
The ground colour ranges from white to bright yellow and greyish brown, and the stripe from bright pink to dark brown, but the species can always be recognised by its resting attitude, with the wings held very steeply over the body. This fairly regular visitor from the continent occurs mainly in late summer in gardens and many other habitats. FP: various low-growing plants.

OBLIQUE STRIPED *Phibalapteryx virgata* fw 10–12mm
The forewings are greyish brown and the darker central band is clearly edged with white on the outside. There is normally a short black streak on each wing-tip. May–August in rough grassland, but rare. FP: bedstraws, especially lady's bedstraw.

OBLIQUE CARPET *Orthonama vittata* fw 12–14mm
The forewings are pale brown with irregular darker lines, including a conspicuous stripe running from the wing-tip to the rear edge of each wing. All wings have tiny black dots on their outer margins. May–September in marshes and other damp habitats. FP: bedstraws. Caterpillar p. 282.

FLAME CARPET *Xanthorhoe designata* fw 12–14mm
The black-edged brick-red central band, with two 'teeth' on its outer edge, distinguishes this from most other carpets, although the extent of the red varies. There is also a short red-brown 'cloak' at the base of the forewings. May–August in many open habitats. FP: ? various cruciferous plants.

RED CARPET *Xanthorhoe munitata* fw 12–15mm
This moth resembles the Flame Carpet but the outer edge of the red band has only 1 bulge and the whole forewing is somewhat paler. June–August on moorland and rough grassland. FP: lady's mantle and bedstraws.

DARK-BARRED TWIN-SPOT CARPET *Xanthorhoe ferrugata* fw *c.* 12mm
The central band on the forewing is usually dark grey or black. Red-banded forms may occur, but the species can always be recognised by the prominent notch near the front of the inner margin of the band. May–August in many habitats. FP: a wide range of herbaceous plants.

RED TWIN-SPOT CARPET *Xanthorhoe spadicearia* fw *c.* 12mm
The dark band on the forewing ranges from red to very dark brown and has fairly distinct white edges, but there is never a deep notch near the front of the inner margin as there is in the Dark-barred Twin-spot Carpet. April–August in many habitats. FP: various low-growing herbs.

LARGE TWIN-SPOT CARPET *Xanthorhoe quadrifasiata* fw *c.* 15mm
Slightly larger than the two previous species and often with indistinct twin spots, this moth ranges from reddish brown to grey-brown and the dark band on the forewing usually has a paler band within it. The outer edge of the band has a strong 'tooth'. June–August. FP: bedstraws and other low-growing herbs.

SILVER-GROUND CARPET *Xanthorhoe montanata* fw 14–18mm
The ground colour is silvery grey, often tinged with brown. The irregular dark band is often broken in the middle; it is brown or grey, usually with a pale blotch near the front, and the very black 'tooth' on its outer margin is conspicuous in the centre of the wing even when the rest of the band is pale. May–August in many habitats. FP: bedstraws and other low-growing herbs.

GARDEN CARPET *Xanthorhoe fluctuata* fw 12–15mm
The ground colour ranges from white to grey and the dark band usually fades away in the rear half of the forewing. The thorax and the wing bases are very dark, forming a short 'cloak' when the moth is at rest. April–October almost everywhere. FP: wild and cultivated crucifers. Caterpillar p. 282.

CHALK CARPET *Scotopteryx bipunctaria* fw 15–18mm
The pale grey forewing is crossed by two slightly darker, wavy bands and there are two small, but conspicuous central black dots. July–August on chalk and limestone grassland, including coastal cliffs. FP: bird's-foot trefoil and related plants.

SHADED BROAD BAR *Scotopteryx chenopodiata* fw 15–20mm
The ground colour of the forewings ranges from dirty yellow to orange-brown. The central band usually consists of several shades of brown and there is normally a dark streak at the wing-tip. The Mallow (*see* p. 176) is larger and has a white zigzag in the outer part of the forewing. June–September in grassy places. FP: clovers and other herbaceous legumes.

LEAD BELLE *Scotopteryx mucronata* fw 15–20mm
The forewing is some shade of grey, often tinged with brown, although northern specimens are completely brown. A dark streaks runs from the wing-tip to a faint white zigzag submarginal line. May–June, mainly on heaths and moors. FP: gorse, broom and related plants. The **July Belle** (*S. luridata*) is almost identical but flies June–August.

ROYAL MANTLE *Catarhoe cuculata* fw 12–15mm
The black and brown 'cloak' at the base of the forewing and the chestnut blotch near the outer edge will always distinguish this moth from superficially similar species. June–July in grassy places on chalk and limestone. FP: bedstraws.

COMMON CARPET *Epirrhoe alternata* fw *c.* 14mm
The central band of the forewing ranges from grey to very black but, unlike that of the Garden Carpet, it nearly always extends right across the wing. The inner white band is quite narrow, and the outer one usually has a fine grey line running through it. May–October almost everywhere. FP: cleavers and bedstraws.

SMALL ARGENT AND SABLE *Epirrhoe tristata* fw *c.* 12mm
This moth resembles the Common Carpet, but the outer margins are just as dark as the central band and the fringes are conspicuously chequered. May–August, often flying by day, mainly on upland moors and grasslands. FP: mainly heath bedstraw (*see also* Argent and Sable p. 182).

WOOD CARPET *Epirrhoe rivata* fw *c.* 15mm
Distinguished from both the Common Carpet and the Galium Carpet by the broader and cleaner outer white band. It also differs from the Galium Carpet in the straight or slightly convex front edge of its forewing. The cross-band and the dark basal area often have a rusty tinge. June–August in and around woods and scrubby habitats. FP: bedstraws.

GALIUM CARPET *Epirrhoe galiata* fw 13–15mm
Distinguished from other carpets by the slightly concave front edge of its forewing. The cross-band ranges from grey to black, often enclosing thin brown lines. The basal area often has a rusty tinge. May–August in grassy places. FP: bedstraws.

PURPLE BAR *Cosmorhoe ocellata* fw 13–15mm
The pattern, with its irregular purplish-brown band on a creamy background, is unlike that of any other species. In common with several close relatives, the insect habitually rests with its abdomen sticking vertically upwards. May–September in grassy and scrubby habitats. FP: bedstraws.

YELLOW SHELL *Camptogramma bilineata* fw 12–15mm
The ground colour ranges from pale yellow to orange-brown, with the darker specimens more common in the north. Numerous fine brown cross-lines are always present and there may be one or two broader brown bands of variable width and density. May–August almost everywhere. FP: dandelion, dock and many other low-growing plants.

THE MALLOW *Larentia clavaria* fw *c.* 20mm
The pointed forewing has a faint zigzag line just inside the outer margin. The dark band is edged with white and its inner edge is deeply notched. September–November in many habitats, including roadsides and gardens. FP: mallows and garden hollyhocks (*see also* Shaded Broad-bar p. 174).

SHOULDER STRIPE *Anticlea badiata* fw 15–18mm
With its short striped or banded 'cloak' and pale central band, this moth is unlikely to be confused with any other species. February–April, mainly in and around hedgerows and woodland margins. FP: wild roses.

THE STREAMER *Anticlea derivata* fw *c.* 15mm
The ground colour is usually silvery grey, often with a purple tinge. The basal part of the forewing is crossed by two black bands. The moth is named for the wavy black line streaming back from the front edge of the forewing. March–May in and around woods and hedgerows. FP: wild roses.

BEAUTIFUL CARPET *Mesoleuca albicillata* fw 15–18mm
The lack of any central dark band immediately distinguishes this moth from other black-and-white species. The Scorched Carpet (*see* p. 192) is smaller and has pale wing-tips. May–August, mainly in open woodland. FP: mainly bramble and raspberry.

DARK SPINACH *Pelurga comitata* fw 15–18mm
The ground colour of the forewing varies from pale yellow to light brown, and the broad central band is composed of several shades of brown with a small central black spot. Both wings have chequered fringes. July–August in many open habitats. FP: goosefoot and orache.

WATER CARPET *Lampropteryx suffumata* fw 15–17mm
There are several superficially similar moths, but this one can usually be identified by its rather shiny forewings and bright white edges to the central dark band. There is a small grey or white patch right at the base of each forewing and the area just beyond the central band is often very white. April–May in damp habitats. FP: bedstraws.

THE PHOENIX *Eulithis prunata* fw 17–20mm
This moth resembles the Water Carpet but is slightly larger and usually browner, and the white markings are less conspicuous. There are also some dark arrowheads near the wing-tip. July–August in woods and gardens. FP: currants and gooseberries.

SMALL PHOENIX *Ecliptopera silaceata* fw 12–17mm
The pattern is very like that of The Phoenix but most specimens are noticeably smaller. The dark central band is often broken in the middle. Several dark arrowheads are prominent in the outer part of the forewing. April–August in many habitats. FP: willowherbs.

STRIPED TWIN-SPOT CARPET *Nebula salicata* fw 12–15mm
The ground colour of this inconspicuous little moth ranges from pale to dark grey and there are numerous light and dark cross-lines, including a dotted white line near the outer margin. The central band is not always distinct from the rest of the wing. May–September, mainly in upland areas. FP: bedstraws.

BARRED YELLOW *Cidaria fulvata* fw 12–14mm
The coloration of the forewing makes this moth unmistakable. June–August in and around woods and hedgerows and also in gardens. FP: wild and cultivated roses.

THE CHEVRON *Eulithis testata* fw 12–20mm
Named for the chevron-shaped central cross-band, this moth can also be recognised by the white 'V' near the wing-tip. The ground colour ranges from orange to brown and paler specimens usually have an orange blotch near the wing-tip. July–September, mainly on heaths and moors. FP: mainly willows and birches.

NORTHERN SPINACH *Eulithis populata* fw 12–18mm
The ground colour ranges from yellow to brown but the dark central band is always visible. The moth resembles the Dark Spinach (*see* p. 176) but the fringes are not chequered and the central band has a double 'tooth' on its outer edge. July–August on heaths and moors. FP: bilberry.

THE SPINACH *Eulithis mellinata* fw 15–18mm
This moth and the Barred Straw have a characteristic resting attitude, with the wings held horizontally above the surface and the hindwings almost completely concealed. The abdomen curls upwards. The wing patterns are similar except that the Barred Straw does not have chequered fringes. June–August, mainly in woods and urban areas. FP: currants.

BARRED STRAW *Eulithis pyraliata* fw 15–18mm
This moth is very like The Spinach but the outer part of the forewing bears a number of small brown arrowheads and the fringes are not chequered. May–September, in and around woods, hedgerows and rough grassland. FP: bedstraws and cleavers.

RED-GREEN CARPET *Chloroclysta siterata* fw *c.* 15mm
The forewings are some shade of green, usually with orange or red marbling and a pale blotch on the front edge, just beyond the dark central band. The hindwing is dark grey. Wooded areas September–May, but dormant in coldest months. FP: various broadleaved trees.

AUTUMN GREEN CARPET *Chloroclysta miata* fw *c.* 16mm
This moth resembles the Red-green Carpet but has no reddish marbling on the forewings. The white cross-lines are more obvious and the hindwings are almost white. November–May in wooded areas, but dormant in coldest months. FP: mainly sallows and birches.

DARK MARBLED CARPET *Chloroclysta citrata* fw 15–20mm
Elaborately marbled with brown, black and white, this extremely variable species is not easily distinguished from the Common Marbled Carpet. The best way to distinguish the species is to look at the underside of the hindwing, where the Dark Marbled Carpet has a sharply bent dark line and the Common Marbled Carpet has a smoothly curved one. July–October in most habitats. FP: various trees and shrubs.

COMMON MARBLED CARPET *Chloroclysta truncata* fw 15–20mm
Although many specimens are intricately marbled and some northern specimens are almost black, a common form has a large brown patch on the forewing and another has a large white patch. May–October almost everywhere. FP: numerous trees and shrubs.

BLUE-BORDERED CARPET *Plemyria rubiginata* fw 12–15mm
The commonest form has a short brown 'cloak' and a brown blotch near the front of each forewing. Some specimens have a complete cross-band on each wing and all individuals have a dark blotch on the outer edge of the forewing. June–August, mainly in damp places. FP: mainly alder and blackthorn.

BROKEN-BARRED CARPET *Electrophaes corylata* fw *c.* 15mm
The ground colour is white, heavily dusted with brown, and the dark areas are brown or black. The jagged central cross-band is strongly constricted and usually broken towards the rear of the wing. May–July, mainly in and around woodlands. FP: hawthorn and various other trees.

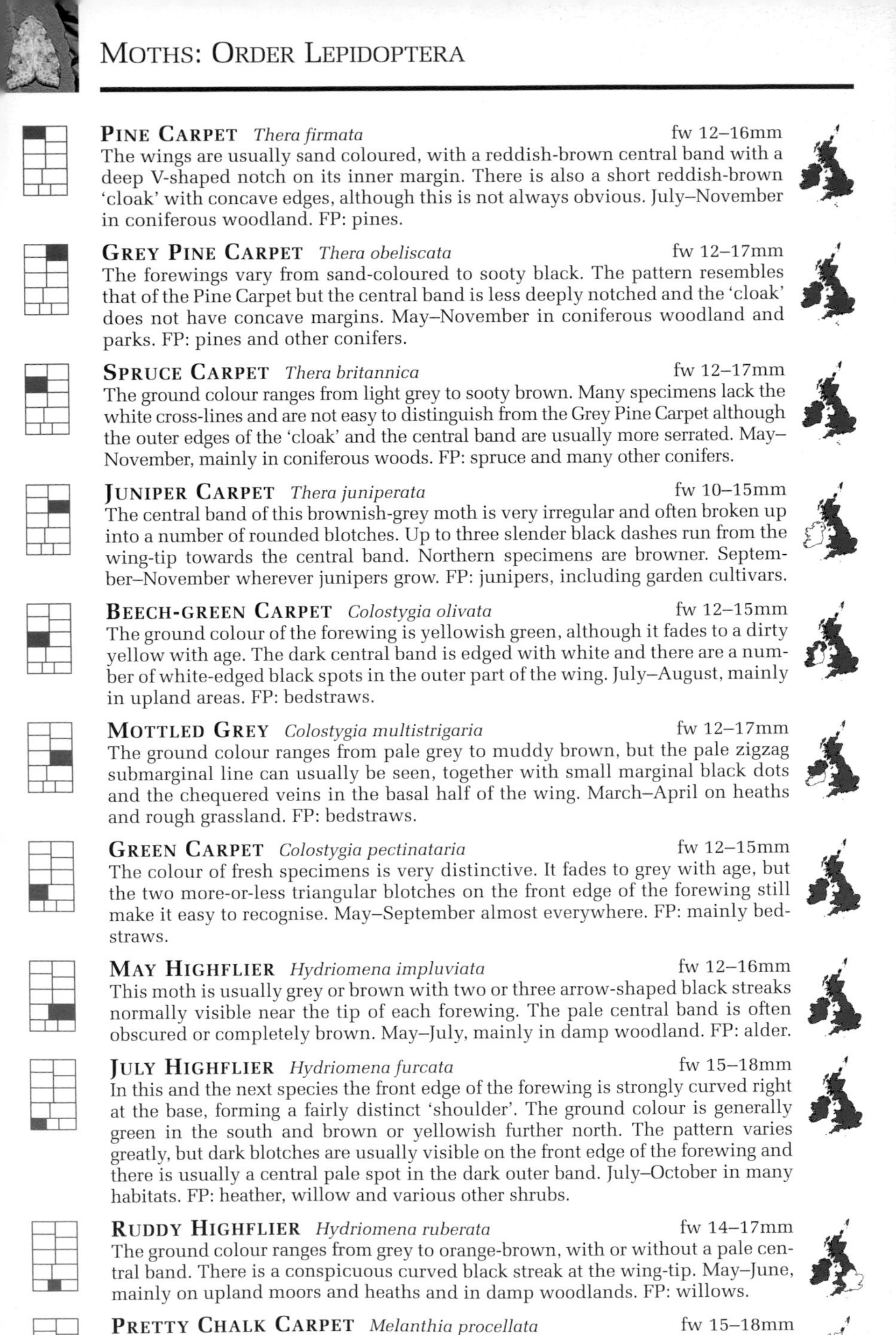

PINE CARPET *Thera firmata* fw 12–16mm
The wings are usually sand coloured, with a reddish-brown central band with a deep V-shaped notch on its inner margin. There is also a short reddish-brown 'cloak' with concave edges, although this is not always obvious. July–November in coniferous woodland. FP: pines.

GREY PINE CARPET *Thera obeliscata* fw 12–17mm
The forewings vary from sand-coloured to sooty black. The pattern resembles that of the Pine Carpet but the central band is less deeply notched and the 'cloak' does not have concave margins. May–November in coniferous woodland and parks. FP: pines and other conifers.

SPRUCE CARPET *Thera britannica* fw 12–17mm
The ground colour ranges from light grey to sooty brown. Many specimens lack the white cross-lines and are not easy to distinguish from the Grey Pine Carpet although the outer edges of the 'cloak' and the central band are usually more serrated. May–November, mainly in coniferous woods. FP: spruce and many other conifers.

JUNIPER CARPET *Thera juniperata* fw 10–15mm
The central band of this brownish-grey moth is very irregular and often broken up into a number of rounded blotches. Up to three slender black dashes run from the wing-tip towards the central band. Northern specimens are browner. September–November wherever junipers grow. FP: junipers, including garden cultivars.

BEECH-GREEN CARPET *Colostygia olivata* fw 12–15mm
The ground colour of the forewing is yellowish green, although it fades to a dirty yellow with age. The dark central band is edged with white and there are a number of white-edged black spots in the outer part of the wing. July–August, mainly in upland areas. FP: bedstraws.

MOTTLED GREY *Colostygia multistrigaria* fw 12–17mm
The ground colour ranges from pale grey to muddy brown, but the pale zigzag submarginal line can usually be seen, together with small marginal black dots and the chequered veins in the basal half of the wing. March–April on heaths and rough grassland. FP: bedstraws.

GREEN CARPET *Colostygia pectinataria* fw 12–15mm
The colour of fresh specimens is very distinctive. It fades to grey with age, but the two more-or-less triangular blotches on the front edge of the forewing still make it easy to recognise. May–September almost everywhere. FP: mainly bedstraws.

MAY HIGHFLIER *Hydriomena impluviata* fw 12–16mm
This moth is usually grey or brown with two or three arrow-shaped black streaks normally visible near the tip of each forewing. The pale central band is often obscured or completely brown. May–July, mainly in damp woodland. FP: alder.

JULY HIGHFLIER *Hydriomena furcata* fw 15–18mm
In this and the next species the front edge of the forewing is strongly curved right at the base, forming a fairly distinct 'shoulder'. The ground colour is generally green in the south and brown or yellowish further north. The pattern varies greatly, but dark blotches are usually visible on the front edge of the forewing and there is usually a central pale spot in the dark outer band. July–October in many habitats. FP: heather, willow and various other shrubs.

RUDDY HIGHFLIER *Hydriomena ruberata* fw 14–17mm
The ground colour ranges from grey to orange-brown, with or without a pale central band. There is a conspicuous curved black streak at the wing-tip. May–June, mainly on upland moors and heaths and in damp woodlands. FP: willows.

PRETTY CHALK CARPET *Melanthia procellata* fw 15–18mm
The sooty black and white pattern, with a dark blotch halfway along the front edge of the forewing and a white one in the outer border, make this easy to recognise. June–August, in and around woods and hedgerows. FP: traveller's-joy.

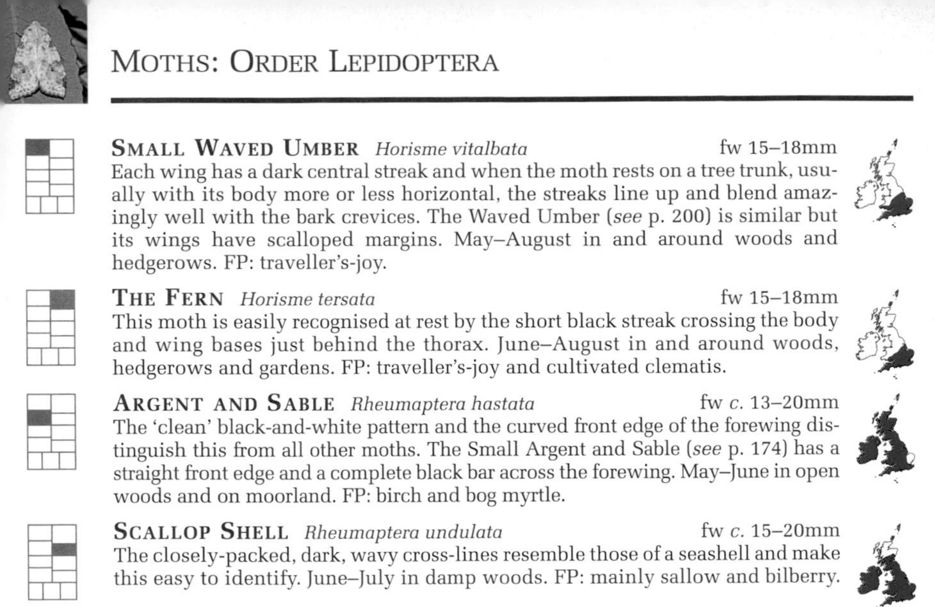

SMALL WAVED UMBER *Horisme vitalbata* fw 15–18mm

Each wing has a dark central streak and when the moth rests on a tree trunk, usually with its body more or less horizontal, the streaks line up and blend amazingly well with the bark crevices. The Waved Umber (*see* p. 200) is similar but its wings have scalloped margins. May–August in and around woods and hedgerows. FP: traveller's-joy.

THE FERN *Horisme tersata* fw 15–18mm

This moth is easily recognised at rest by the short black streak crossing the body and wing bases just behind the thorax. June–August in and around woods, hedgerows and gardens. FP: traveller's-joy and cultivated clematis.

ARGENT AND SABLE *Rheumaptera hastata* fw *c.* 13–20mm

The 'clean' black-and-white pattern and the curved front edge of the forewing distinguish this from all other moths. The Small Argent and Sable (*see* p. 174) has a straight front edge and a complete black bar across the forewing. May–June in open woods and on moorland. FP: birch and bog myrtle.

SCALLOP SHELL *Rheumaptera undulata* fw *c.* 15–20mm

The closely-packed, dark, wavy cross-lines resemble those of a seashell and make this easy to identify. June–July in damp woods. FP: mainly sallow and bilberry.

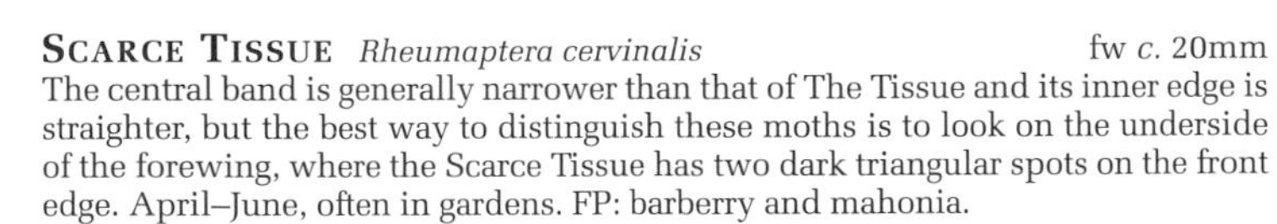

SCARCE TISSUE *Rheumaptera cervinalis* fw *c.* 20mm

The central band is generally narrower than that of The Tissue and its inner edge is straighter, but the best way to distinguish these moths is to look on the underside of the forewing, where the Scarce Tissue has two dark triangular spots on the front edge. April–June, often in gardens. FP: barberry and mahonia.

THE TISSUE *Triphosa dubitata* fw *c.* 20mm

Although most easily distinguished from the Scarce Tissue by looking at the underside (*see above*), this moth has more obviously scalloped wing margins and also has different flight times. Woodland margins and scrub from August–October and again in spring after hibernation. FP: buckthorn and alder buckthorn.

DARK UMBER *Philereme transversata* fw 17–20mm

The wings range from mid- to dark brown, with scalloped edges and lots of wavy cross-lines. The central band bends sharply just behind the front edge of the forewing and its outer edge bears a short 2-pronged spur. July–August in wooded areas. FP: buckthorn and alder buckthorn. Caterpillar p.282.

BROWN SCALLOP *Philereme vetulata* fw *c.* 15mm

This moth resembles a pale Dark Umber but has a less obvious central band on the forewing. There is also a small but conspicuous black dot near the centre of the wing. June–August in woods and scrubby places. FP: buckthorn.

CLOAKED CARPET *Euphyia biangulata* fw *c.* 15mm

The dark central band has a very conspicuous 2-pronged spur on its outer edge and this, together with the greenish tinge in the basal area, distinguishes it from all other superficially similar species. June–August in damp woods and hedgerows. FP: stitchworts.

SHARP-ANGLED CARPET *Euphyia unangulata* fw 12–15mm

This moth is not unlike the Cloaked Carpet, but the central band has just 1 sharp tooth. June–August in wooded areas. FP: ? chickweeds and stitchworts. The similar **White-banded Carpet** (*Spargania luctuata*) has a very dark outer margin to the wing.

WINTER MOTH *Operophtera brumata* fw 12–15mm

The female has vestigial wings and is most often found on tree trunks. The drab greyish-brown male, with a variably developed darker band across the forewing, is often found on lighted windows. October–January almost everywhere. FP: most deciduous trees and some conifers.

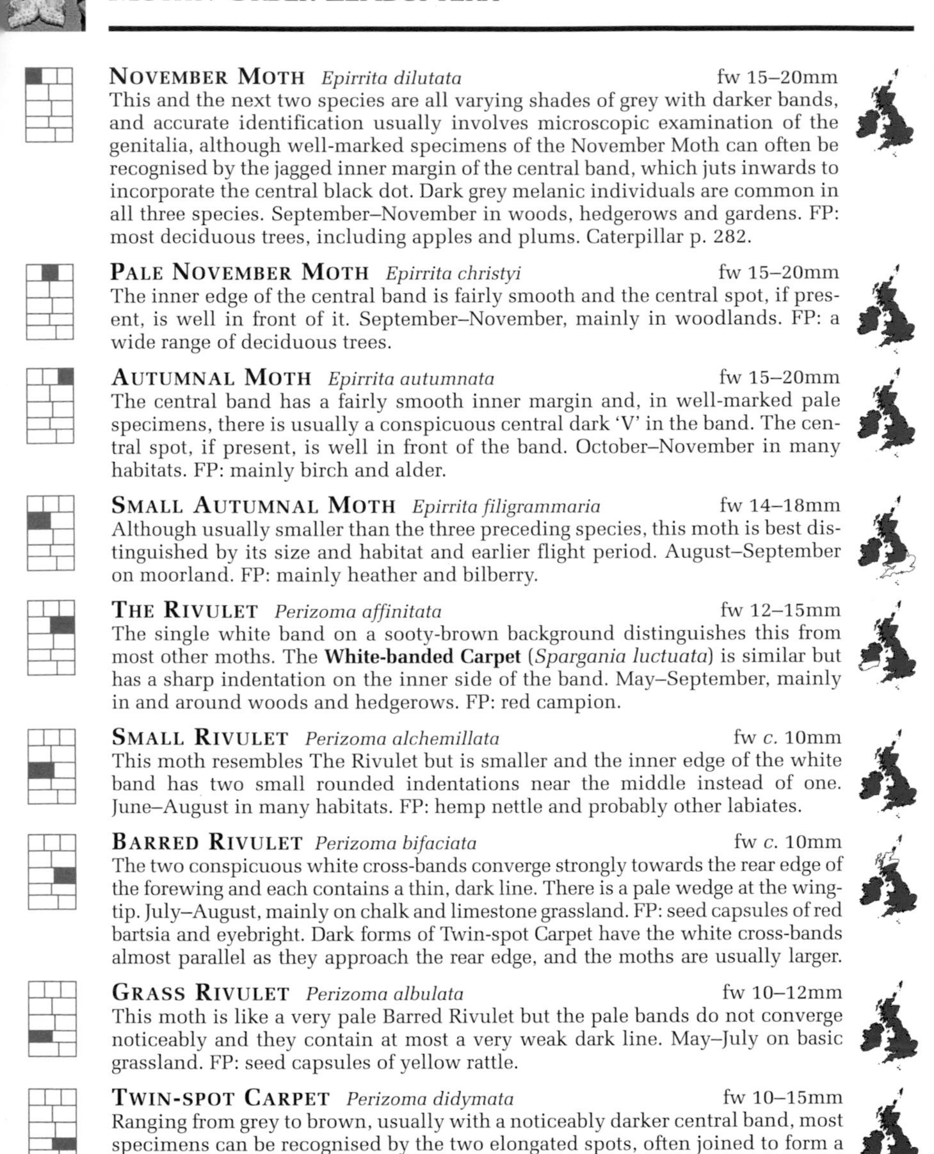

NOVEMBER MOTH *Epirrita dilutata* fw 15–20mm
This and the next two species are all varying shades of grey with darker bands, and accurate identification usually involves microscopic examination of the genitalia, although well-marked specimens of the November Moth can often be recognised by the jagged inner margin of the central band, which juts inwards to incorporate the central black dot. Dark grey melanic individuals are common in all three species. September–November in woods, hedgerows and gardens. FP: most deciduous trees, including apples and plums. Caterpillar p. 282.

PALE NOVEMBER MOTH *Epirrita christyi* fw 15–20mm
The inner edge of the central band is fairly smooth and the central spot, if present, is well in front of it. September–November, mainly in woodlands. FP: a wide range of deciduous trees.

AUTUMNAL MOTH *Epirrita autumnata* fw 15–20mm
The central band has a fairly smooth inner margin and, in well-marked pale specimens, there is usually a conspicuous central dark 'V' in the band. The central spot, if present, is well in front of the band. October–November in many habitats. FP: mainly birch and alder.

SMALL AUTUMNAL MOTH *Epirrita filigrammaria* fw 14–18mm
Although usually smaller than the three preceding species, this moth is best distinguished by its size and habitat and earlier flight period. August–September on moorland. FP: mainly heather and bilberry.

THE RIVULET *Perizoma affinitata* fw 12–15mm
The single white band on a sooty-brown background distinguishes this from most other moths. The **White-banded Carpet** (*Spargania luctuata*) is similar but has a sharp indentation on the inner side of the band. May–September, mainly in and around woods and hedgerows. FP: red campion.

SMALL RIVULET *Perizoma alchemillata* fw *c.* 10mm
This moth resembles The Rivulet but is smaller and the inner edge of the white band has two small rounded indentations near the middle instead of one. June–August in many habitats. FP: hemp nettle and probably other labiates.

BARRED RIVULET *Perizoma bifaciata* fw *c.* 10mm
The two conspicuous white cross-bands converge strongly towards the rear edge of the forewing and each contains a thin, dark line. There is a pale wedge at the wingtip. July–August, mainly on chalk and limestone grassland. FP: seed capsules of red bartsia and eyebright. Dark forms of Twin-spot Carpet have the white cross-bands almost parallel as they approach the rear edge, and the moths are usually larger.

GRASS RIVULET *Perizoma albulata* fw 10–12mm
This moth is like a very pale Barred Rivulet but the pale bands do not converge noticeably and they contain at most a very weak dark line. May–July on basic grassland. FP: seed capsules of yellow rattle.

TWIN-SPOT CARPET *Perizoma didymata* fw 10–15mm
Ranging from grey to brown, usually with a noticeably darker central band, most specimens can be recognised by the two elongated spots, often joined to form a 'U', near the outer edge of the forewing. June–August, often diurnal, mostly in uplands. FP: numerous low-growing herbs and shrubs.

SANDY CARPET *Perizoma flavofasciata* fw 10–14mm
The wavy white cross-lines on a sandy background make this quite easy to recognise. The twin lobes projecting into the outer white band are also distinctive. June–August in many open habitats. FP: campions, especially red campion.

MARSH CARPET *Perizoma sagittata* fw 13–18mm
The dark central band on the forewing is occasionally broken near the middle, but this moth is unlikely to be mistaken for any other. June–August in and around marshes and other damp, grassy habitats. FP: meadow rue.

The 50 or so British pug moths are all small, predominantly grey or brown moths that generally rest with their forewings spread so that the front edges form a straight line perpendicular to the body. The hindwings are sometimes completely concealed beneath the forewings. It is often necessary to examine the genitalia for a positive identification and only some of the commoner and more easily identified species are described here. Many of the species exhibit melanism and these dark individuals are very difficult to identify.

TOADFLAX PUG *Eupithecia linariata* fw *c.* 10mm
The brown and sooty-grey forewings are fairly diagnostic. The outer edge of the central band sweeps fairly smoothly into the front margin. July–August on grassland and waste places and sometimes in gardens. FP: toadflax and occasionally garden antirrhinums.

FOXGLOVE PUG *Eupithecia pulchellata* fw 10–12mm
This moth resembles the Toadflax Pug but is usually paler and the outer edge of the grey central band has a distinct kink close to the front edge. There is also a larger dark patch at the base of the forewing, where the Toadflax Pug has just a black bar. May–August in many open habitats, including gardens. FP: foxglove flowers.

MARBLED PUG *Eupithecia irriguata* fw *c.* 10mm
The forewings are largely white, with sooty-brown blotches on the front edge and a conspicuous elongated black spot near the middle. The thorax is dark. The Bordered Pug (*see* p. 188) is larger and has a white thorax and black abdomen. April–June, mainly in woodland. FP: oak.

MOTTLED PUG *Eupithecia exiguata* fw *c.* 12mm
This moth can usually be identified by the row of small black darts or wedges in the outer part of the forewing. The submarginal band beyond them is broken into three dark blotches. April–July almost anywhere. FP: a wide range of deciduous trees and shrubs.

NETTED PUG *Eupithecia venosata* fw 10–15mm
Although the forewings range from creamy grey to light brown and the markings vary in intensity, this moth is unlikely to be mistaken for any other. May–June in many open habitats. FP: the seed capsules of campions, especially bladder campion.

LIME-SPECK PUG *Eupithecia centaureata* fw 10–12mm
The large black blotch on the forewing means that this pug is not likely to be confused with any other, although the resting insect is easily mistaken for a bird-dropping. April–October in almost any open habitat. FP: the flowers of a wide range of herbaceous plants. Caterpillar p. 282.

WORMWOOD PUG *Eupithecia absinthiata* fw *c.* 12mm
The reddish-brown forewing has several dark spots on the front edge and another near the centre, as well as a white spot at the rear corner. There is a black band at the front of the abdomen. June–July in many habitats, including waste ground and gardens. FP: the flowers of wormwood and many other composites.

CURRANT PUG *Eupithecia assimilata* fw 9–12mm
This is like the Wormwood Pug but generally smaller and the white spot at the rear corner of the forewing is more obvious. There is also a white spot at the rear of the hindwing. May–August in woods, hedgerows and gardens. FP: currants and hops.

COMMON PUG *Eupithecia vulgata* fw 10–12mm
The wings range from grey to deep brown, usually with numerous wavy cross-lines that bend sharply near the front edge. There is a subterminal line composed of small white lunules, but no obvious central black spot. The abdomen is bordered with black dots. May–August almost anywhere. FP: a wide range of trees, shrubs and herbaceous plants.

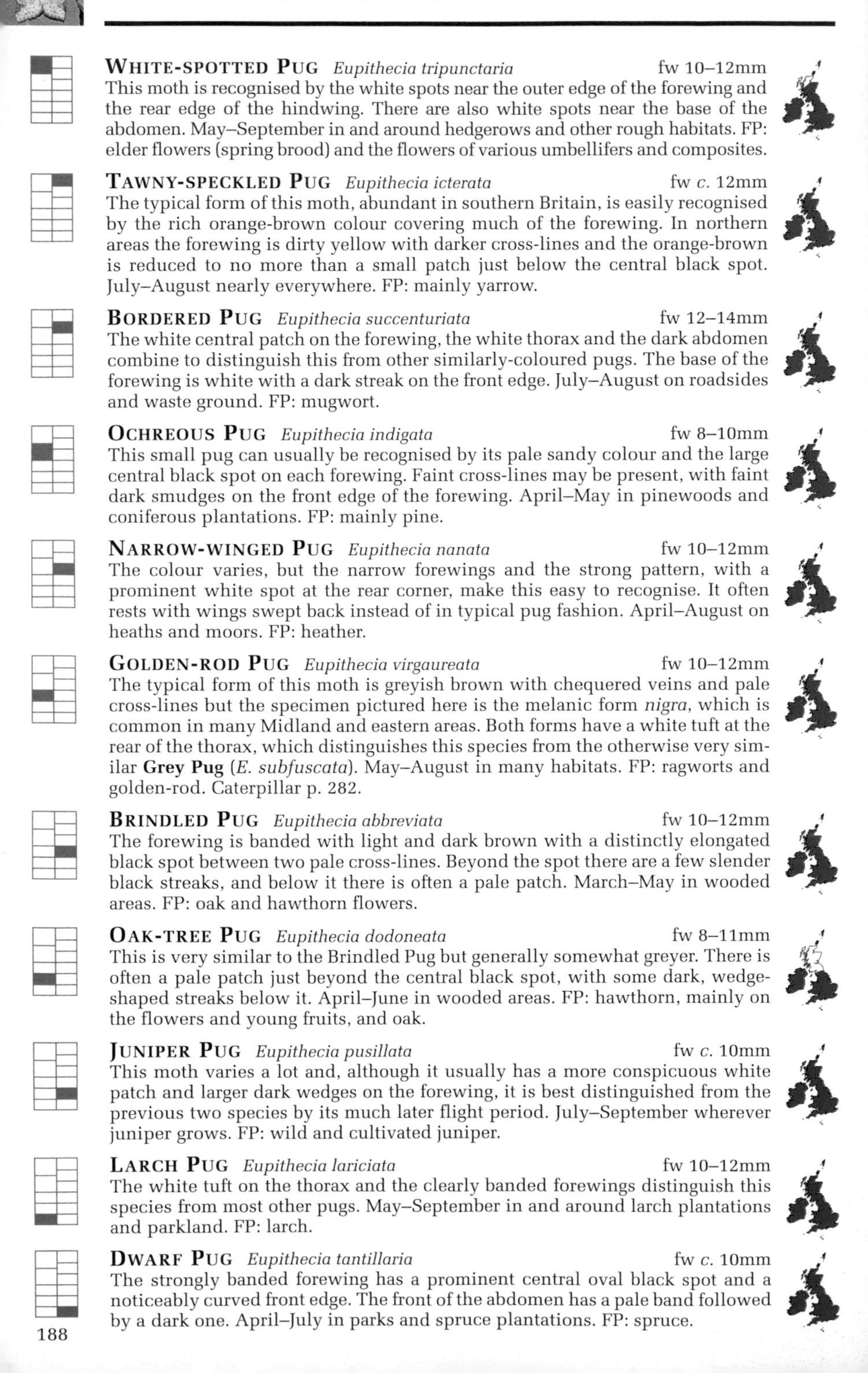

WHITE-SPOTTED PUG *Eupithecia tripunctaria* fw 10–12mm
This moth is recognised by the white spots near the outer edge of the forewing and the rear edge of the hindwing. There are also white spots near the base of the abdomen. May–September in and around hedgerows and other rough habitats. FP: elder flowers (spring brood) and the flowers of various umbellifers and composites.

TAWNY-SPECKLED PUG *Eupithecia icterata* fw *c.* 12mm
The typical form of this moth, abundant in southern Britain, is easily recognised by the rich orange-brown colour covering much of the forewing. In northern areas the forewing is dirty yellow with darker cross-lines and the orange-brown is reduced to no more than a small patch just below the central black spot. July–August nearly everywhere. FP: mainly yarrow.

BORDERED PUG *Eupithecia succenturiata* fw 12–14mm
The white central patch on the forewing, the white thorax and the dark abdomen combine to distinguish this from other similarly-coloured pugs. The base of the forewing is white with a dark streak on the front edge. July–August on roadsides and waste ground. FP: mugwort.

OCHREOUS PUG *Eupithecia indigata* fw 8–10mm
This small pug can usually be recognised by its pale sandy colour and the large central black spot on each forewing. Faint cross-lines may be present, with faint dark smudges on the front edge of the forewing. April–May in pinewoods and coniferous plantations. FP: mainly pine.

NARROW-WINGED PUG *Eupithecia nanata* fw 10–12mm
The colour varies, but the narrow forewings and the strong pattern, with a prominent white spot at the rear corner, make this easy to recognise. It often rests with wings swept back instead of in typical pug fashion. April–August on heaths and moors. FP: heather.

GOLDEN-ROD PUG *Eupithecia virgaureata* fw 10–12mm
The typical form of this moth is greyish brown with chequered veins and pale cross-lines but the specimen pictured here is the melanic form *nigra*, which is common in many Midland and eastern areas. Both forms have a white tuft at the rear of the thorax, which distinguishes this species from the otherwise very similar **Grey Pug** (*E. subfuscata*). May–August in many habitats. FP: ragworts and golden-rod. Caterpillar p. 282.

BRINDLED PUG *Eupithecia abbreviata* fw 10–12mm
The forewing is banded with light and dark brown with a distinctly elongated black spot between two pale cross-lines. Beyond the spot there are a few slender black streaks, and below it there is often a pale patch. March–May in wooded areas. FP: oak and hawthorn flowers.

OAK-TREE PUG *Eupithecia dodoneata* fw 8–11mm
This is very similar to the Brindled Pug but generally somewhat greyer. There is often a pale patch just beyond the central black spot, with some dark, wedge-shaped streaks below it. April–June in wooded areas. FP: hawthorn, mainly on the flowers and young fruits, and oak.

JUNIPER PUG *Eupithecia pusillata* fw *c.* 10mm
This moth varies a lot and, although it usually has a more conspicuous white patch and larger dark wedges on the forewing, it is best distinguished from the previous two species by its much later flight period. July–September wherever juniper grows. FP: wild and cultivated juniper.

LARCH PUG *Eupithecia lariciata* fw 10–12mm
The white tuft on the thorax and the clearly banded forewings distinguish this species from most other pugs. May–September in and around larch plantations and parkland. FP: larch.

DWARF PUG *Eupithecia tantillaria* fw *c.* 10mm
The strongly banded forewing has a prominent central oval black spot and a noticeably curved front edge. The front of the abdomen has a pale band followed by a dark one. April–July in parks and spruce plantations. FP: spruce.

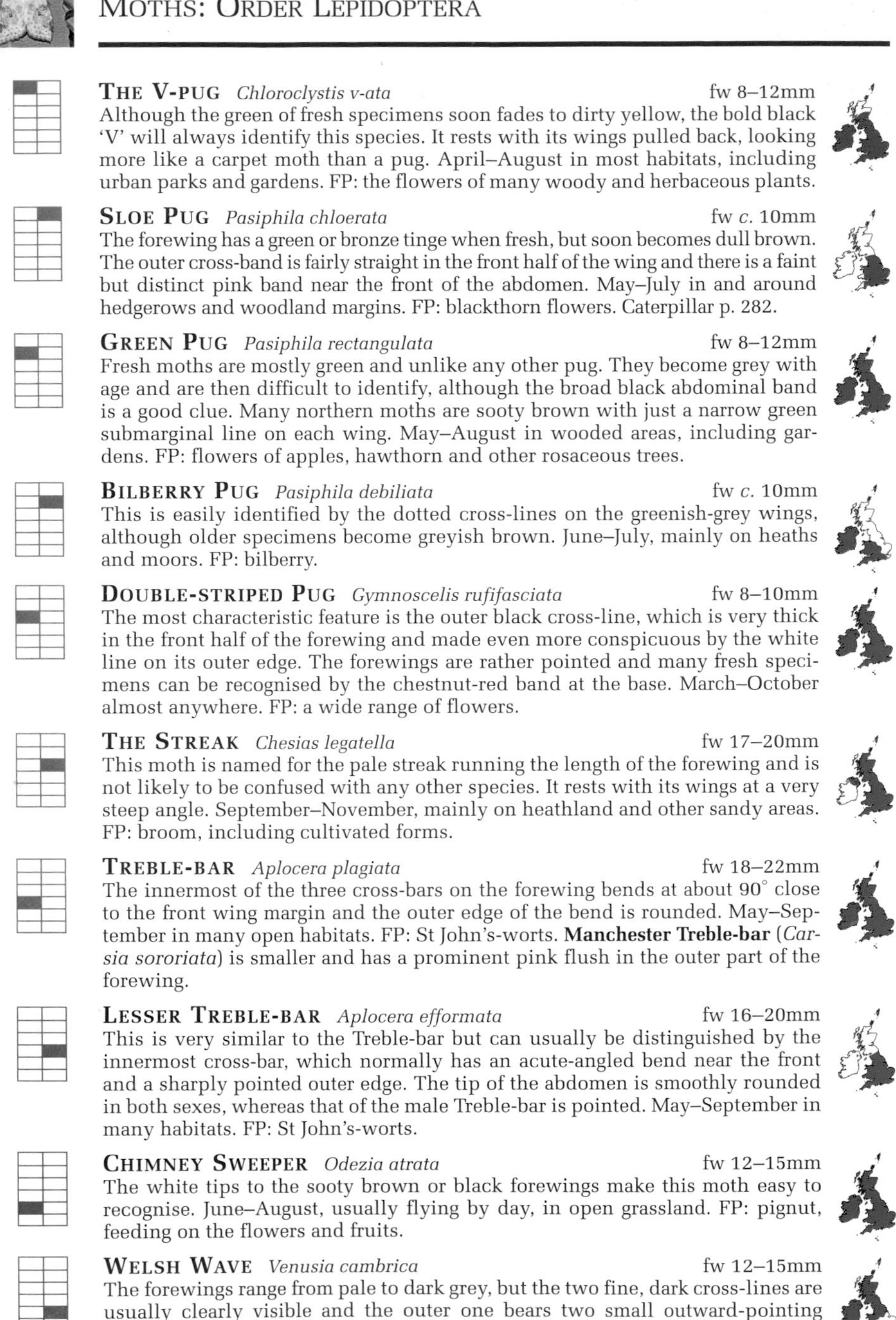

THE V-PUG *Chloroclystis v-ata* fw 8–12mm
Although the green of fresh specimens soon fades to dirty yellow, the bold black 'V' will always identify this species. It rests with its wings pulled back, looking more like a carpet moth than a pug. April–August in most habitats, including urban parks and gardens. FP: the flowers of many woody and herbaceous plants.

SLOE PUG *Pasiphila chloerata* fw *c.* 10mm
The forewing has a green or bronze tinge when fresh, but soon becomes dull brown. The outer cross-band is fairly straight in the front half of the wing and there is a faint but distinct pink band near the front of the abdomen. May–July in and around hedgerows and woodland margins. FP: blackthorn flowers. Caterpillar p. 282.

GREEN PUG *Pasiphila rectangulata* fw 8–12mm
Fresh moths are mostly green and unlike any other pug. They become grey with age and are then difficult to identify, although the broad black abdominal band is a good clue. Many northern moths are sooty brown with just a narrow green submarginal line on each wing. May–August in wooded areas, including gardens. FP: flowers of apples, hawthorn and other rosaceous trees.

BILBERRY PUG *Pasiphila debiliata* fw *c.* 10mm
This is easily identified by the dotted cross-lines on the greenish-grey wings, although older specimens become greyish brown. June–July, mainly on heaths and moors. FP: bilberry.

DOUBLE-STRIPED PUG *Gymnoscelis rufifasciata* fw 8–10mm
The most characteristic feature is the outer black cross-line, which is very thick in the front half of the forewing and made even more conspicuous by the white line on its outer edge. The forewings are rather pointed and many fresh specimens can be recognised by the chestnut-red band at the base. March–October almost anywhere. FP: a wide range of flowers.

THE STREAK *Chesias legatella* fw 17–20mm
This moth is named for the pale streak running the length of the forewing and is not likely to be confused with any other species. It rests with its wings at a very steep angle. September–November, mainly on heathland and other sandy areas. FP: broom, including cultivated forms.

TREBLE-BAR *Aplocera plagiata* fw 18–22mm
The innermost of the three cross-bars on the forewing bends at about 90° close to the front wing margin and the outer edge of the bend is rounded. May–September in many open habitats. FP: St John's-worts. **Manchester Treble-bar** (*Carsia sororiata*) is smaller and has a prominent pink flush in the outer part of the forewing.

LESSER TREBLE-BAR *Aplocera efformata* fw 16–20mm
This is very similar to the Treble-bar but can usually be distinguished by the innermost cross-bar, which normally has an acute-angled bend near the front and a sharply pointed outer edge. The tip of the abdomen is smoothly rounded in both sexes, whereas that of the male Treble-bar is pointed. May–September in many habitats. FP: St John's-worts.

CHIMNEY SWEEPER *Odezia atrata* fw 12–15mm
The white tips to the sooty brown or black forewings make this moth easy to recognise. June–August, usually flying by day, in open grassland. FP: pignut, feeding on the flowers and fruits.

WELSH WAVE *Venusia cambrica* fw 12–15mm
The forewings range from pale to dark grey, but the two fine, dark cross-lines are usually clearly visible and the outer one bears two small outward-pointing spikes. June–August, mainly on moorland. FP: mainly rowan.

DINGY SHELL *Euchoeca nebulata* fw 9–12mm
This moth habitually rests with its wings closed vertically above the body in butterfly-fashion. Upper and lower surfaces are both sand-coloured, with indistinct cross-lines and chequered fringes. May–August, in and around damp woodland. FP: alder.

Small White Wave *Asthena albulata* fw *c.* 10mm
The many wavy brown cross-lines on the pure white background make it unlikely that this moth will be confused with any other. The Satin Wave (*see* p. 170) is similar but has fewer cross-lines and no marginal dots. May–September in and around woodland. FP: mainly hazel and birch.

Small Yellow Wave *Hydrelia flammeolaria* fw *c.* 10mm
The 'stretched-out' resting position and the overall yellow appearance make this little moth very easy to recognise. The Dingy Shell (*see* p. 190) has fewer and straighter cross-lines and a very different resting position. May–July, mainly in wooded areas. FP: maples and alder.

Yellow-barred Brindle *Acasis viretata* fw 10–15mm
The pale areas are green in fresh specimens, but soon fade to yellowish. The central part of the forewing may be crossed by a more-or-less solid black band or, as pictured here, by a collection of wavy cross-lines. The veins in the outer part of the wing often bear conspicuous black dashes. May–September in woods and hedgerows. FP: dogwood, hawthorn, and various other shrubs.

The Magpie *Abraxas grossulariata* fw 18–25mm
This is an unmistakable moth: although the dark spots may be reduced, the yellow band is rarely missing from the forewing. June–August in many habitats, including moorland. FP: a wide range of shrubs, including heathers: a common pest of gooseberries. Caterpillar p. 282.

Clouded Magpie *Abraxas sylvata* fw 18–23mm
The coloration is similar to that of The Magpie but the markings are much less clear. The base of the forewing is dark. May–July, mainly in woodland. FP: elms.

Clouded Border *Lomaspilis marginata* fw 10–14mm
The sooty-brown blotches around the wing margins immediately identify this moth, which may also have a complete dark cross-band on the forewing. May–September, mainly in wooded areas. FP: mainly poplars and willows.

Scorched Carpet *Ligdia adustata* fw 12–14mm
The black and white pattern, with a dark 'cloak' at the bases of the forewings, makes this very easy to recognise, although difficult to spot when at rest on lichen-covered bark and similar surfaces. The Beautiful Carpet (*see* p. 176) is larger and lacks the clear white wing-tips. April–September in and around woods and hedgerows. FP: spindle. Caterpillar p. 282.

Peacock Moth *Macaria notata* fw *c.* 15mm
The forewing pattern and the angled hindwings distinguish this and the next species from all our other moths, but the two species are not easy to separate. In the Peacock, the narrow submarginal line on the hindwing is generally complete. May–August in wooded areas. FP: mainly birch.

Sharp-angled Peacock *Macaria alternata* fw *c.* 15mm
This moth is very similar to the Peacock Moth but the submarginal line on the hindwing is normally dotted. The outermost brown patch on the front edge tends to be narrower than in the Peacock Moth and often somewhat triangular. May–August, mainly in wooded areas. FP: mainly sallow and blackthorn.

Tawny-barred Angle *Macaria liturata* fw 14–18mm
This moth is similar in shape to the last two species but lacks the central black spot on the forewing. The cross-lines are more obvious and the rusty colour of the outer one should identify the species quite easily. A dark brown form with broad orange cross-bands is common in some areas. May–September, mainly in coniferous woods. FP: pine and spruce.

The V-moth *Macaria wauaria* fw 14–18mm
The conspicuous central 'V' mark on the forewing immediately identifies this moth. July–August, most common in gardens. FP: currants and gooseberry.

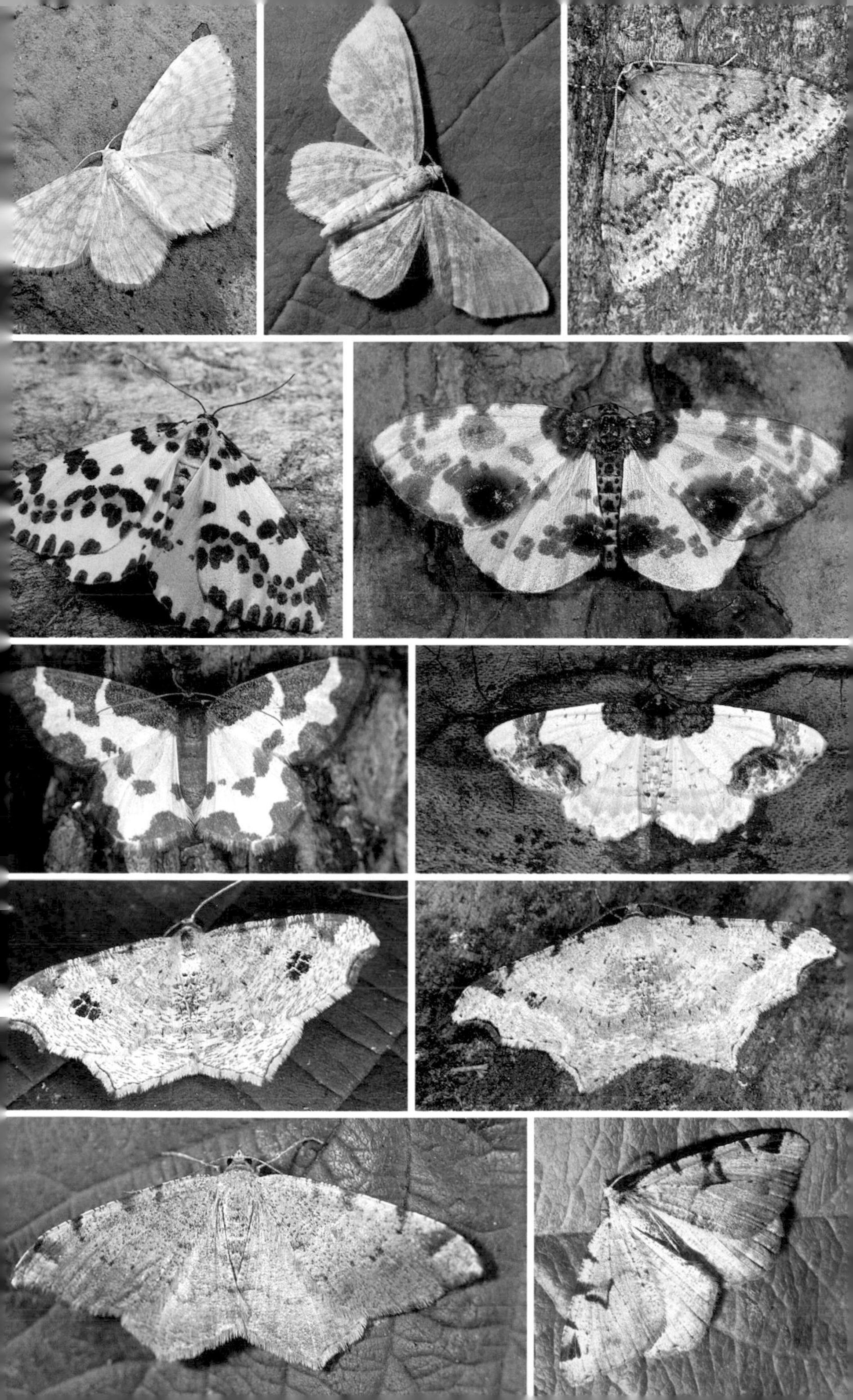

LATTICED HEATH *Chiasmia clathrata* fw 10–15mm
The ground colour varies from white to deep cream and the brown lattice, formed by the dark veins and cross-lines, varies in thickness, but the moth is unlikely to be confused with many others. The Common Heath (*see* p. 202) has similar colours but lacks the dark veins and the males have feathery antennae. May–September in open habitats. FP: clovers and other legumes.

BROWN SILVER-LINE *Petrophora chlorosata* fw 15–18mm
The ground colour varies from greyish brown to mid-brown, often with a pinkish tinge. The two cross-lines on the forewing are normally silvery white with brown edges, although the inner line may show little or no silver. There may be a very faint white submarginal line as well. April–June in woods and on heathland. FP: bracken.

BARRED UMBER *Plagodis pulveraria* fw 17–20mm
The ground colour ranges from pale pinkish brown to purplish brown but is always heavily speckled. The central cross-band bulges strongly on its outer edge, which is often quite jagged. May–June, mainly in woodland. FP: mainly hazel and birch.

SCORCHED WING *Plagodis dolabraria* fw 15–20mm
The 'scorch marks' on the forewing make this moth very easy to recognise. Males rest with the adomen curled upwards. May–June in wooded areas. FP: mainly oak, but many other deciduous trees may be used.

HORSE CHESTNUT *Pachycnemia hippocastanaria* fw *c.* 15mm
The resting position, with the wings overlapping and wrapped around the body, is very unusual for a geometer moth. The forewings vary from light grey to brown and the area between the two cross-lines may be noticeably darker than the rest of the wing. The inner cross-line is sharply bent in the middle. April–August on heathland. FP: heath and ling. The moth has no connection with horse chestnut trees other than having a coloration matching that of the bark.

BRIMSTONE MOTH *Opisthograptis luteolata* fw 15–20mm
This attractive moth cannot possibly be confused with any other species. Specimens of the autumn brood are noticeably smaller than those flying earlier in the year. April–October, almost everywhere but especially around hedgerows. FP: blackthorn, hawthorn and other rosaceous trees and shrubs. Caterpillar p. 284.

BORDERED BEAUTY *Epione repandaria* fw 12–16mm
The greyish border of the forewing peters out at the wing-tip. The colours vary a little but the moth cannot be mistaken for anything other than the very rare **Dark Bordered Beauty** (*E. vespertaria*) of northern areas. The latter species differs in that the male's dark border does not peter out at the wing-tip and the female lacks the orange veins. July–September, mainly in damp woodland. FP: mainly sallows.

SPECKLED YELLOW *Pseudopanthera macularia* fw 12–15mm
The brown blotches vary in size but this day-flying moth is quite unmistakable. May–July in open woods and scrubby areas. FP: mainly wood sage: possibly other labiates.

SCALLOPED HAZEL *Odontopera bidentata* fw 20–25mm
The ground colour ranges from dirty white to deep brown and the outer edges of both wings are strongly toothed. The area between the wavy cross-lines may be the same colour as the rest of the forewing or it may form a darker band. May–July, wherever there are trees. FP: wide range of deciduous trees and conifers.

SCALLOPED OAK *Crocallis elinguaria* fw 18–22mm
The ground colour varies from primrose-yellow to orange and is often heavily speckled, but the darker central band, tapering towards the rear and containing a prominent dark spot, is always visible. June–September in many habitats, including town parks and gardens. FP: wide range of deciduous trees and shrubs. Caterpillar p. 284.

LILAC BEAUTY *Apeira syringaria* fw *c.* 20mm
The unusual resting position of this moth, with the front edge of the forewing raised and folded, is unlike that of any other species and it gives the insect a remarkable resemblance to a dead leaf (*see* p. 41). The female is less orange than the male pictured here, but both sexes have the distinctive lilac patches at the front of the forewing. June–September in woods and hedgerows. FP: mainly honeysuckle. Caterpillar p. 284.

LARGE THORN *Ennomos autumnaria* fw 20–30mm
The female is larger than the male, shown here, and the rusty spotting is much paler. The moth, in common with its congeners, rests with its wings partly raised. August–October, mainly in woodland. FP: numerous deciduous trees.

AUGUST THORN *Ennomos quercinaria* fw *c.* 20mm
The female is much paler than the male, pictured here. Both sexes have a rusty tinge and some males are quite brown. The outer cross-line usually has a distinct 'wobble' close to the front margin and the inner one bends sharply inwards at the margin. August–September in wooded habitats. FP: mainly oak.

CANARY-SHOULDERED THORN *Ennomos alniaria* fw 15–20mm
The bright yellow thoracic hair distinguishes this moth from the other thorns. The wings are often heavily dusted with brown, and the outer cross-line runs into the front margin without a kink. July–October in wooded habitats. FP: birch, lime and some other deciduous trees.

DUSKY THORN *Ennomos fuscantaria* fw 17–20mm
The dark outer part of the forewing, contrasting strongly with the pale area between the cross-lines, distinguishes this species. Dark specimens of the August Thorn can be distinguished by the kinked outer cross-line. July–October in many habitats, including parks and gardens. FP: ash and privet.

SEPTEMBER THORN *Ennomos erosaria* fw 17–20mm
The lack of rusty spotting distinguishes this moth from the previous 4 species. The outer cross-line is also less curved than in these other species. July–October in wooded areas, including gardens. FP: various deciduous trees, especially oak and birch.

EARLY THORN *Selenia dentaria* fw 15–25mm
This thorn rests with its wings closed vertically over its body like those of butterflies. The upper and lower surfaces are similar in colour, although the upperside of the hindwing is more or less unmarked. Spring insects are larger and generally darker than those of the summer brood pictured here and females (A) of both broods are usually paler than the males (B). March–September nearly everywhere. FP: a wide range of deciduous trees and shrubs. Caterpillar p. 284.

LUNAR THORN *Selenia lunularia* fw 15–22mm
The hindwings of this moth are much more deeply incised than those of the other thorns'. The ground colour ranges from greyish yellow to purplish brown, but in any one individual the upper and lower surfaces are more or less alike. The moth normally rests with its wings open and partly raised. May–August in wooded areas. FP: various deciduous trees. Caterpillar p. 284.

PURPLE THORN *Selenia tetralunaria* fw 17–25mm
Summer insects are smaller and often much browner than the spring brood pictured here, but are easily distinguished from the Lunar Thorn by the black spot near the rear edge of the upperside of each hindwing. April–October, mainly in wooded areas. FP: many deciduous trees.

FEATHERED THORN *Colotois pennaria* fw 18–23mm
The male (A), with extremely feathery antennae, is usually some shade of orange and often heavily speckled with brown. The female (B) is often dull brown. There is usually a conspicuous black or white spot near the wing-tip in both sexes, or it may be white with a black rim. The outer cross-line is almost straight, but both lines are commonly obscured. September–December in most wooded areas. FP: wide range of deciduous trees.

SWALLOWTAILED MOTH *Ourapteryx sambucaria* fw 20–30mm
Named for the small 'tail' on each hindwing, this moth is quite unlike any other British species. June–October in many habitats, including town parks and gardens, flying with a ghostly appearance and commonly resting on lighted windows. FP: numerous trees and shrubs, including hawthorn and ivy. Caterpillar p. 284

ORANGE MOTH *Angerona prunaria* fw 20–30mm
The male is usually bright orange and the female is dull yellow, both with a variable amount of brown flecks, but both sexes can exist in a brown form in which the normal ground colour is restricted to a central band on each wing. All forms have chequered margins. June–July, mainly in woodland. FP: wide range of non-coniferous trees and shrubs.

SMALL BRINDLED BEAUTY *Apocheima hispidaria* fw *c.* 15mm
Only the male is winged, with two dark cross-lines near the middle of the forewing, although the inner one may fade before reaching the front margin. The antennae are reddish brown. February–March in wooded areas. FP: mainly oak.

PALE BRINDLED BEAUTY *Phigalia pilosaria* fw 20–25mm
Only the male is winged, with broader wings than those of the Small Brindled Beauty. The antennae are brown. January–March in wooded areas, including parkland. FP: wide range of deciduous trees and shrubs. Caterpillar p. 284.

BRINDLED BEAUTY *Lycia hirtaria* fw 20–25mm
Both sexes of this very hairy moth are fully winged and the wings are commonly heavily dusted with gold. The width of the dark cross-lines varies a good deal. As with the two previous species, forms with unmarked sooty-brown wings are common. March–May in many habitats. FP: many deciduous trees and shrubs.

OAK BEAUTY *Biston strataria* fw 15–30mm
Although the brown bands vary in width and the pale areas are sometimes very densely speckled with brown or grey, this moth is not likely to be confused with any other. February–April, mainly in woodland. FP: a variety of deciduous trees, but especially oak. Caterpillar p. 284.

BELTED BEAUTY *Lycia zonaria* fw 12–15mm
The male of this rare moth is easily recognised by the silver and grey pattern on the wings and by the orange bands on the abdomen. The female is wingless but still has the orange bands. March–April, mainly in sandy coastal habitats. FP: mostly low-growing legumes, especially bird's-foot trefoil.

PEPPERED MOTH *Biston betularia* fw 20–30mm
The 'normal' form (A) of this moth has white wings heavily peppered with black, while the melanic form (B) is black with just a white spot at the base of the forewing. Both forms are now quite common in both rural and urban areas. May–September in woods, parks and gardens. FP: wide range of deciduous trees and shrubs, including garden raspberries and currants. Caterpillar p. 284.

DOTTED BORDER *Agriopis marginaria* fw 15–20mm
The male forewing includes various shades of brown but the black dots around the outer edge and the sharply-angled outer cross-line are usually clearly visible. The short-winged female cannot fly. January–April, mainly in wooded areas. FP: numerous deciduous trees, shrubs and heather. Caterpillar p. 284.

SCARCE UMBER *Agriopis aurantiaria* fw 17–20mm
The male resembles the Dotted Border but is more golden and the outer cross-line on the forewing is less sharply angled. The outer margins are only weakly dotted. The female has very small wings. October–December in wooded areas. FP: numerous deciduous trees and shrubs.

SPRING USHER *Agriopis leucophaearia* fw *c.* 15mm
The two cross-lines on the male's forewing are usually clearly marked, the inner strongly but smoothly curved and the outer rather wavy. The area between is usually paler than the rest of the wing, but some individuals have completely brown wings. The female has only vestigial wings. January–March in wooded areas, including parks and mature gardens. FP: mainly oak.

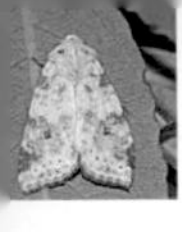

WAVED UMBER *Menophra abruptaria* fw *c.* 20mm

The ground colour ranges from off-white to reddish brown, and the hindwings are scalloped. When resting on trees with its body horizontal the dark streaks on the wings blend with the bark fissures. April–August in woods, parks and gardens. FP: mainly privet, but probably ash and lilac also. The Small Waved Umber (*see* p. 182) does not have scalloped margins. Caterpillar p. 284.

MOTTLED UMBER *Erannis defoliaria* fw 18–25mm

The male's forewings range from off-white to mid-brown, usually heavily dusted with darker scales and having a prominent black spot near the centre. The very irregular dark band is rarely absent from the outer part of the wing. The female has only the tiniest stumps for wings. October–January nearly everywhere. FP: numerous deciduous trees. Caterpillar p. 284.

WILLOW BEAUTY *Peribatodes rhomboidaria* fw 15–25mm

The wings are grey, often with a dusting of brown. The 2 central cross-lines on the forewing converge strongly towards the rear and there is often a dark spot where they meet the rear margin. The outer line bends sharply near the front of the wing but is then almost straight, with dark dots where it crosses the veins. June–October almost everywhere. FP: a wide range of shrubs and trees, including conifers.

SATIN BEAUTY *Deileptenia ribeata* fw 20–25mm

This moth is often very similar to the Willow Beauty, although usually browner and the central cross-lines are less distinct and less convergent. June–August in coniferous woods and plantations. FP: mainly conifers, including yew.

MOTTLED BEAUTY *Alcis repandata* fw 20–25mm

This is an *extremely* variable moth, ranging from pale grey to dark brown, but it can usually be recognised by the smooth curves of the bold outer cross-line on the forewing, especially the strong S-shaped curve near the front. There is often a dark smudge or ring about halfway along the front edge of the wing. The hindwings are more strongly scalloped than those of similar species. June–August nearly everywhere. FP: numerous trees and shrubs, including conifers.

DOTTED CARPET *Alcis jubata* fw *c.* 15mm

The grey colour and the prominent black dot on the forewing distinguish this from a few similar moths; the most distinctive feature is the sharp bend at each end of the outer cross-line. June–September in mature woodlands and plantations. FP: arboreal lichens.

GREAT OAK BEAUTY *Hypomecis roboraria* fw 20–30mm

Its size distinguishes this moth from most other geometers. The central cross-lines converge towards the rear, where they often link up to form a dark blotch. June–July in mature oak woods, but uncommon. FP: oaks. Caterpillar p. 286.

PALE OAK BEAUTY *Hypomecis punctinalis* fw 20–25mm

The wings are usually pale grey, with a slightly darker or browner patch at the base of the forewing. The cross-lines converge slightly towards the rear, but do not form a dark blotch. There is generally a small central ring in the hindwing. April–October, mainly in woodland. FP: oaks, birch and various other deciduous trees.

BRUSSELS LACE *Cleorodes lichenaria* fw 15–18mm

The ground colour is basically pale grey, but is usually heavily dusted with green and black. It differs from other superficially similar species in having a bold, irregular outer cross-line on the hindwing as well as on the forewing, where it bulges strongly outwards in the front half. June–October in woods and dense scrub. FP: arboreal lichens. Caterpillar p. 286.

GREY BIRCH *Aethalura punctulata* fw 12–15mm

Ranging from light to dark grey, this moth can be distinguished from similarly-coloured species by the 4 dark spots on the front of the forewing. An irregular cross-line may run back from the outermost spot and sometimes from the innermost one as well, but these lines rarely make it right across the wing. May–June in birch woods. FP: birch and probably alder.

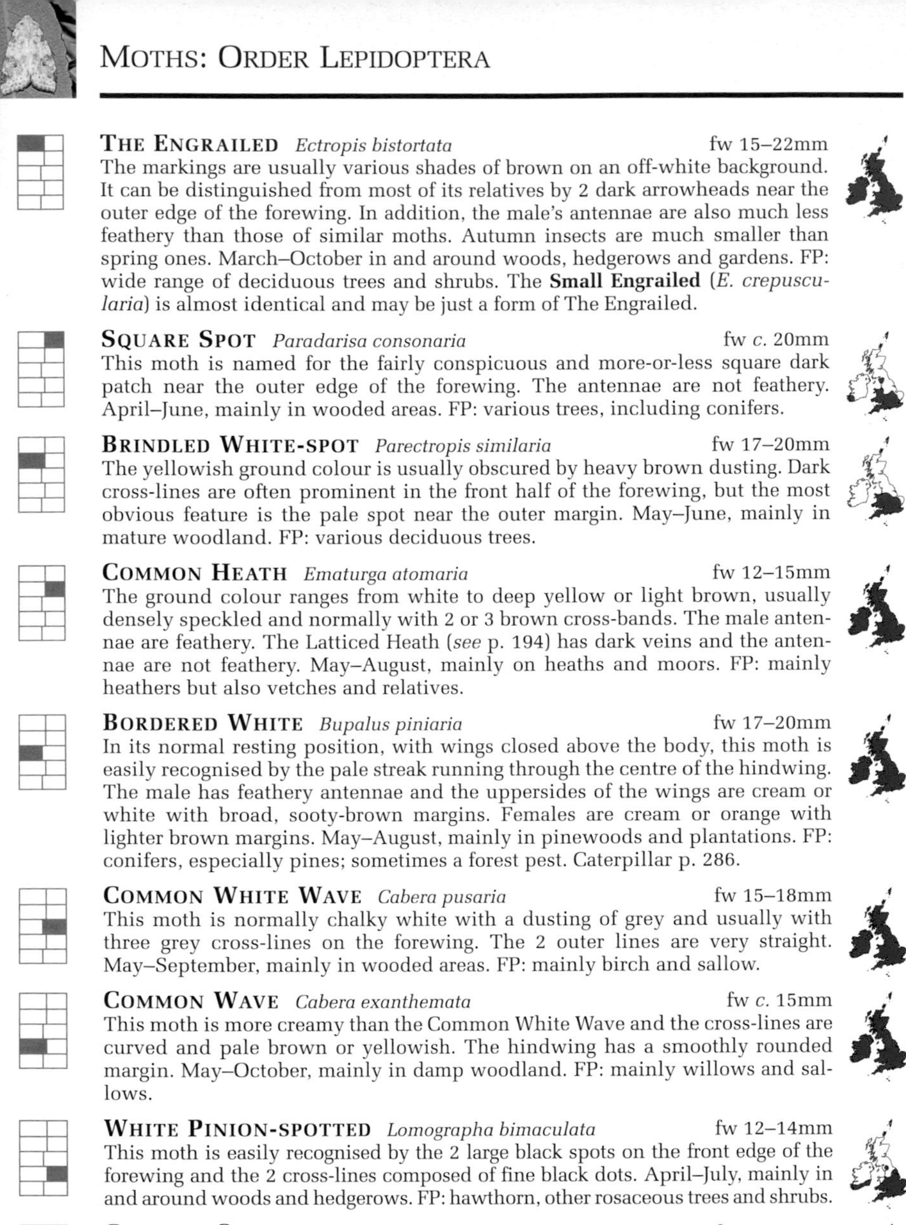

THE ENGRAILED *Ectropis bistortata* fw 15–22mm
The markings are usually various shades of brown on an off-white background. It can be distinguished from most of its relatives by 2 dark arrowheads near the outer edge of the forewing. In addition, the male's antennae are also much less feathery than those of similar moths. Autumn insects are much smaller than spring ones. March–October in and around woods, hedgerows and gardens. FP: wide range of deciduous trees and shrubs. The **Small Engrailed** (*E. crepuscularia*) is almost identical and may be just a form of The Engrailed.

SQUARE SPOT *Paradarisa consonaria* fw *c.* 20mm
This moth is named for the fairly conspicuous and more-or-less square dark patch near the outer edge of the forewing. The antennae are not feathery. April–June, mainly in wooded areas. FP: various trees, including conifers.

BRINDLED WHITE-SPOT *Parectropis similaria* fw 17–20mm
The yellowish ground colour is usually obscured by heavy brown dusting. Dark cross-lines are often prominent in the front half of the forewing, but the most obvious feature is the pale spot near the outer margin. May–June, mainly in mature woodland. FP: various deciduous trees.

COMMON HEATH *Ematurga atomaria* fw 12–15mm
The ground colour ranges from white to deep yellow or light brown, usually densely speckled and normally with 2 or 3 brown cross-bands. The male antennae are feathery. The Latticed Heath (*see* p. 194) has dark veins and the antennae are not feathery. May–August, mainly on heaths and moors. FP: mainly heathers but also vetches and relatives.

BORDERED WHITE *Bupalus piniaria* fw 17–20mm
In its normal resting position, with wings closed above the body, this moth is easily recognised by the pale streak running through the centre of the hindwing. The male has feathery antennae and the uppersides of the wings are cream or white with broad, sooty-brown margins. Females are cream or orange with lighter brown margins. May–August, mainly in pinewoods and plantations. FP: conifers, especially pines; sometimes a forest pest. Caterpillar p. 286.

COMMON WHITE WAVE *Cabera pusaria* fw 15–18mm
This moth is normally chalky white with a dusting of grey and usually with three grey cross-lines on the forewing. The 2 outer lines are very straight. May–September, mainly in wooded areas. FP: mainly birch and sallow.

COMMON WAVE *Cabera exanthemata* fw *c.* 15mm
This moth is more creamy than the Common White Wave and the cross-lines are curved and pale brown or yellowish. The hindwing has a smoothly rounded margin. May–October, mainly in damp woodland. FP: mainly willows and sallows.

WHITE PINION-SPOTTED *Lomographa bimaculata* fw 12–14mm
This moth is easily recognised by the 2 large black spots on the front edge of the forewing and the 2 cross-lines composed of fine black dots. April–July, mainly in and around woods and hedgerows. FP: hawthorn, other rosaceous trees and shrubs.

CLOUDED SILVER *Lomographa temerata* fw 12–15mm
A white line wriggling through the sooty smudge on the outer part of the forewing will usually identify this moth, although the smudge may be rather faint in females. A black central dot is usually very conspicuous in the forewing, with another dark mark on the rear edge. May–July almost anywhere with trees or shrubs, including town parks and gardens. FP: hawthorn and other rosaceous trees and shrubs, including cultivated plums and apples.

BLACK VEINED MOTH *Siona lineata* fw *c.* 20mm
This rare moth is sometimes mistaken for a white butterfly when disturbed by day. It also feeds with its wings up like a butterfly. Creamy white at first, it becomes purer white with age and the black veins show up more clearly. May–July in rough grassland. FP: marjoram and other lime-loving grassland herbs.

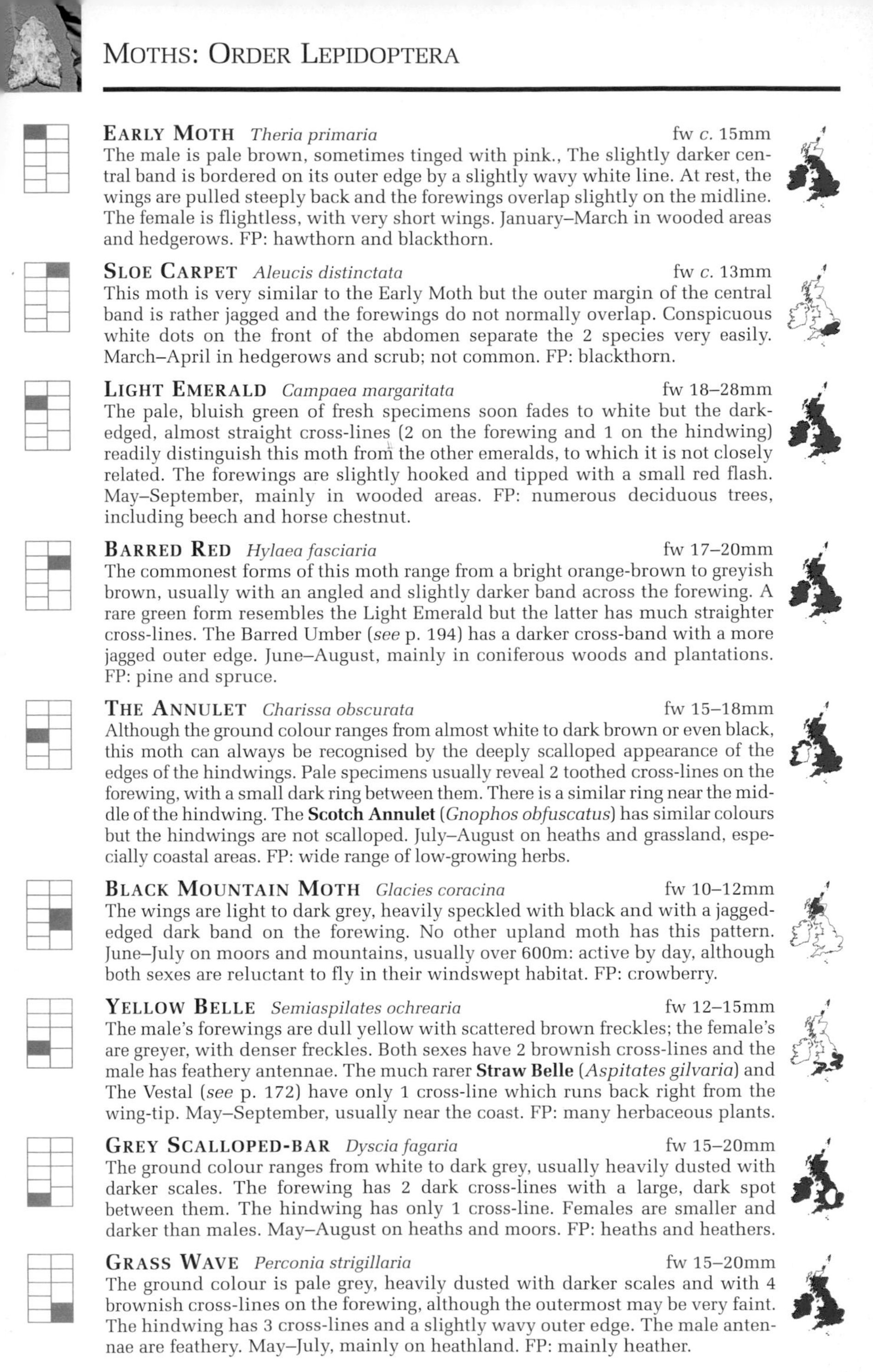

EARLY MOTH *Theria primaria* fw *c.* 15mm

The male is pale brown, sometimes tinged with pink., The slightly darker central band is bordered on its outer edge by a slightly wavy white line. At rest, the wings are pulled steeply back and the forewings overlap slightly on the midline. The female is flightless, with very short wings. January–March in wooded areas and hedgerows. FP: hawthorn and blackthorn.

SLOE CARPET *Aleucis distinctata* fw *c.* 13mm

This moth is very similar to the Early Moth but the outer margin of the central band is rather jagged and the forewings do not normally overlap. Conspicuous white dots on the front of the abdomen separate the 2 species very easily. March–April in hedgerows and scrub; not common. FP: blackthorn.

LIGHT EMERALD *Campaea margaritata* fw 18–28mm

The pale, bluish green of fresh specimens soon fades to white but the dark-edged, almost straight cross-lines (2 on the forewing and 1 on the hindwing) readily distinguish this moth from the other emeralds, to which it is not closely related. The forewings are slightly hooked and tipped with a small red flash. May–September, mainly in wooded areas. FP: numerous deciduous trees, including beech and horse chestnut.

BARRED RED *Hylaea fasciaria* fw 17–20mm

The commonest forms of this moth range from a bright orange-brown to greyish brown, usually with an angled and slightly darker band across the forewing. A rare green form resembles the Light Emerald but the latter has much straighter cross-lines. The Barred Umber (*see* p. 194) has a darker cross-band with a more jagged outer edge. June–August, mainly in coniferous woods and plantations. FP: pine and spruce.

THE ANNULET *Charissa obscurata* fw 15–18mm

Although the ground colour ranges from almost white to dark brown or even black, this moth can always be recognised by the deeply scalloped appearance of the edges of the hindwings. Pale specimens usually reveal 2 toothed cross-lines on the forewing, with a small dark ring between them. There is a similar ring near the middle of the hindwing. The **Scotch Annulet** (*Gnophos obfuscatus*) has similar colours but the hindwings are not scalloped. July–August on heaths and grassland, especially coastal areas. FP: wide range of low-growing herbs.

BLACK MOUNTAIN MOTH *Glacies coracina* fw 10–12mm

The wings are light to dark grey, heavily speckled with black and with a jagged-edged dark band on the forewing. No other upland moth has this pattern. June–July on moors and mountains, usually over 600m: active by day, although both sexes are reluctant to fly in their windswept habitat. FP: crowberry.

YELLOW BELLE *Semiaspilates ochrearia* fw 12–15mm

The male's forewings are dull yellow with scattered brown freckles; the female's are greyer, with denser freckles. Both sexes have 2 brownish cross-lines and the male has feathery antennae. The much rarer **Straw Belle** (*Aspitates gilvaria*) and The Vestal (*see* p. 172) have only 1 cross-line which runs back right from the wing-tip. May–September, usually near the coast. FP: many herbaceous plants.

GREY SCALLOPED-BAR *Dyscia fagaria* fw 15–20mm

The ground colour ranges from white to dark grey, usually heavily dusted with darker scales. The forewing has 2 dark cross-lines with a large, dark spot between them. The hindwing has only 1 cross-line. Females are smaller and darker than males. May–August on heaths and moors. FP: heaths and heathers.

GRASS WAVE *Perconia strigillaria* fw 15–20mm

The ground colour is pale grey, heavily dusted with darker scales and with 4 brownish cross-lines on the forewing, although the outermost may be very faint. The hindwing has 3 cross-lines and a slightly wavy outer edge. The male antennae are feathery. May–July, mainly on heathland. FP: mainly heather.

HAWKMOTHS: FAMILY SPHINGIDAE

These stout-bodied moths generally have narrow, pointed forewings and most of them fly rapidly. At rest, the wings are commonly held flat and swept sharply back like delta-winged aircraft. Most hawkmoths have a long proboscis and feed while hovering in front of flowers, although some species lack a proboscis and do not feed in the adult stage. The caterpillars of nearly all species have a curved horn at the rear and those of most British species are green with oblique camouflaging stripes (*see* pp. 286–8). There are 9 resident British species, but several others may arrive from continental Europe or even from Africa during the summer.

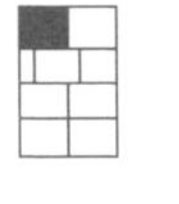

CONVOLVULUS HAWKMOTH *Agrius convolvuli* fw 40–55mm

All wings are dull grey with a dusting of white and usually marked with a few dark streaks. The body is also grey, but the abdomen has pink and black bands on the sides. July–November, usually at dusk, an irregular visitor to the British Isles; probes many garden flowers with a proboscis as long as the body – so long that it has its own sheath in the pupal stage, curving from the body like a jug handle. FP: various bindweeds. Privet Hawkmoth has pink-banded hindwings; Pine Hawkmoth has no pink at all.

PINE HAWKMOTH *Hyloicus pinastri* fw 30–40mm

Wings and body are dull brownish grey, dusted with white. Look for the chequered fringes and the three or four black streaks near the middle of the forewing. May–August, feeding on honeysuckle and other night-scented flowers, and resting on pine trunks by day. FP: pine and spruce. Convolvulus and Privet Hawkmoths are larger, with pink stripes on the abdomen. Caterpillar p. 286.

PRIVET HAWKMOTH *Sphinx ligustri* fw 45–55mm

The light and dark brown forewings are pulled tightly back along the sides of the body at rest, so the moth is easily mistaken for a broken twig or a piece of bark. The hindwings and abdomen carry pink and black stripes. June–July in many habitats and not uncommon in suburban areas. FP: mainly privet, but ash and lilac also eaten. Convolvulus and Pine Hawkmoths are both much greyer. Caterpillar p. 286.

DEATH'S-HEAD HAWKMOTH *Acherontia atropos* fw 50–60mm

The skull-like pattern on the thorax immediately identifies this moth, the largest of European hawkmoths. The hindwing is golden with dark brown stripes. If handled, the insect can emit a loud squeak by forcing air through its proboscis. Flies, rather jerkily, April–October but, having a very short proboscis, it cannot feed on the wing. It occasionally takes nectar, but more often drinks sap oozing from trees. It also enters beehives and takes honey, possibly deceiving the bees into accepting it by emitting a squeak said to be similar to that of the queen bee. The thoracic pattern may be an additional protective device, for it has a fairly strong resemblance to the face of a large bee. A rare visitor from Africa and SW Asia. FP: mainly potato, but nightshades, tomato, and related plants are also eaten. Caterpillar p. 286.

OLEANDER HAWKMOTH *Daphnis nerii* fw 35–55mm

Its coloration makes this moth quite unmistakable. June–October; a rare visitor from Africa, but has been found in many parts of the British Isles. FP: oleander is the principal food-plant, but caterpillars also eat periwinkle and vine leaves.

LIME HAWKMOTH *Mimas tiliae* fw 30–35mm

The ground colour of the wings may be green, pale brown, pink or orange, and the dark green band on the forewing is often complete, but the moth is easily recognised by its rather 'ragged' forewings. At rest, it is often mistaken for a dead leaf. May–July, but does not feed. FP: various deciduous trees, but mainly lime. Caterpillar p. 286.

EYED HAWKMOTH *Smerinthus ocellata* fw 35–40mm

At rest, the moth blends well with the bark on which it usually rests (A), but when disturbed it raises its forewings and displays the large eye-spots on its hindwings (B). At the same time it may heave its body up and down and this is more than enough to scare off birds and other enemies. Flies May–September, but does not feed. FP: mainly willows and apple. Caterpillar p. 286.

POPLAR HAWKMOTH *Laothoe populi* fw 35–40mm
The unusual resting position, with the hindwings protruding beyond the front edge of the forewings, gives the moth the appearance of a dead leaf and readily distinguishes it from other hawkmoths. All 4 wings are greyish brown, usually tinged with pink, and each hindwing bears a brick-coloured patch that is revealed when the moth is alarmed, as with the Eyed Hawkmoth. May–June; a small 2nd brood may fly August–September. Adults do not feed. FP: poplars and willows. Caterpillar p. 286.

BROAD-BORDERED BEE HAWKMOTH *Hemaris fuciformis* fw *c.* 20mm
This day-flying moth looks very much like a bumble-bee because most of the scales fall from its wings during its first flight, leaving just the brown borders. It darts rapidly from flower to flower May–June, but has become rare in recent years. FP: bedstraws and honeysuckle. Narrow-bordered Bee Hawkmoth has narrower borders, especially on the hindwings. Caterpillar p. 286.

NARROW-BORDERED BEE HAWKMOTH *Hemaris tityus* fw *c.* 20mm
This day-flying species behaves just like its Broad-bordered relative, but has narrower and blacker wing borders, which are detectable even in flight. It lacks the prominent dark spot near the centre of the forewing and its abdominal bands are black rather than chestnut. May–June, usually in woodlands in S Britain but often on bogs and moors elsewhere. It has become rare in recent years. FP: devil's-bit scabious and field scabious. Caterpillar p. 286.

HUMMINGBIRD HAWKMOTH *Macroglossum stellatarum* fw 20–25mm
This day-flying moth is usually seen as a brownish blur, making a humming sound as it hovers in front of flowers and probes them for nectar with its long tongue. It also hovers in front of sunny walls to absorb reflected warmth. The hindwings are largely orange. At night it can be found nestling in warm crevices. A summer visitor to the Britain; flies mainly May–September and frequently visits gardens. It may occasionally survive mild winters in the S. FP: bedstraws. Caterpillar p. 288.

SPURGE HAWKMOTH *Hyles euphorbiae* fw 30–35mm
The forewings vary a good deal in the amount of pink. Look for the narrow brownish triangle running to the wing-tip. The hindwings are pink and black. Flies May–September; rare summer visitor. FP: various spurges. Bedstraw Hawkmoth is very similar but the forewing has a continuous brown front margin.

BEDSTRAW HAWKMOTH *Hyles gallii* fw 30–35mm
The continuous brown front edge to the forewing distinguishes this from the Spurge Hawkmoth. Both species have pink and black hindwings. Flies May–July; occasional summer visitor. FP: bedstraws and willowherbs.

STRIPED HAWKMOTH *Hyles livornica* fw c. 35mm
The white veins on the forewings readily identify this species. The hindwings are pink and black, just like in the Spurge Hawkmoth. Flies throughout the summer; a sporadic summer visitor. FP: bedstraws and fuchsias, and sometimes vines.

ELEPHANT HAWKMOTH *Deilephila elpenor* fw 30–35mm
The pink and green coloration of this moth is unmistakable, although the green may turn yellowish with age. Flies May–August, often visiting honeysuckle at dusk. FP: willowherbs and bedstraws, and also garden fuchsias. The insect gets its name for the trunk-like front end of the caterpillar (*see* p. 288).

SMALL ELEPHANT HAWKMOTH *Deilephila porcellus* fw 20–25mm
Much smaller and yellower than the Elephant Hawkmoth, this species lacks the conspicuous pink stripe down the centre of the abdomen. Flies May–July and is particularly attracted to honeysuckle and rhododendron flowers. FP: bedstraws and sometimes willowherbs. The caterpillar (*see* p. 288) resembles that of the Elephant Hawkmoth but has no horn on the rear.

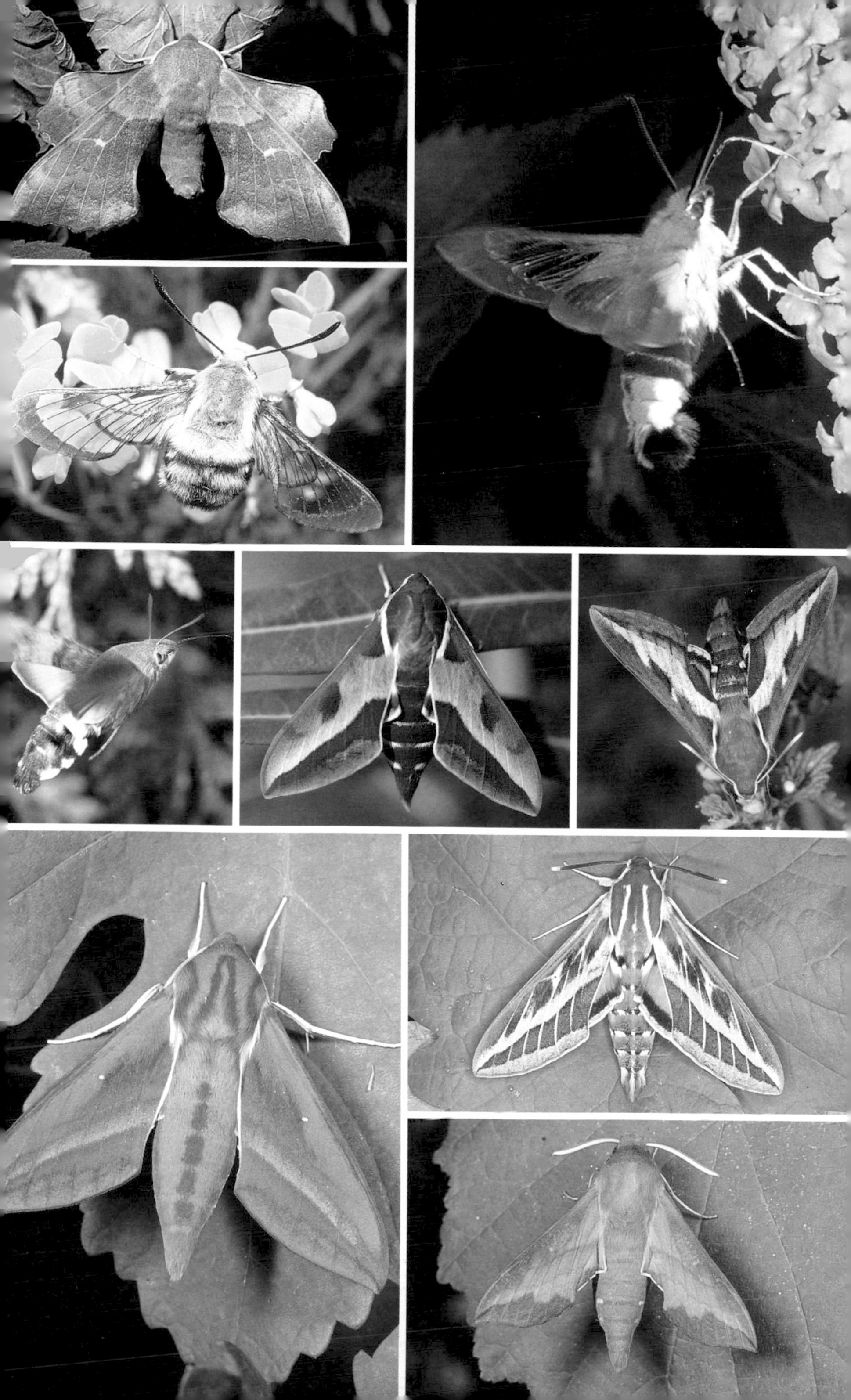

KITTENS AND PROMINENTS: FAMILY NOTODONTIDAE

The moths in this family are mostly grey or brown, with relatively stout, hairy bodies. Apart from the venation, there is little to suggest that they are all related. The prominents get their name for a small tuft of scales that sticks up from the midline when the insects are at rest. The caterpillars usually have little or no hair, although many bear fleshy outgrowths on the back. They often rest with both front and rear ends raised. There are 21 resident British species.

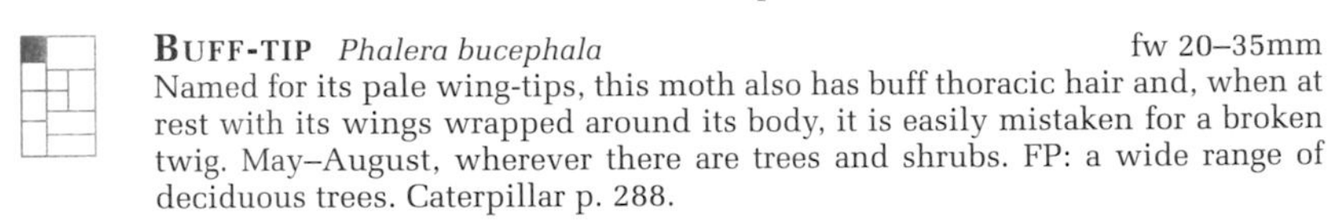

BUFF-TIP *Phalera bucephala* fw 20–35mm

Named for its pale wing-tips, this moth also has buff thoracic hair and, when at rest with its wings wrapped around its body, it is easily mistaken for a broken twig. May–August, wherever there are trees and shrubs. FP: a wide range of deciduous trees. Caterpillar p. 288.

PUSS MOTH *Cerura vinula* fw 30–40mm

This moth is easily recognised by the black spots on the thorax and the pattern of narrow arches in the outer part of the forewing. The Leopard Moth (*see* p. 156) looks similar at rest, but has spotted wings. May–July in gardens and other lightly wooded habitats. FP: willows and poplars. Caterpillar p. 288.

POPLAR KITTEN *Furcula bifida* fw 16–22mm

The outer margin of the central grey band is gently curved and is very black, especially in the front half. The inner margin of this band is almost straight and also very black. May–July in parks, gardens and damp woodlands. FP: poplars.

SALLOW KITTEN *Furcula furcula* fw 14–18mm

This moth is like a small Poplar Kitten, with females of both species being larger than the males. The outer margin of the central grey band is irregularly toothed and only thinly lined with black. May–August, mainly in lightly wooded areas. FP: willows, especially goat willow.

ALDER KITTEN *Furcula bicuspis* fw 16–20mm

The very dark central band, usually sharply constricted in the centre, separates this from the 2 previous species. May–July, mostly in damp woodland. FP: alder and birch.

LOBSTER MOTH *Stauropus fagi* fw 25–35mm

Both wings range from greyish brown to rust-coloured, often dusted with yellow. The forewing usually has a pale patch at the base and a red smudge on the rear edge. At rest, with the hindwing protruding beyond the front edge of the forewing, the moth is easily mistaken for a dead leaf. May–July in mature woodland. FP: mainly beech and oak. The moth is named for its extraordinary caterpillar (*see* p. 288).

GREAT PROMINENT *Peridea anceps* fw 23–33mm

The mottled brown forewings, often tinged with yellow or green, lack the white markings of the Lobster Moth. The hindwings are off-white apart from the greyish-brown front edge protruding beyond the forewing at rest. April–June in wooded areas. FP: oak. Caterpillar p. 290.

IRON PROMINENT *Notodonta dromedarius* fw 18–25mm

Rust-coloured streaks on the dark grey forewings give this moth its name, although the streaks are sometimes rather faint in N areas. May–August in wooded areas, including hedgerows and gardens. FP: mainly birch and alder. Caterpillar p. 288.

PEBBLE PROMINENT *Eligmodonta ziczac* fw 17–25mm

This moth is named for the brown or grey pebble-like blotch in the outer part of the forewing, separated from the pale central region by a black crescent. May–August in a wide variety of habitats, including town parks and gardens. FP: poplars and willows. Caterpillar p. 288.

PALE PROMINENT *Pterostoma palpina* fw 18–25mm

The long palps readily identify this moth, which is easily mistaken for a sliver of wood or bark. Only the male has the twin tufts of hairs projecting beyond the wings at rest. May–August in many habitats, including town parks and gardens. FP: poplars and willows. Caterpillar p. 288.

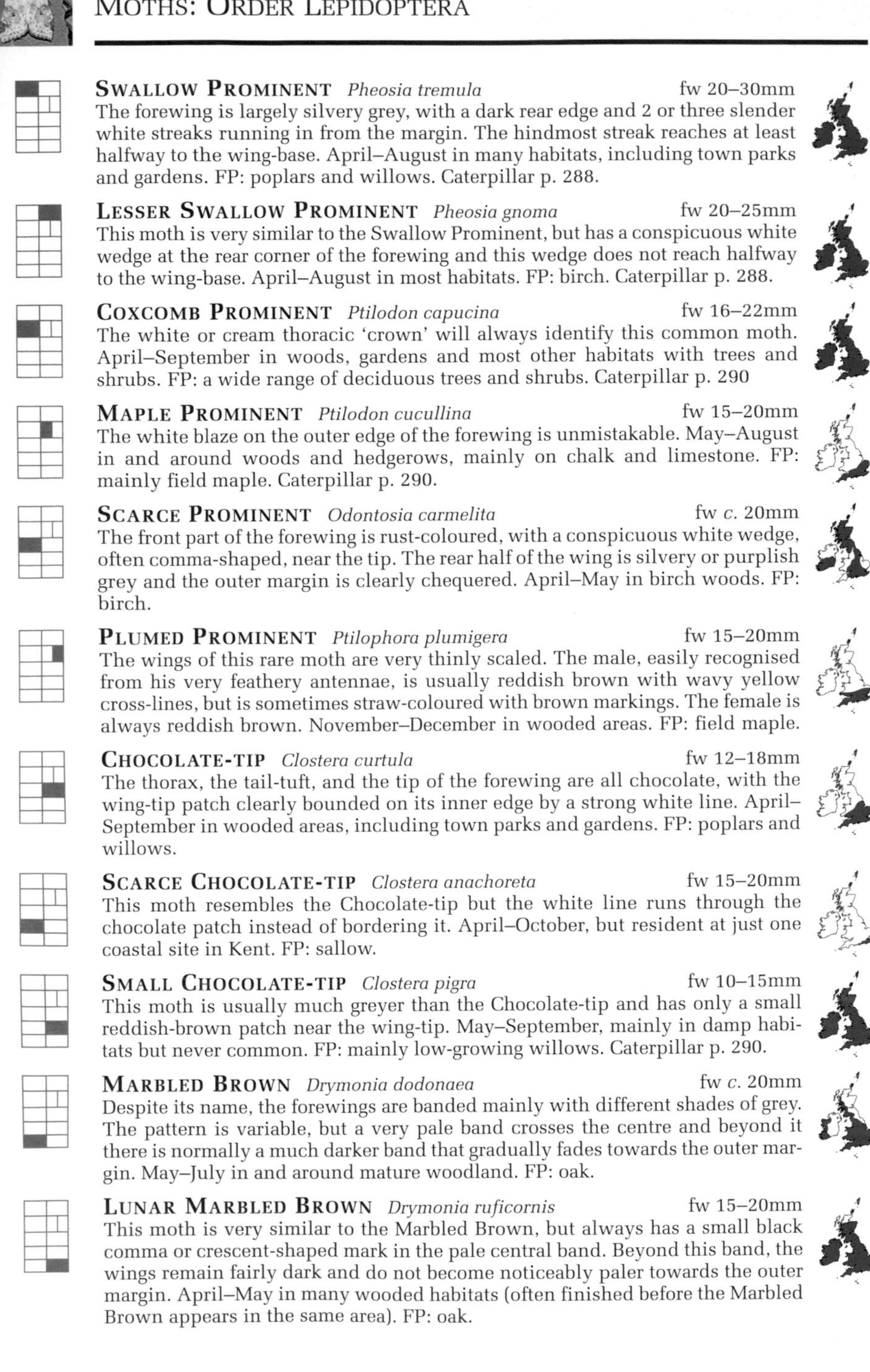

SWALLOW PROMINENT *Pheosia tremula* fw 20–30mm
The forewing is largely silvery grey, with a dark rear edge and 2 or three slender white streaks running in from the margin. The hindmost streak reaches at least halfway to the wing-base. April–August in many habitats, including town parks and gardens. FP: poplars and willows. Caterpillar p. 288.

LESSER SWALLOW PROMINENT *Pheosia gnoma* fw 20–25mm
This moth is very similar to the Swallow Prominent, but has a conspicuous white wedge at the rear corner of the forewing and this wedge does not reach halfway to the wing-base. April–August in most habitats. FP: birch. Caterpillar p. 288.

COXCOMB PROMINENT *Ptilodon capucina* fw 16–22mm
The white or cream thoracic 'crown' will always identify this common moth. April–September in woods, gardens and most other habitats with trees and shrubs. FP: a wide range of deciduous trees and shrubs. Caterpillar p. 290

MAPLE PROMINENT *Ptilodon cucullina* fw 15–20mm
The white blaze on the outer edge of the forewing is unmistakable. May–August in and around woods and hedgerows, mainly on chalk and limestone. FP: mainly field maple. Caterpillar p. 290.

SCARCE PROMINENT *Odontosia carmelita* fw *c.* 20mm
The front part of the forewing is rust-coloured, with a conspicuous white wedge, often comma-shaped, near the tip. The rear half of the wing is silvery or purplish grey and the outer margin is clearly chequered. April–May in birch woods. FP: birch.

PLUMED PROMINENT *Ptilophora plumigera* fw 15–20mm
The wings of this rare moth are very thinly scaled. The male, easily recognised from his very feathery antennae, is usually reddish brown with wavy yellow cross-lines, but is sometimes straw-coloured with brown markings. The female is always reddish brown. November–December in wooded areas. FP: field maple.

CHOCOLATE-TIP *Clostera curtula* fw 12–18mm
The thorax, the tail-tuft, and the tip of the forewing are all chocolate, with the wing-tip patch clearly bounded on its inner edge by a strong white line. April–September in wooded areas, including town parks and gardens. FP: poplars and willows.

SCARCE CHOCOLATE-TIP *Clostera anachoreta* fw 15–20mm
This moth resembles the Chocolate-tip but the white line runs through the chocolate patch instead of bordering it. April–October, but resident at just one coastal site in Kent. FP: sallow.

SMALL CHOCOLATE-TIP *Clostera pigra* fw 10–15mm
This moth is usually much greyer than the Chocolate-tip and has only a small reddish-brown patch near the wing-tip. May–September, mainly in damp habitats but never common. FP: mainly low-growing willows. Caterpillar p. 290.

MARBLED BROWN *Drymonia dodonaea* fw *c.* 20mm
Despite its name, the forewings are banded mainly with different shades of grey. The pattern is variable, but a very pale band crosses the centre and beyond it there is normally a much darker band that gradually fades towards the outer margin. May–July in and around mature woodland. FP: oak.

LUNAR MARBLED BROWN *Drymonia ruficornis* fw 15–20mm
This moth is very similar to the Marbled Brown, but always has a small black comma or crescent-shaped mark in the pale central band. Beyond this band, the wings remain fairly dark and do not become noticeably paler towards the outer margin. April–May in many wooded habitats (often finished before the Marbled Brown appears in the same area). FP: oak.

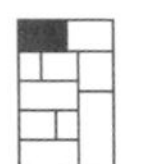

FIGURE-OF-EIGHT *Diloba caeruleocephala* fw 5–20mm

The greyish-brown forewings are dominated by the 2 white or cream marks that give the moth its name. Both marks vary and they often join up, but it is the inner one that is most like an '8'. The male antennae are feathery. The Figure-of-Eighty Moth (*see* p. 164) has finer markings and it is the outer one that resembles an '8'. September–November in and around woods and hedgerows: common in gardens. FP: hawthorn, blackthorn and many other rosaceous trees and shrubs. Caterpillar p. 290. (NB: many books place this moth in the Family Noctuidae.)

TUSSOCK MOTHS: FAMILY LYMANTRIIDAE

These moths are all rather hairy and the females of most species have tufts of detachable hairs at the rear which they use to conceal their egg clusters. The males, which are usually smaller than the females, have very feathery antennae. The adult moths have no proboscis and do not feed. Most of the caterpillars bear dense tufts or tussocks of often very colourful hairs (*see* p. 49). The hairs of adults and larvae are usually barbed and make the insects unpleasant or even painful to handle.

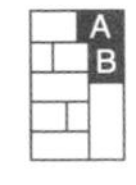

THE VAPOURER *Orgyia antiqua* fw 12–18mm

Only the male (A) has functional wings. The female's (B) wings are reduced to minute flaps and she rarely moves from her cocoon after emergence, and usually lays her eggs on the cocoon itself. July–October, the male flying mainly by day in most habitats, including town parks and streets. FP: almost any deciduous tree or shrub. Caterpillar p. 290.

PALE TUSSOCK *Calliteara pudibunda* fw 20–30mm

The female is paler than the male pictured here and, although the cross-lines are usually quite clear, the area between them is little or no darker than the rest of the wing. Both this and the Dark Tussock rest with their front legs stretched out straight in front of the body. May–June in most habitats, including town parks and gardens. FP: a wide range of trees and shrubs. Caterpillar p. 290.

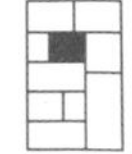

DARK TUSSOCK *Dicallomera fascelina* fw 18–30mm

This moth is usually darker than the Pale Tussock, but best distinguished by the yellow or orange decoration on the cross-lines. July–August, mainly on heathland and in coastal habitats. FP: heather and other low-growing shrubs.

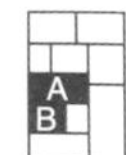

BROWN-TAIL *Euproctis chrysorrhoea* fw 15–20mm

The wings are generally pure white in both sexes, but the moth can be recognised by the brown abdominal hairs. When disturbed, both this and the Yellow-tail may roll over and pretend to be dead (B) (*see* p. 39). July–October in scrubby habitats, mainly near the coast but apparently increasing its range. FP: bramble, blackthorn, and many other deciduous trees and shrubs. Caterpillar p. 290

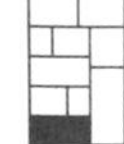

YELLOW-TAIL *Euproctis similis* fw 15–23mm

The female's wings are usually pure white, but the male normally has a dark smudge at the rear corner of the forewing. The abdomen has a tuft of golden hairs at the tip. July–August almost everywhere. FP: wide range of deciduous trees and shrubs, but mainly hawthorn and blackthorn. Caterpillar p. 290.

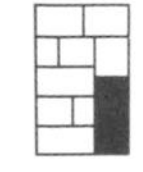

WHITE SATIN *Leucoma salicis* fw 18–30mm

This moth is usually larger than the other pure white species, but is best distinguished by the shiny or silky texture of the wings, the white abdominal hairs, and the black-and-white ringed legs. July–August, mainly in damp, wooded areas. FP: poplars and willows. The **Black-V Moth** (*Arctornis l-nigrum*), a rare visitor, has a small black 'V' or 'L' on the forewing and rests with its wings much flatter.

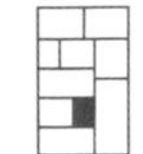

BLACK ARCHES *Lymantria monacha* fw 18–30mm

Although the black pattern varies a good deal, this moth is not likely to be mistaken for any other. The pink-banded abdomen will confirm the identity of the occasional dark specimen. July–August, in and around woodland. FP: mainly oak.

TIGERS, ERMINES AND FOOTMEN: FAMILY ARCTIIDAE

Tiger moths are named for their bold and often striped patterns. The forewings provide good camouflage at rest, while the more colourful hindwings advertise the insects' unpalatable nature (*see* p. 47). Ermines are named for their predominantly white wings, while the neat appearance of the slender-bodied footmen is supposed to be reminiscent of well-groomed servants. At first sight, the three groups appear to have nothing in common, but the underlying venation is similar and all have very hairy caterpillars. There are 29 resident British species.

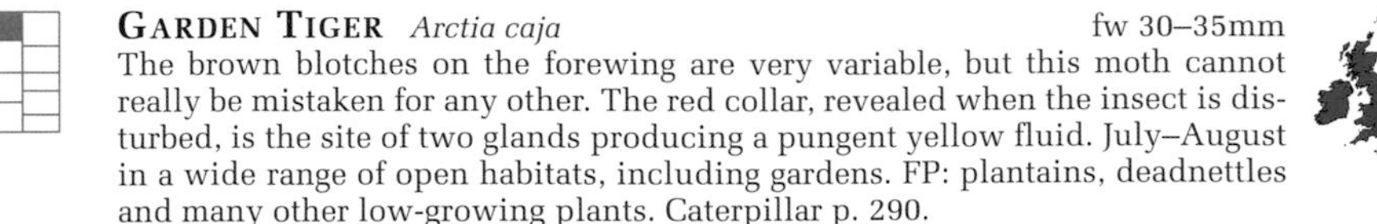

GARDEN TIGER *Arctia caja* fw 30–35mm

The brown blotches on the forewing are very variable, but this moth cannot really be mistaken for any other. The red collar, revealed when the insect is disturbed, is the site of two glands producing a pungent yellow fluid. July–August in a wide range of open habitats, including gardens. FP: plantains, deadnettles and many other low-growing plants. Caterpillar p. 290.

CREAM-SPOT TIGER *Arctia villica* fw 25–30mm

The cream-spotted black forewings are very characteristic. The hindwings are yellow with variable black spotting and there is also a conspicuous white 'shoulder flash' on the thorax. May–July in open habitats, but most common in coastal areas and on heathland. FP: a wide range of low-growing herbaceous plants. The Scarlet Tiger sometimes has yellowish hindwings, but it has no white at the base of the forewing and no white 'shoulder flashes'.

JERSEY TIGER *Euplagia quadripunctaria* fw *c.* 30mm

The diagonal stripes on the forewing, including a conspicuous 'Y' or 'V' in the outer half, are unlike those of any other tiger moth. The hindwings range from bright red to yellow, with black spots in the outer half. July–September, often day-flying, in gardens and other open habitats, especially near the coast. FP: hemp agrimony, deadnettles, stinging nettle, bramble and many other low-growing plants.

SCARLET TIGER *Callimorpha dominula* fw *c.* 25mm

The ground colour of the forewing is black with a blue or green sheen. The white and yellow blotches vary but there is never a pale patch at the base of the wing. The hindwing is either red or yellow. June–July, night- and day-flying, mainly in damp habitats. FP: hemp agrimony, bramble, sallow and many other herbs and shrubs. Caterpillar p. 290.

WOOD TIGER *Parasemia plantaginis* fw 15–20mm

The pale areas of the forewing are white or pale yellow, and those of the hindwing range from white to red. The thorax has cream or yellow 'shoulder flashes'. May–July on heaths and scrubby grassland, and also in open woodland. The male flies by day but the female is usually nocturnal. FP: a wide range of low-growing plants.

RUBY TIGER *Phragmatobia fuliginosa* fw 15–20mm

The forewings are brick-red and quite thinly scaled, especially in the centre. The hindwings range from pink with sooty borders to almost completely grey. April–September; common in most open habitats including gardens. FP: dandelions, plantains and many other low-growing herbs and shrubs.

CLOUDED BUFF *Diacrisia sannio* fw 17–22mm

The male (A), with his pink-fringed yellow forewings and white hindwings, is quite unmistakable. The slightly smaller female (B) has golden-orange wings, with heavy black markings on the hindwings. June–August, mainly on heathland and rough grassland. FP: heathers and many other low-growing plants.

THE CINNABAR *Tyria jacobaeae* fw *c.* 20mm

This moth, with its brilliant red hindwings, cannot be confused with any other. May–August in open grassy places, especially on light and sandy soils; often flies by day. FP: ragwort, which it has been used to control in some places. Caterpillar p. 292.

WHITE ERMINE *Spilosoma lubricipeda* fw 17–24mm
The black spots vary in size and number and are sometimes very sparse. There is always a black spot near the centre of the hindwing. The ground colour is occasionally pale cream and then the moth may be confused with the Buff Ermine. Weakly-spotted specimens may resemble the female Muslin Ermine, but are easily distinguished by the yellow abdomen. May–September in most habitats, including town parks and gardens. FP: docks, dandelions and most other low-growing herbs. Caterpillar p. 292. The rare **Water Ermine** (*S. urticae*) is almost spotless and never has a black central spot in the hindwing.

BUFF ERMINE *Spilosoma luteum* fw *c.* 20mm
The wings are cream or yellowish, the forewing with a fairly conspicuous line of elongated black spots running from the wing-tip to the centre of the rear edge. Cream specimens of the White Ermine never have such a prominent line. Other spots vary and may coalesce to form streaks. May–August in most habitats. FP: a wide variety of herbs and shrubs. Caterpillar p. 292.

MUSLIN ERMINE *Diaphora mendica* fw 15–20mm
The male (A) is easily identified in mainland Britain by his greyish-brown wings with scattered black spots. The female (B) resembles the White Ermine but is more thinly scaled and the abdomen is white instead of yellow. Irish males are creamy white and resemble the females. May–June, the male flying at night and the female by day. Most open habitats. FP: docks, plantains and many other low-growing herbs.

COMMON FOOTMAN *Eilema lurideola* fw *c.* 15mm
The pale yellow streak at the front of the forewing tapers strongly and usually fails to reach the wing-tip. The moth rests with its wings only slightly rolled around the body. July–August, in most habitats, including orchards and gardens. FP: lichens.

SCARCE FOOTMAN *Eilema complana* fw 15–18mm
The yellow streak at the front of the forewing reaches the wing-tip and is of constant width. The wings are wrapped tightly round the body at rest, giving the insect a very slender appearance. July–August and, despite its name, quite common in most habitats. FP: lichens.

DINGY FOOTMAN *Eilema griseola* fw 15–18mm
The front edge of the forewing is strongly arched instead of almost straight and the yellow streak often fades towards the tip. The wings are browner than those of the 2 preceding species and they are laid almost flat at rest. June–August, mainly in damp woods. FP: lichens.

BUFF FOOTMAN *Eilema depressa* fw 15–17mm
Resembles the Dingy Footman in shape, but the thorax is yellowish instead of grey and the male has conspicuous dark yellow fringes on the outer edges of the wings. Only the female has a yellow streak running the length of the forewing. July–August, mainly in mature woodland. FP: lichens, mainly arboreal species.

ORANGE FOOTMAN *Eilema sororcula* fw *c.* 15mm
The plain orange wings are unlike those of any other moth. May–June in woodland and dense scrub; uncommon. FP: arboreal lichens.

FOUR-SPOTTED FOOTMAN *Lithosia quadra* fw 20–25mm
This, our largest footman, is an unmistakable insect, although only the female (A) has the 4 spots. Both sexes rest with their wings wrapped tightly around the body and the male (B), a good deal smaller than his mate, is easily mistaken for a broken twig. July–September, rare in mature woodland. FP: lichens.

FOUR-DOTTED FOOTMAN *Cybosia mesomella* fw *c.* 15mm
The forewings are usually white, with pale yellow borders. Some specimens are pale yellow all over, but all have a small black dot halfway along the front and rear edges. June–August, on heaths and damp grassland, and also woodland clearings. FP: lichens.

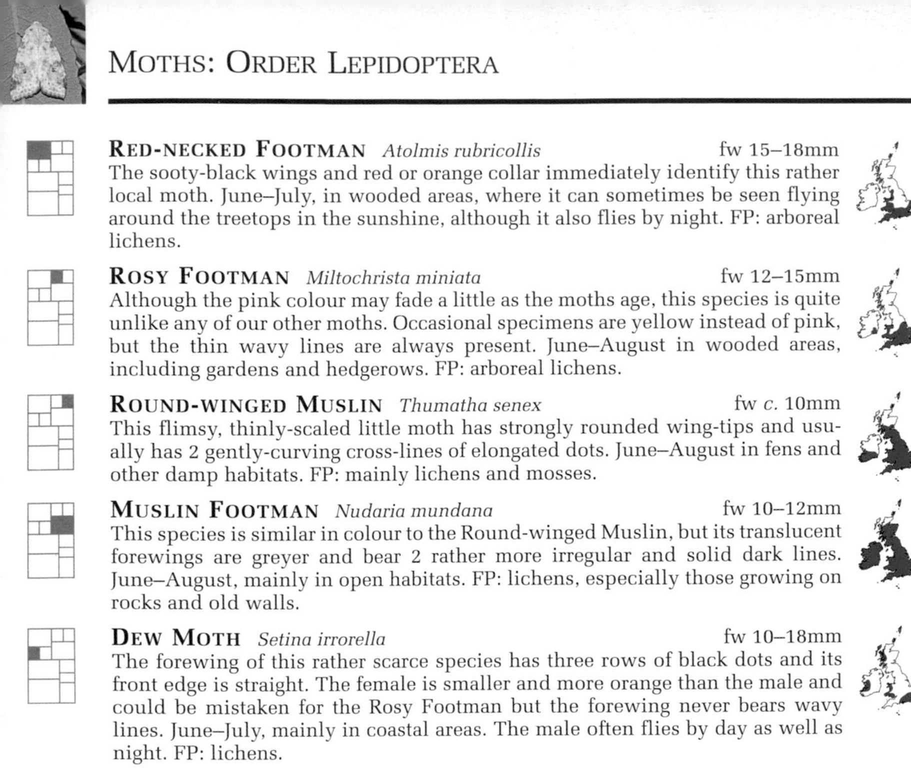

RED-NECKED FOOTMAN *Atolmis rubricollis* fw 15–18mm
The sooty-black wings and red or orange collar immediately identify this rather local moth. June–July, in wooded areas, where it can sometimes be seen flying around the treetops in the sunshine, although it also flies by night. FP: arboreal lichens.

ROSY FOOTMAN *Miltochrista miniata* fw 12–15mm
Although the pink colour may fade a little as the moths age, this species is quite unlike any of our other moths. Occasional specimens are yellow instead of pink, but the thin wavy lines are always present. June–August in wooded areas, including gardens and hedgerows. FP: arboreal lichens.

ROUND-WINGED MUSLIN *Thumatha senex* fw *c.* 10mm
This flimsy, thinly-scaled little moth has strongly rounded wing-tips and usually has 2 gently-curving cross-lines of elongated dots. June–August in fens and other damp habitats. FP: mainly lichens and mosses.

MUSLIN FOOTMAN *Nudaria mundana* fw 10–12mm
This species is similar in colour to the Round-winged Muslin, but its translucent forewings are greyer and bear 2 rather more irregular and solid dark lines. June–August, mainly in open habitats. FP: lichens, especially those growing on rocks and old walls.

DEW MOTH *Setina irrorella* fw 10–18mm
The forewing of this rather scarce species has three rows of black dots and its front edge is straight. The female is smaller and more orange than the male and could be mistaken for the Rosy Footman but the forewing never bears wavy lines. June–July, mainly in coastal areas. The male often flies by day as well as night. FP: lichens.

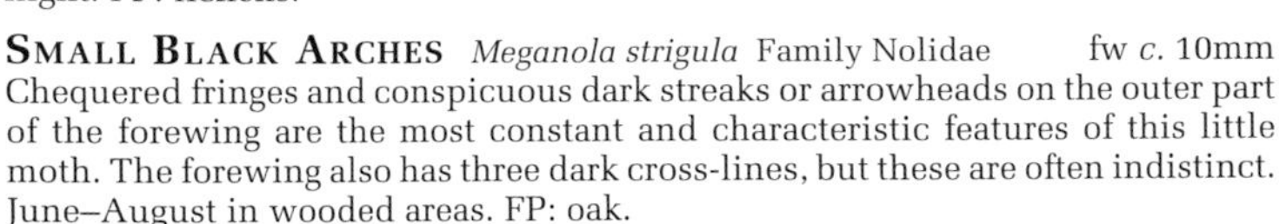

SMALL BLACK ARCHES *Meganola strigula* Family Nolidae fw *c.* 10mm
Chequered fringes and conspicuous dark streaks or arrowheads on the outer part of the forewing are the most constant and characteristic features of this little moth. The forewing also has three dark cross-lines, but these are often indistinct. June–August in wooded areas. FP: oak.

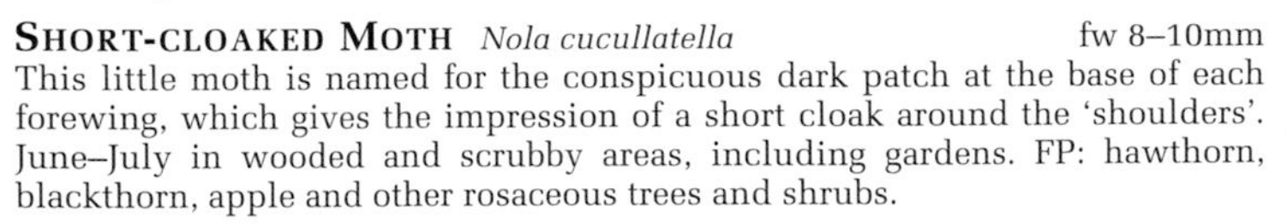

SHORT-CLOAKED MOTH *Nola cucullatella* fw 8–10mm
This little moth is named for the conspicuous dark patch at the base of each forewing, which gives the impression of a short cloak around the 'shoulders'. June–July in wooded and scrubby areas, including gardens. FP: hawthorn, blackthorn, apple and other rosaceous trees and shrubs.

KENT BLACK ARCHES *Meganola albula* fw 10–12mm
This species is much whiter than its relatives in this small family, although the greyish-brown markings vary a good deal in extent. The front edge of the forewing is usually noticeably chequered. As in other family members, the basal and central areas of the forewings bear upright tufts of scales, although these are often missing from older specimens. June–August in open habitats, mostly near the coast. FP: bramble, wild strawberry and related plants.

EMPEROR MOTH *Saturnia pavonia* Family Saturniidae fw 30–40mm
Seen from the front, in the way that most birds would see it, this moth bears a strong resemblance to a cat's face and undoubtedly scares off potential predators (*see* p. 42). All 4 wings carry eye-spots. The female is larger and greyer than the male , and her antennae are much simpler. April–June in all kinds of open country. Males (A) fly rapidly by day, but females (B) fly mainly at dusk. FP: heather, bramble, blackthorn and many other shrubs. Caterpillar p. 280.

KENTISH GLORY *Endromis versicolora* Family Endromidae fw 25–35mm
The pattern of this moth is unmistakable, although the female is greyer and larger than the male shown here. Restricted to open woodland in the Scottish Highlands. Flies April–May; the males by day and the females at night. FP: birch. Caterpillar p. 280.

NOCTUID MOTHS: FAMILY NOCTUIDAE

These are generally sturdy moths, mostly with brown or grey forewings. Most have a more-or-less kidney-shaped spot or ring called the reniform stigma in the outer part of the forewing, and a rounded or oval orbicular stigma nearer to the middle. Many also have a bullet-shaped claviform stigma behind the other 2. At rest, the forewings normally overlap and are held either tent-like or flat over the body. The hindwings are mostly brown or grey, although some species flash brightly coloured hindwings in flight (*see* p. 224). The larvae (*see* pp. 292–6) are usually plump and fleshy and rarely have much hair. There are nearly 400 resident British species.

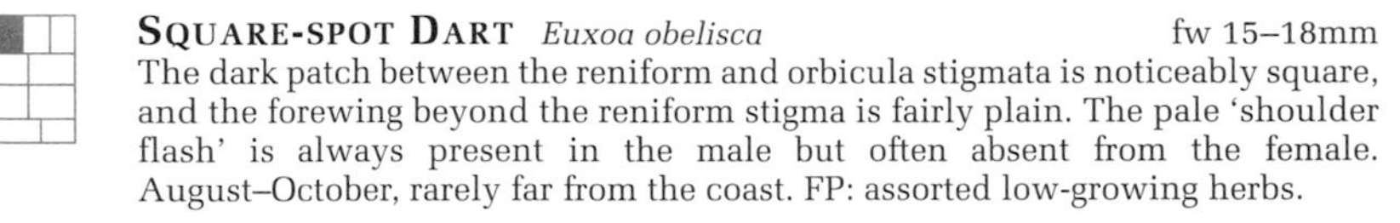

SQUARE-SPOT DART *Euxoa obelisca* fw 15–18mm
The dark patch between the reniform and orbicula stigmata is noticeably square, and the forewing beyond the reniform stigma is fairly plain. The pale 'shoulder flash' is always present in the male but often absent from the female. August–October, rarely far from the coast. FP: assorted low-growing herbs.

WHITE-LINE DART *Euxoa tritici* fw 12–18mm
The forewings range from light brown to almost black. The patch between the stigmata is like that of the Square-spot Dart but the outer part of the wing usually carries a number of dark 'arrowheads'. The moth is named for the white streak running parallel to the front edge, although this is not present in all specimens. July–September on heaths and other open habitats. FP: bedstraws and other low-growing herbs.

GARDEN DART *Euxoa nigricans* fw 15–20mm
The ground colour ranges from fawn to dark brown. Usually visible are 2 pale, wavy cross-lines, the outer curving strongly around the outside of the reniform stigma. June–October, mostly on cultivated land. FP: wide range of low-growing herbaceous plants.

LIGHT FEATHERED RUSTIC *Agrotis cinerea* fw 12–18mm
The forewings range from pale grey to deep brown, with 2 or three thin, wavy cross-lines. The reniform stigma may exist as a dark spot, but the other stigmata are usually missing. May–July on calcareous grassland. FP: mainly wild thyme.

ARCHER'S DART *Agrotis vestigialis* fw 14–18mm
The forewings range from silvery grey to brown but the moth is easily recognised by the large, bullet-like claviform stigma and the row of short, dark wedges in the outer part of the wing. The pale 'shoulder pads' are also quite distinctive. July–September, mainly in coastal areas and sandy heaths. FP: bedstraws and other low-growing plants.

TURNIP MOTH *Agrotis segetum* fw 15–20mm
The forewing is some shade of brown, often with extensive pale patches, and the 3 stigmata often have pale centres. The hindwing is very white, with or without dark veins. April–November, most common on cultivated land. FP: roots of many wild and cultivated plants.

HEART AND CLUB *Agrotis clavis* fw 15–20mm
The forewings range from pale greyish-brown to almost black, with the claviform stigma broad and blunt and often pale-centred. Pale 'shoulder pads' are usually conspicuous. May–August in many open habitats, including gardens. FP: numerous low-growing herbaceous plants.

HEART AND DART *Agrotis exclamationis* fw 15–20mm
This moth is readily distinguished from the Heart and Club by its black collar. The forewings are usually plain brown, with little decoration other than the solid, dark stigmata, the claviform stigma being relatively long and pointed. May–September almost everywhere, especially common on cultivated land. FP: a wide range of wild and cultivated herbaceous plants.

SHUTTLE-SHAPED DART *Agrotis puta* fw 12–15mm
Named for the pale, elongate orbicular stigma, with a dark central streak. The forewings are pale brown in the male and dark brown in the female. The thorax is always pale. March–October in most open habitats. FP: many wild and cultivated herbaceous plants.

DARK SWORD-GRASS *Agrotis ipsilon* fw 15–25mm
The distribution of light and dark brown on the forewing varies but the moth is easily recognised by the black dart on the outer edge of the reniform stigma and the two smaller inward-pointing darts near the wing-tip. The wings are partly wrapped around the body at rest. This common immigrant can be found in most habitats, mainly in late summer. FP: many low-growing plants.

THE FLAME *Axylia putris* fw *c.* 15mm
The forewings are largely straw-coloured, with dark reniform and orbicular stigmata and a dark brown front margin. At rest, with the wings wrapped tightly around the body, the moth looks very like a broken twig or reed. June–September in many open habitats. FP: a wide range of low-growing herbs. The similarly coloured Shuttle-shaped Dart (*see* p. 222) has a pale thorax.

FLAME SHOULDER *Ochropleura plectra* fw 12–15mm
The pale front edge to the otherwise chestnut-brown forewing makes this moth very easy to identify. April–September almost everywhere, including gardens. FP: a wide range of low-growing herbaceous plants.

LARGE YELLOW UNDERWING *Noctua pronuba* fw 20–25mm
The forewing ranges from light to very dark brown and the male may be extensively marbled with light and dark bands. The black dot near the wing-tip does not touch the margin. The yellow hindwing has a narrow dark border and no other markings. June–November, nearly everywhere. FP: many wild and cultivated plants. Caterpillar p. 292.

LUNAR YELLOW UNDERWING *Noctua orbona* fw 17–20mm
This moth is like the Large Yellow Underwing but the black mark meets the front wing-margin and the hindwing bears a dark crescent. June–September, mainly on heathland and rough grassland: uncommon. FP: mainly grasses.

LESSER YELLOW UNDERWING *Noctua comes* fw 15–20mm
The ground colour of the forewings ranges from light brown to dark brown or almost grey and the stigmata may be light or dark, but there is never a discrete black spot near the wing-tip. The hindwing has a dark crescent. June–October; common almost everywhere. FP: polyphagous, on shrubs as well as herbaceous plants.

BROAD-BORDERED YELLOW UNDERWING *Noctua fimbriata* fw 20–30mm
The forewings are generally orange or light brown, often heavily dusted with grey and/or green. The reniform and orbicular stigmata have pale outlines and there is a pale cross-band beyond them. The hindwing has a very broad black border. The wings are slightly rolled around the body at rest, whereas those of the other yellow underwings are laid flat. July–September, mainly in wooded areas. FP: many herbaceous plants and low-growing shrubs.

LESSER BROAD-BORDERED YELLOW UNDERWING *N. janthe* fw 15–20mm
The forewings are deep brown, with a wine-red front margin in the outer half and often tinged with purple all over. Fresh specimens have a light green collar, although this becomes brown with age. The hindwing has a broad black border and a dark smudge at the base. July–September nearly everywhere. FP: numerous herbaceous plants and low shrubs.

LEAST YELLOW UNDERWING *Noctua interjecta* fw 14–18mm
The forewing is usually reddish brown, often dusted with darker scales and bearing 1 or more thin cross-lines, but without conspicuous stigmata. The black edging on the hindwing continues along the inner margin towards the base. June–August in many open habitats. FP: grasses and various other plants.

DOUBLE DART *Graphiphora augur* fw *c.* 20mm
The forewings range from greyish brown to reddish brown and the stigmata are represented simply by plain black outlines. These are sometimes indistinct, and then the moth resembles the Mouse Moth (*see* p. 250). June–August in wooded areas, including parks and gardens. FP: blackthorn, hawthorn and other shrubs, as well as docks and other herbaceous plants.

COUSIN GERMAN *Protolampra sobrina* fw *c.* 15mm
Most specimens can be recognised by the silvery-grey basal half of the forewing and the purplish-brown outer half, although some individuals lack the grey area. There is an irregular central cross-line, but the stigmata are inconspicuous. July–September in upland birch woods; rare. FP: heather, bilberry, and birch.

AUTUMNAL RUSTIC *Eugnorisma glareosa* fw *c.* 15mm
Although the ground colour varies a good deal from pale grey to brown, this moth is easy to recognise by the black anvil-shaped spot between the reniform and orbicular stigmata. August–October in many open habitats. FP: bedstraws, heathers, and many other low-growing plants.

TRUE LOVER'S KNOT *Lycophotia porphyrea* fw 12–15mm
The forewings are varying shades of brown but the prominent white or white-edged stigmata and the almost semicircular cross-line just beyond the reniform stigma distinguish this from other similar moths. June–August on heaths and moors. FP: heathers.

PEARLY UNDERWING *Peridroma saucia* fw 20–23mm
Named for its pearly-white hindwings, this common immigrant has variable amounts of light brown on the forewings but can often be recognised by the grey crest running along the thorax. Most of the year, almost anywhere, but most common in late summer and autumn. FP: numerous wild and cultivated herbaceous plants.

INGRAILED CLAY *Diarsia mendica* fw 12–18mm
As its scientific name suggests, this is a very variable moth, with forewings ranging from pale grey or straw-coloured to dark brown and from almost plain to heavily marbled. There is often a dark square between the reniform and orbicular stigmata and a dark triangle on the inner edge of the latter, and most specimens have a small black dot just below it. June–August in many habitats. FP: a wide range of herbaceous and woody plants, including heathers and bramble.

BARRED CHESTNUT *Diarsia dahlii* fw *c.* 16mm
The forewings range from light to dark brown, with a strongly-arched front edge and usually banded with purplish brown. The stigmata are ringed with white but otherwise much the same colour as the rest of the wing and not readily seen. August–September on heaths and in light woodland. FP: birch, brambles and numerous other shrubs.

PURPLE CLAY *Diarsia brunnea* fw 15–20mm
Named for the strong purple tinge to its rich brown wings, this moth usually has a conspicuous straw-coloured reniform stigma and there is a dark, more or less square spot between this and the orbicular stigma. June–August in wooded areas. FP: many herbs and low-growing shrubs.

SMALL SQUARE-SPOT *Diarsia rubi* fw 12–15mm
The pale to mid-brown forewings often have a pink tinge. There is a light brown patch between the orbicular and reniform stigmata, both of which normally have pale outlines. The claviform stigma is reduced to a small dark spot or dash between the orbicular stigma and the rear margin. May–September nearly everywhere. FP: many wild and cultivated herbaceous plants.

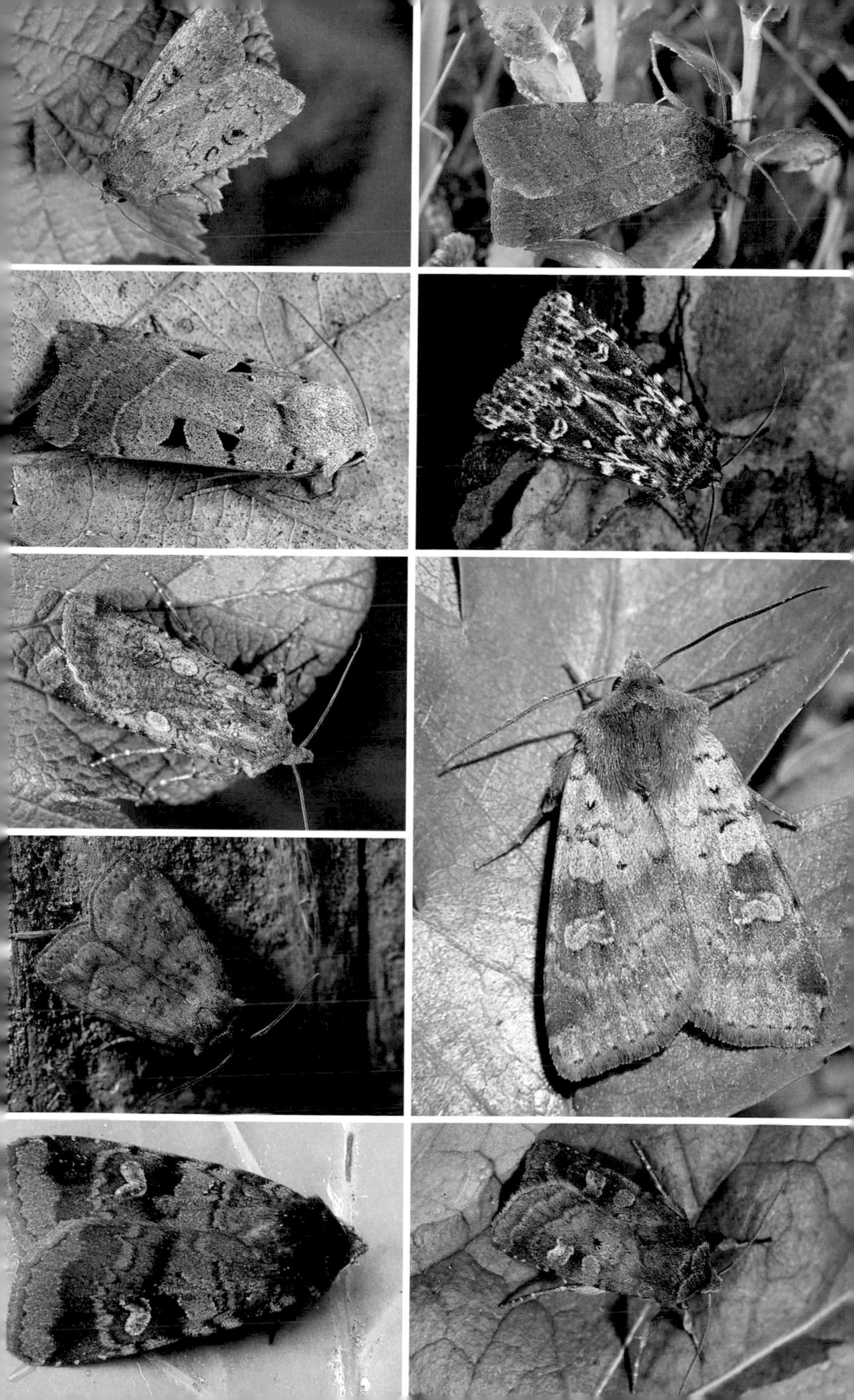

SETACEOUS HEBREW CHARACTER *Xestia c-nigrum* fw 15–20mm
The pale triangular mark sitting in a black 'cradle' at the front of the dark brown and often purplish forewing distinguishes this from all other moths. The wings are laid flat at rest. The Hebrew Character (*see* p. 236) has a similar 'cradle' but the pale spot is less pronounced and the wings are held tented at rest. May–October nearly everywhere; especially common on cultivated land. FP: deadnettles and many other wild and cultivated herbaceous plants.

TRIPLE-SPOTTED CLAY *Xestia ditrapezium* fw 17–20mm
The forewings are usually rust-coloured, often with a purplish tinge. The outer arm of the black 'cradle' is quite square and the pale area between the arms is much the same colour as the rest of the wing. There is a prominent black mark near the wing-tip. The hindwings are yellowish white. June–August, mainly in damp woodland. FP: various herbs and shrubs.

DOUBLE SQUARE-SPOT *Xestia triangulum* fw 17–20mm
This is very similar to the Triple-spotted Clay but the forewings are usually much paler and greyer and the hindwings are dull greyish-brown. June–August in and around woods and hedgerows, including garden hedges. FP: numerous herbs and shrubs.

DOTTED CLAY *Xestia baja* fw 17–20mm
The forewings range from yellowish brown to rusty brown, often with a greyish tinge. The stigmata are inconspicuous, but the moth can always be recognised by the two black dots near the wing-tip. July–September, mainly in and around woods. FP: various herbs and shrubs.

SQUARE-SPOTTED CLAY *Xestia rhomboidea* fw 17–20mm
The forewings are dingy brown with a very dark, more or less square mark between the inconspicuous stigmata. There is an irregular dark brown cross-band in the outer part of the wing but there is no black streak near the tip. July–August, mainly in mature woodland; uncommon. FP: various herbs and shrubs.

SIX-STRIPED RUSTIC *Xestia sexstrigata* fw *c.* 16mm
The dark veins and fine cross-lines on the light brown forewings make this moth easy to recognise. July–August in many habitats, especially damp places. FP: bedstraws and many other herbs.

SQUARE-SPOT RUSTIC *Xestia xanthographa* fw 14–18mm
The ground colour of the forewings is always some shade of brown, on which the rather square, pale reniform stigma for which the moth is named is usually very obvious. July–October on grassland and cultivated land. FP: mainly grasses.

HEATH RUSTIC *Xestia agathina* fw *c.* 15mm
This greyish-brown or reddish-brown moth is easily recognised by the black wedge enclosing the white orbicular stigma in the centre of the forewing, and also by the pale flash at the base of the forewing. August–September on heaths and moors. FP: heather.

THE GOTHIC *Naenia typica* fw 16–22mm
The white veins and white cross-lines distinguish this from most other moths. The 1st cross-line is more or less straight and the hindwings are grey. The Feathered Gothic (*see* p. 234) has dark cross-lines. June–July nearly everywhere. FP: many wild and cultivated plants.

GREEN ARCHES *Anaplectoides prasina* fw 20–25mm
The pale patch beyond the reniform stigma will usually identify this moth, even when the fine green of the freshly-emerged insect has faded to yellowish brown. June–July in deciduous woodland. FP: a wide range of woodland herbs and shrubs.

GREAT BROCADE *Eurois occulta* fw *c.* 25mm
This is the only large grey-winged moth to rest with its wings folded flat over the body. Resident moths, found only in Scotland, are sooty black with a few light markings, but immigrants are much paler. July–September, breeding in boggy areas; immigrants occur in all habitats. FP: mainly bog myrtle.

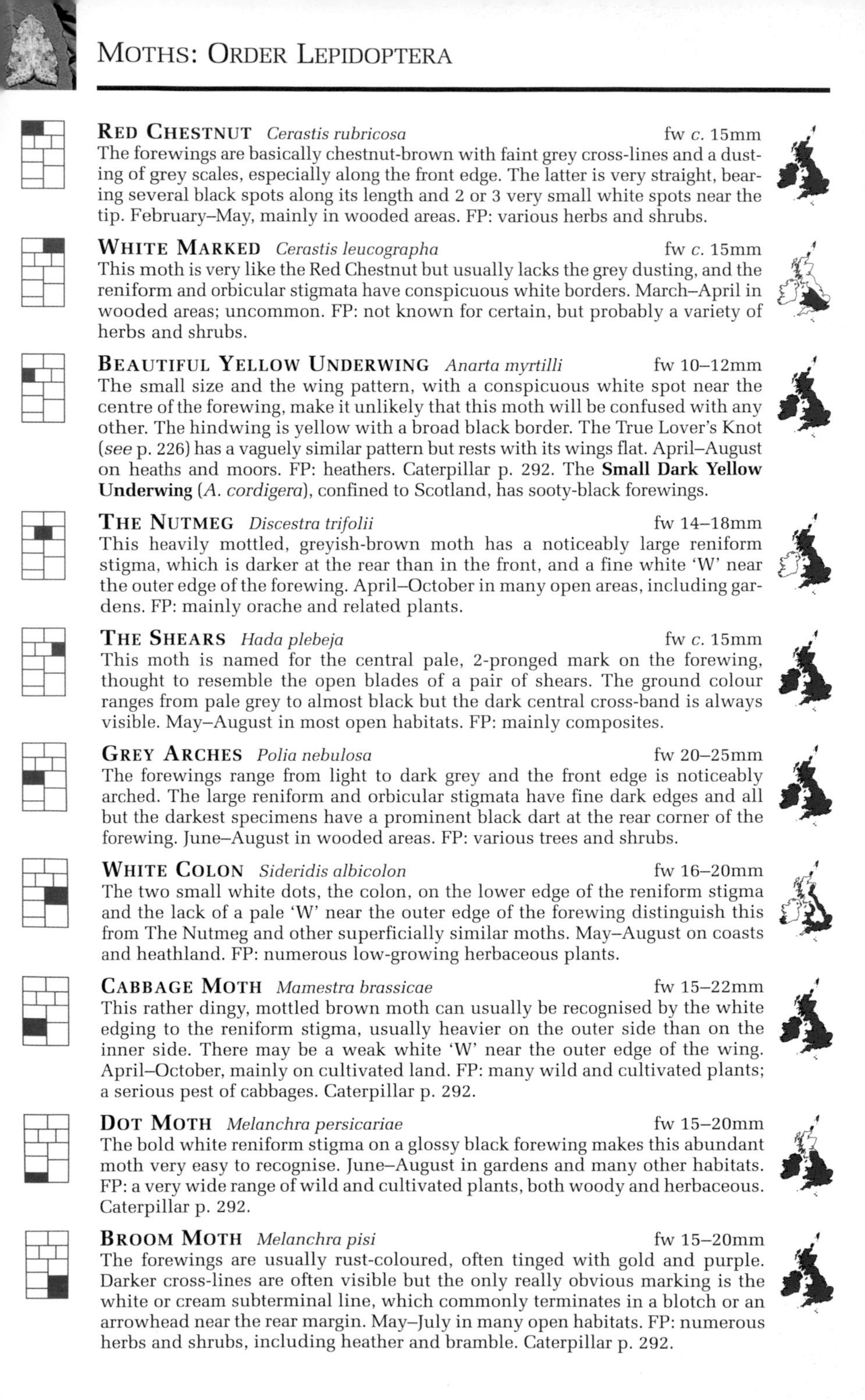

RED CHESTNUT *Cerastis rubricosa* fw *c.* 15mm
The forewings are basically chestnut-brown with faint grey cross-lines and a dusting of grey scales, especially along the front edge. The latter is very straight, bearing several black spots along its length and 2 or 3 very small white spots near the tip. February–May, mainly in wooded areas. FP: various herbs and shrubs.

WHITE MARKED *Cerastis leucographa* fw *c.* 15mm
This moth is very like the Red Chestnut but usually lacks the grey dusting, and the reniform and orbicular stigmata have conspicuous white borders. March–April in wooded areas; uncommon. FP: not known for certain, but probably a variety of herbs and shrubs.

BEAUTIFUL YELLOW UNDERWING *Anarta myrtilli* fw 10–12mm
The small size and the wing pattern, with a conspicuous white spot near the centre of the forewing, make it unlikely that this moth will be confused with any other. The hindwing is yellow with a broad black border. The True Lover's Knot (*see* p. 226) has a vaguely similar pattern but rests with its wings flat. April–August on heaths and moors. FP: heathers. Caterpillar p. 292. The **Small Dark Yellow Underwing** (*A. cordigera*), confined to Scotland, has sooty-black forewings.

THE NUTMEG *Discestra trifolii* fw 14–18mm
This heavily mottled, greyish-brown moth has a noticeably large reniform stigma, which is darker at the rear than in the front, and a fine white 'W' near the outer edge of the forewing. April–October in many open areas, including gardens. FP: mainly orache and related plants.

THE SHEARS *Hada plebeja* fw *c.* 15mm
This moth is named for the central pale, 2-pronged mark on the forewing, thought to resemble the open blades of a pair of shears. The ground colour ranges from pale grey to almost black but the dark central cross-band is always visible. May–August in most open habitats. FP: mainly composites.

GREY ARCHES *Polia nebulosa* fw 20–25mm
The forewings range from light to dark grey and the front edge is noticeably arched. The large reniform and orbicular stigmata have fine dark edges and all but the darkest specimens have a prominent black dart at the rear corner of the forewing. June–August in wooded areas. FP: various trees and shrubs.

WHITE COLON *Sideridis albicolon* fw 16–20mm
The two small white dots, the colon, on the lower edge of the reniform stigma and the lack of a pale 'W' near the outer edge of the forewing distinguish this from The Nutmeg and other superficially similar moths. May–August on coasts and heathland. FP: numerous low-growing herbaceous plants.

CABBAGE MOTH *Mamestra brassicae* fw 15–22mm
This rather dingy, mottled brown moth can usually be recognised by the white edging to the reniform stigma, usually heavier on the outer side than on the inner side. There may be a weak white 'W' near the outer edge of the wing. April–October, mainly on cultivated land. FP: many wild and cultivated plants; a serious pest of cabbages. Caterpillar p. 292.

DOT MOTH *Melanchra persicariae* fw 15–20mm
The bold white reniform stigma on a glossy black forewing makes this abundant moth very easy to recognise. June–August in gardens and many other habitats. FP: a very wide range of wild and cultivated plants, both woody and herbaceous. Caterpillar p. 292.

BROOM MOTH *Melanchra pisi* fw 15–20mm
The forewings are usually rust-coloured, often tinged with gold and purple. Darker cross-lines are often visible but the only really obvious marking is the white or cream subterminal line, which commonly terminates in a blotch or an arrowhead near the rear margin. May–July in many open habitats. FP: numerous herbs and shrubs, including heather and bramble. Caterpillar p. 292.

GLAUCOUS SHEARS *Papestra biren* fw 15–18mm
Looking rather like a grey version of the Dot Moth, this moth can be recognised by its 3 black-edged, white stigmata. The claviform stigma is broad and almost as big as the orbicular stigma immediately above it. May–August on moorland. FP: heather, bilberry and other low-growing plants.

BEAUTIFUL BROCADE *Lacanobia contigua* fw 15–20mm
This well-named moth can always be recognised by the broad, pale grey band, often tinged with pink, running diagonally back from the orbicular stigma. June–July on heaths and moors and in open woodland. FP: many trees and shrubs, especially birch and sallow.

LIGHT BROCADE *Lacanobia w-latinum* fw *c.* 20mm
Although resembling the Beautiful Brocade in colour, this moth lacks the pale diagonal band. The pale orbicular and reniform stigmata stand out clearly on the brown central cross-band, which contrasts strongly with the grey band beyond it. May–August, mainly on calcareous grassland. FP: many low-growing shrubs and herbaceous plants.

PALE-SHOULDERED BROCADE *Lacanobia thalassina* fw 15–20mm
The forewing pattern resembles that of the Light Brocade but is much browner, especially in the outer part. The pale 'shoulder flash' is quite prominent, as is the black line joining the two pale cross-lines. May–August, mainly in wooded areas. FP: numerous trees and shrubs.

DOG'S-TOOTH *Lacanobia suasa* fw 15–20mm
The typical form is greyish brown with a central thick black claviform stigma (the 'dog's tooth') on the forewing and a white 'W' or two white spikes near the outer edge. There is also a brick-coloured form in which the tooth is not much darker than the rest of the wing. May–September in most open habitats, especially damp places. FP: many herbaceous plants.

BRIGHT-LINE BROWN-EYE *Lacanobia oleracea* fw 15–20mm
This moth is named for the clear white submarginal line and the light brown or orange reniform stigma. The submarginal line has two spikes or a conspicuous 'W' in the centre, and the orbicular stigma may be a white spot or a ring. May–September nearly everywhere; very common on cultivated land. FP: numerous wild and cultivated herbs and shrubs.

BROAD-BARRED WHITE *Hecatera bicolorata* fw 12–15mm
No other moth has such a clear-cut black band on grey wings. May–August, mainly on rough grassland. FP: hawkweeds, sowthistles and related composites, feeding mainly on the flowers.

THE CAMPION *Hadena rivularis* fw *c.* 15mm
The lower edge of the reniform stigma extends inwards to meet the base of the elongate orbicular stigma so that the two marks form a white-edged 'V'. The purple tinge in the central part of the forewing fades with age. May–September in most open, grassy habitats. FP: seed capsules of campions.

THE LYCHNIS *Hadena bicruris* fw 14–17mm
This moth is very like The Campion but the reniform and orbicular stigmata usually remain separate and there is no purple tinge on the forewings. May–September in most open habitats. FP: seed capsules of campions and related plants, including cultivated sweet williams.

VARIED CORONET *Hadena compta* fw 12–15mm
The unbroken central white band distinguishes this moth from its relatives and most other similar black and white species. June–September, mainly in parks and gardens. FP: seed capsules of sweet william and, less often, bladder campion.

MARBLED CORONET *Hadena confusa* fw *c.* 15mm
This is very like the Varied Coronet but the white cross-band is broken in the centre and there is a prominent white patch at the wing-tip, although in many N and W areas the white is replaced by a dirty yellow. May–August on open grassland, especially coastal areas. FP: seed capsules of campions.

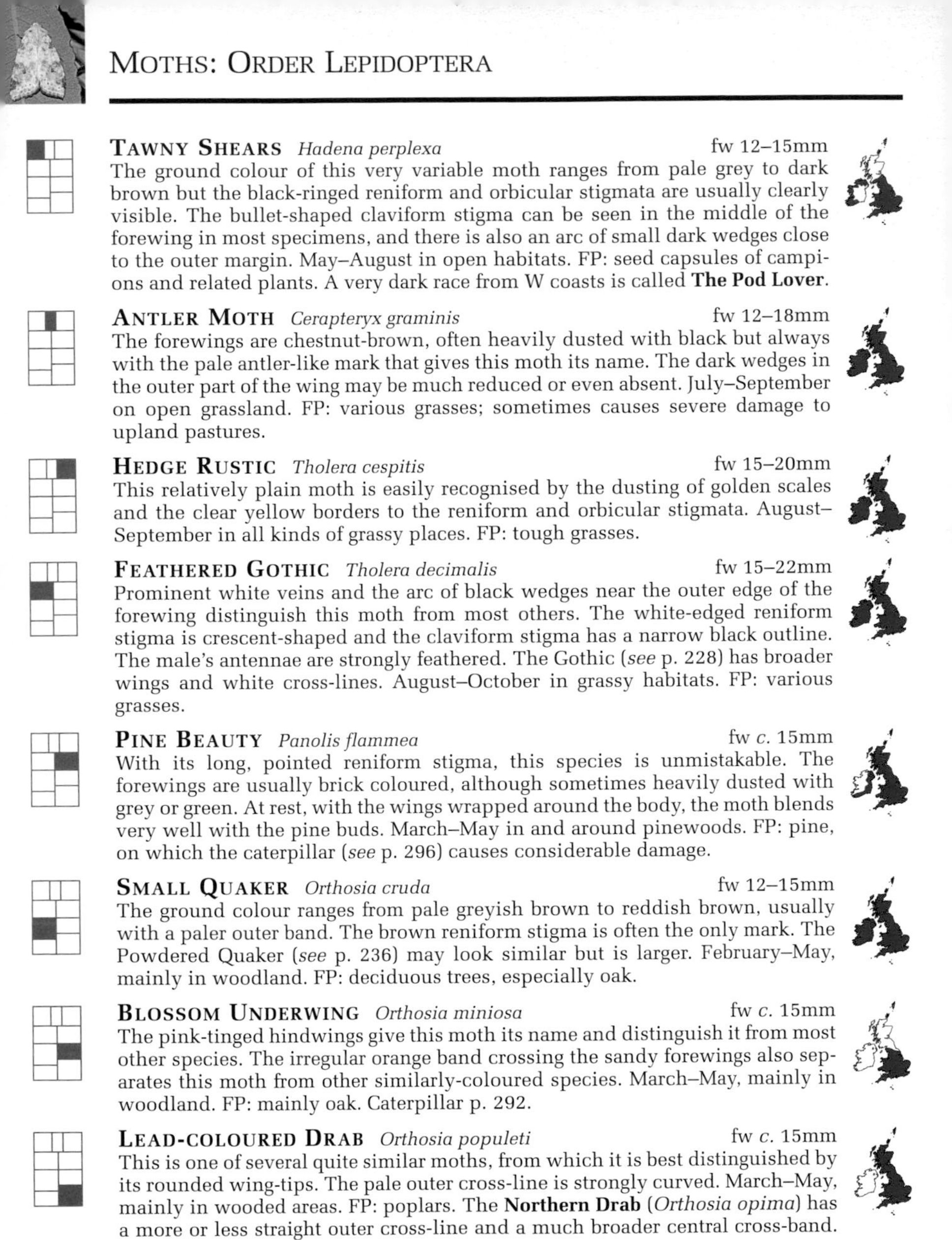

TAWNY SHEARS *Hadena perplexa* fw 12–15mm
The ground colour of this very variable moth ranges from pale grey to dark brown but the black-ringed reniform and orbicular stigmata are usually clearly visible. The bullet-shaped claviform stigma can be seen in the middle of the forewing in most specimens, and there is also an arc of small dark wedges close to the outer margin. May–August in open habitats. FP: seed capsules of campions and related plants. A very dark race from W coasts is called **The Pod Lover**.

ANTLER MOTH *Cerapteryx graminis* fw 12–18mm
The forewings are chestnut-brown, often heavily dusted with black but always with the pale antler-like mark that gives this moth its name. The dark wedges in the outer part of the wing may be much reduced or even absent. July–September on open grassland. FP: various grasses; sometimes causes severe damage to upland pastures.

HEDGE RUSTIC *Tholera cespitis* fw 15–20mm
This relatively plain moth is easily recognised by the dusting of golden scales and the clear yellow borders to the reniform and orbicular stigmata. August–September in all kinds of grassy places. FP: tough grasses.

FEATHERED GOTHIC *Tholera decimalis* fw 15–22mm
Prominent white veins and the arc of black wedges near the outer edge of the forewing distinguish this moth from most others. The white-edged reniform stigma is crescent-shaped and the claviform stigma has a narrow black outline. The male's antennae are strongly feathered. The Gothic (*see* p. 228) has broader wings and white cross-lines. August–October in grassy habitats. FP: various grasses.

PINE BEAUTY *Panolis flammea* fw *c.* 15mm
With its long, pointed reniform stigma, this species is unmistakable. The forewings are usually brick coloured, although sometimes heavily dusted with grey or green. At rest, with the wings wrapped around the body, the moth blends very well with the pine buds. March–May in and around pinewoods. FP: pine, on which the caterpillar (*see* p. 296) causes considerable damage.

SMALL QUAKER *Orthosia cruda* fw 12–15mm
The ground colour ranges from pale greyish brown to reddish brown, usually with a paler outer band. The brown reniform stigma is often the only mark. The Powdered Quaker (*see* p. 236) may look similar but is larger. February–May, mainly in woodland. FP: deciduous trees, especially oak.

BLOSSOM UNDERWING *Orthosia miniosa* fw *c.* 15mm
The pink-tinged hindwings give this moth its name and distinguish it from most other species. The irregular orange band crossing the sandy forewings also separates this moth from other similarly-coloured species. March–May, mainly in woodland. FP: mainly oak. Caterpillar p. 292.

LEAD-COLOURED DRAB *Orthosia populeti* fw *c.* 15mm
This is one of several quite similar moths, from which it is best distinguished by its rounded wing-tips. The pale outer cross-line is strongly curved. March–May, mainly in wooded areas. FP: poplars. The **Northern Drab** (*Orthosia opima*) has a more or less straight outer cross-line and a much broader central cross-band.

CLOUDED DRAB *Orthosia incerta* fw 15–20mm
The ground colour ranges widely, from pale grey to purplish brown, and the stigmata are not always obvious. The wavy outer cross-line, with sharp bends fore and aft, distinguishes this from the previous species. March–May, almost anywhere with trees. FP: many deciduous trees. Caterpillar p. 292.

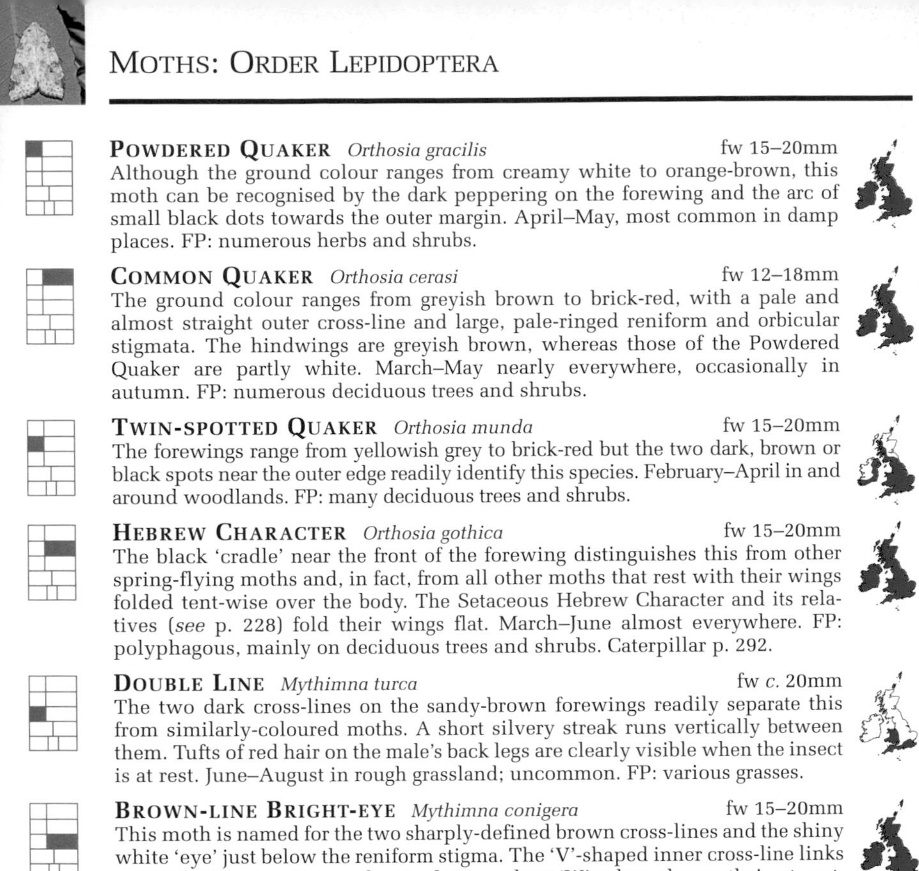

POWDERED QUAKER *Orthosia gracilis* fw 15–20mm
Although the ground colour ranges from creamy white to orange-brown, this moth can be recognised by the dark peppering on the forewing and the arc of small black dots towards the outer margin. April–May, most common in damp places. FP: numerous herbs and shrubs.

COMMON QUAKER *Orthosia cerasi* fw 12–18mm
The ground colour ranges from greyish brown to brick-red, with a pale and almost straight outer cross-line and large, pale-ringed reniform and orbicular stigmata. The hindwings are greyish brown, whereas those of the Powdered Quaker are partly white. March–May nearly everywhere, occasionally in autumn. FP: numerous deciduous trees and shrubs.

TWIN-SPOTTED QUAKER *Orthosia munda* fw 15–20mm
The forewings range from yellowish grey to brick-red but the two dark, brown or black spots near the outer edge readily identify this species. February–April in and around woodlands. FP: many deciduous trees and shrubs.

HEBREW CHARACTER *Orthosia gothica* fw 15–20mm
The black 'cradle' near the front of the forewing distinguishes this from other spring-flying moths and, in fact, from all other moths that rest with their wings folded tent-wise over the body. The Setaceous Hebrew Character and its relatives (*see* p. 228) fold their wings flat. March–June almost everywhere. FP: polyphagous, mainly on deciduous trees and shrubs. Caterpillar p. 292.

DOUBLE LINE *Mythimna turca* fw *c.* 20mm
The two dark cross-lines on the sandy-brown forewings readily separate this from similarly-coloured moths. A short silvery streak runs vertically between them. Tufts of red hair on the male's back legs are clearly visible when the insect is at rest. June–August in rough grassland; uncommon. FP: various grasses.

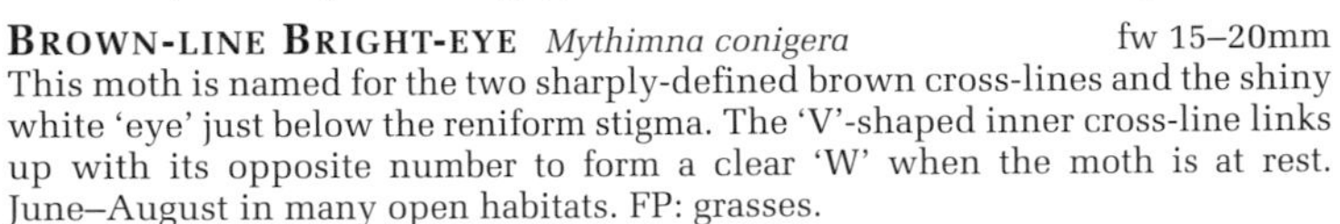

BROWN-LINE BRIGHT-EYE *Mythimna conigera* fw 15–20mm
This moth is named for the two sharply-defined brown cross-lines and the shiny white 'eye' just below the reniform stigma. The 'V'-shaped inner cross-line links up with its opposite number to form a clear 'W' when the moth is at rest. June–August in many open habitats. FP: grasses.

THE CLAY *Mythimna ferrago* fw 15–18mm
♂ The forewings range from straw-coloured to rusty red, with an oval white spot at the base of the reniform stigma and an arc of small black dots in the outer part of the wing. Males are easily recognised by a black 'V' under the abdomen. June–August in all open habitats. FP: grasses.

WHITE POINT *Mythimna albipuncta* fw 15–17mm
This species resembles The Clay but is usually brighter and the white spot is rounder and 'cleaner'. There may be an arc of tiny black and white dots in the outer part of the forewing. May–November in coastal areas, mainly as immigrants. FP: grasses.

STRIPED WAINSCOT *Mythimna pudorina* fw 15–20mm
This moth differs from similar wainscots in that the dark central streak has no accompanying white line. The forewing is often tinged with pink and the hindwing is dull brown. The Southern Wainscot (*see* p. 238) has dotted cross-lines. June–July in watery habitats. FP: reeds and other grasses.

SMOKY WAINSCOT *Mythimna impura* fw 15–18mm
The central white vein has a dark line along its rear edge and there are usually several small black dots in the outer part of the forewing. The hindwing is quite grey, not largely white as in most other wainscots. June–October on grasslands. FP: various grasses.

COMMON WAINSCOT *Mythimna pallens* fw *c.* 15mm
The forewings range from pale straw-coloured to brick-red, with pale veins. There may be a number of black dots in the outer part of the wing. The hindwing is usually pure white. June–October in most grassy places, including parks and gardens. FP: various grasses.

SOUTHERN WAINSCOT *Mythimna straminea* fw 15–18mm
The forewings are pale grey, often tinged with pink, and a dark streak runs along the rear edge of the central white vein. There is an arc of small black dots on the forewing *and* the hindwing. July–August in wet places. FP: reeds.

MATHEW'S WAINSCOT *Mythimna favicolor* fw 15–18mm
The forewings are generally much yellower than those of other wainscots, with little or no white on the veins. There is an arc of tiny black dots in the outer region. June–September on coastal salt marshes. FP: grasses.

SHORE WAINSCOT *Mythimna litoralis* fw 15–18mm
The heavy white streak, with a brown line on both sides, distinguishes this from other wainscots. The hindwing is pure white. June–October on sandy coasts. FP: marram grass.

L-ALBUM WAINSCOT *Mythimna l-album* fw *c.* 15mm
The long white central 'L' on the forewing readily identifies this moth. July–November on coastal grasslands, often as an immigrant. FP: marram and perhaps other grasses.

SHOULDER-STRIPED WAINSCOT *Mythimna comma* fw 15–20mm
The very pale front edge of the forewing and the dense black streak running out from the base help to distinguish this from other wainscots. May–July in grassy places, most common in damp habitats. FP: various grasses.

FLAME WAINSCOT *Mythimna flammea* fw 15–18mm
The strongly-arched front edge and pointed tip of the forewing distinguish this from all our other wainscots. There is often a reddish 'glow' around the central streak, which is never present in the superficially similar Silky Wainscot (*see* p. 260). May–July in watery habitats. FP: reeds.

THE WORMWOOD *Cucullia absinthii* fw 15–20mm
The black spots near the front and outer edges of the forewing distinguish this moth from its relatives, all of which have prominent crests and rest with their narrow wings pulled tightly back along the body. July–August, mainly on disturbed land. FP: wormwood and mugwort.

THE SHARK *Cucullia umbratica* fw 20–25mm
The grey forewings are tinged with brown, especially near the front edge, but otherwise the only obvious markings are thin black streaks, which stop abruptly at the outer fringes. June–August in many open and disturbed habitats. FP: sowthistles and related composites.

CHAMOMILE SHARK *Cucullia chamomillae* fw *c.* 20mm
This moth is very like The Shark but the black streaks continue into the marginal fringes on the outer edge of the forewing. April–May, mainly on arable land and other disturbed places. FP: scentless mayweed and related plants.

THE STARWORT *Cucullia asteris* fw *c.* 20mm
The rust-coloured stripe along the front edge of the forewing, enclosing the pale-ringed reniform and orbicular stigmata, distinguishes this from related species. June–August, mainly in coastal regions. FP: mainly sea aster, but also cultivated Michaelmas daisies.

MULLEIN MOTH *Shargacucullia verbasci* fw 20–25mm
This common moth resembles The Starwort but the forewing is browner and there are no stigmata. The outer margins are scalloped, and when at rest in its typical position the moth looks just like a sliver of bark. April–May in gardens and waste or disturbed ground. FP: mainly wild and cultivated mulleins, the caterpillars (*see* p. 294) often completely destroying the flower spikes.

TOADFLAX BROCADE *Calophasia lunula* fw *c.* 15mm
The banded outer margin of the forewing and the black-ringed white stigmata make this moth easy to identify. May–September in disturbed coastal areas, where some individuals are probably immigrants. FP: purple and common toadflax.

Minor Shoulder-knot *Brachylomia viminalis* fw 12–15mm
There is a short black streak at the base of the forewing and another runs diagonally forward from it. Most specimens are light grey with a darker central band. Others are dark grey, but the pale reniform stigma is normally obvious and the black streaks are visible in all but the darkest specimens. July–August in damp woods and other wet habitats. FP: willows.

The Sprawler *Asteroscopus sphinx* fw 17–22mm
The forewings are grey or warm brown, with a long black basal streak, several other dark streaks, and a conspicuous white zigzag submarginal line. October–December in wooded areas. FP: numerous deciduous trees and shrubs. Caterpillar p. 294.

Brindled Ochre *Dasypolia templi* fw 18–25mm
Size and colour are enough to separate this moth from most others, although the density of the brown or grey dusting varies. The reniform stigma and the pale submarginal line are often the only obvious markings. August–October and again in spring after hibernation, mainly on coastal grassland. FP: umbellifers.

Deep-brown Dart *Aporophyla lutulenta* fw 15–18mm
The forewings are often entirely sooty brown, with no obvious markings, although a slightly darker cross-band may be visible in the centre. September–October in most open habitats. FP: numerous herbs and shrubs. The **Northern Deep-brown Dart** (*A. lueneburgensis*) is *very* similar and may be conspecific.

Black Rustic *Aporophyla nigra* fw 17–20mm
The glossy, sooty brown or black forewings, with a pale outer edge to the reniform stigma, are unlike those of any other moth. August–October in most open habitats. FP: many herbs and low-growing shrubs.

Golden-rod Brindle *Lithomoia solidaginis* fw *c.* 20mm
This poorly-named moth, which has nothing to do with golden-rod, is quite easily identified by the large, pale reniform stigma and the black wedges in the outer part of the wing. August–October on moorland. FP: heather, bilberry, and other moorland plants.

Tawny Pinion *Lithophane semibrunnea* fw 15–20mm
The dark thoracic crest and prominent 'shoulder pads' distinguish this from most other moths. A dark streak runs in towards the centre from the rear corner of the forewing. October–November and again in spring after hibernation, mainly in open woodland. FP: ash. The **Pale Pinion** (*L. hepatica*) has a pale crest and no black streak. *See also* Light Arches (*see* p. 252).

Grey Shoulder-knot *Lithophane ornitopus* fw 16–20mm
The bold, branching, antler-like black streak at the base of the forewing distinguishes this moth from superficially similar species. The lower half of the reniform stigma is often orange or pale brown. August–November and again in spring after hibernation, in wooded areas. FP: oaks.

Blair's Shoulder-knot *Lithophane leautieri* fw 17–20mm
The black streaks vary but there is always a long, straight one at the base of the forewing and this readily distinguishes the species from the Grey Shoulder-knot. September–November, mainly in parks and gardens. FP: Lawson cypress and related trees. Caterpillar p. 294.

Red Sword-grass *Xylena vetusta* fw 25–30mm
The longitudinal division of the forewing into a pale front half and a rich-brown rear half is not found in any other moth, although this division is not easy to see when the wings are wrapped around the body at rest. September–November and again in spring after hibernation on moors and rough pasture. FP: rushes, sedges and low-growing shrubs.

Sword-grass *Xylena exsoleta* fw 25–30mm
This looks very like the Red Sword-grass but lacks the rich brown on the rear half of the forewing and has paler feet. Mullein Moth (*see* p. 238) has a pale crest. September–October and again in spring, mainly on upland pastures. FP: many woody and herbaceous plants. Caterpillar p. 294.

EARLY GREY *Xylocampa areola* fw 15–18mm
The ground colour ranges from light to dark grey and may have a pink tinge, but the moth is easily recognised by the oval mark linking the lower edges of the reniform and orbicular stigmata. February–May in woods, hedgerows and gardens. FP: honeysuckle.

GREEN-BRINDLED CRESCENT *Allophyes oxyacanthae* fw 17–20mm
The typical form of this moth has light brown wings dusted, often quite heavily, with shiny green scales. There is also a dark brown form, with similar markings but without green scales. Both forms have a clear white crescent near the rear corner of the forewing. September–November in woods, hedgerows and gardens. FP: various rosaceous trees and shrubs. Caterpillar p. 294.

MERVEILLE-DU-JOUR *Dichona aprilina* fw 18–24mm
The extent and density of the black markings vary a good deal but the moth is unlikely to be confused with any other apart from the Scarce Merveille-du-jour (*see* p. 246), which is smaller and has heavier black markings in the outer part of the forewing. August–October, mainly in woodland. FP: oaks.

BRINDLED GREEN *Dryobotodes eremita* fw *c.* 15mm
The forewings are basically pale to dark green, marbled with black and white and often tinged with reddish brown in the outer half. The pale orbicular stigma is connected to another pale patch in the centre of the wing and the rear of this patch is normally drawn out to point to the rear corner of the wing. August–October, mainly woods and hedgerows. FP: oak.

DARK BROCADE *Blepharita adusta* fw *c.* 20mm
The forewings are mid- to dark brown, often with a strong reddish tinge. A solid black bar usually joins the two central cross-lines below the orbicular stigma and the middle part of the pale submarginal line usually forms a 'W'. May–August, mainly in open country. FP: many woody and herbaceous plants, including heather.

LARGE RANUNCULUS *Polymixis flavicincta* fw *c.* 20mm
The grey-green forewings are usually heavily dusted with orange, especially in the outer parts, although specimens from the south-west may have little or no orange. September–October, mainly on disturbed ground, including gardens. FP: numerous wild and cultivated herbs.

FEATHERED RANUNCULUS *Eumichtis lichenea* fw 15–18mm
Males are easily separated from the Large Ranunculus by their feathery antennae. Females are less easy to distinguish but the orbicular stigma is usually more obvious and the smaller size is a good guide. The orange patches on the forewings are not always well developed. August–October, usually in coastal areas. FP: wide range of herbaceous plants.

GREY CHI *Antitype chi* fw 15–20mm
This moth is named after the central dark mark on the forewing, said to resemble the Greek letter chi (χ), although the ends are occasionally missing and the mark is reduced to a simple dash. The ground colour ranges from very pale grey to dark greenish grey. August–October in open areas, including town parks and gardens. FP: numerous herbaceous and woody plants.

THE SATELLITE *Eupsilia transversa* fw 17–20mm
This moth is named for the two 'satellite' dots beside the large reniform stigma. These marks are usually brilliant white or yellow and they stand out clearly on the rich brown, scalloped forewings, although many individuals have orange marks. October–April, mainly in wooded habitats; dormant in the coldest weather. FP: numerous deciduous trees and shrubs.

THE SUSPECTED *Parastichtis suspecta* fw *c.* 15mm
This is a very variable moth, with forewings ranging from dull greyish brown to a rich chestnut, often tinged with purple and mottled with brick-red. The reniform and orbicular stigmata usually have clear pale edges and there is usually a pale band on the outer margin, although some specimens are virtually unmarked. June–August, mainly in damp woodland. FP: mainly birch.

THE CHESTNUT *Conistra vaccinii* fw *c.* 15mm
The forewings range from light orange to rich chestnut, often marbled with yellow and with a large dark spot in the rear half of the reniform stigma. At rest, the wings are folded almost flat with a large overlap and the moth appears almost rectangular from above. September–May, mainly in woodland; dormant in the coldest weather. FP: various deciduous trees.

DARK CHESTNUT *Conistra ligula* fw 12–15mm
This moth is very similar to The Chestnut but usually darker. The outer edge of the forewing is straighter, giving the resting moth an even more truncated appearance when seen from above. October–March in many habitats, but often dormant until January. FP: various trees, shrubs and herbaceous plants.

DOTTED CHESTNUT *Conistra rubiginea* fw 15–17mm
The forewings range from yellow to chestnut-brown, with dense brown or black spotting, including a large spot in the rear half of the reniform stigma. Some of the sallows (*see* p. 246) have similar coloration but a more triangular outline at rest. September–November and again in spring after hibernation, mainly in wooded areas. FP: many deciduous trees and shrubs.

THE BRICK *Agrochola circellaris* fw 15–20mm
The ground colour resembles that of The Chestnut, with a blackish spot in the rear half of the reniform stigma, but the resting outline is more triangular. A greyish cross-band towards the outer margin has a thin red line on its outer edge. August–December in wooded habitats, including gardens. FP: deciduous trees, especially poplars, sallows and ash.

RED-LINE QUAKER *Agrochola lota* fw 15–18mm
This moth is named for the strong red submarginal line on the forewing, although the line is edged on the outside with an equally strong yellow line. The reniform and orbicular stigmata are also ringed with red and the rear half of the reniform is very black. September–December in many habitats. FP: willows.

YELLOW-LINE QUAKER *Agrochola macilenta* fw *c.* 15mm
This moth is just like a brown version of the Red-line Quaker although the stigmata are not outlined in red and the red of the submarginal line is less intense. The reniform stigma is often the same colour as the rest of the wing. September–December in many habitats. FP: a wide range of trees, shrubs and herbaceous plants.

FLOUNCED CHESTNUT *Agrochola helvola* fw 15–18mm
Although the colours vary a little, the wide central cross-band always stands out slightly paler than the rest of the forewing. There is usually a sharply angled darker line through the middle of this band and the outer edge is marked with dark dots. September–October, mainly in deciduous woodland. FP: numerous deciduous trees, heather, and bilberry.

BROWN-SPOT PINION *Agrochola litura* fw 14–18mm
The solid black mark near the wing-tip and short black cross-line right at the base separate this from otherwise similar moths. The basal half of the forewing is often much paler than the rest. August–November in many habitats. FP: numerous herbaceous plants when young, often turning to trees and shrubs later.

BEADED CHESTNUT *Agrochola lychnidis* fw 14–18mm
The ground colour of the forewing ranges from yellowish brown to a very deep brown and there are several black marks along the front edge. The orbicular stigma is dark and slit-like and stands out clearly on most specimens. The reniform stigma is also quite narrow but not always darker than the rest of the wing. September–November in most habitats. FP: numerous herbaceous plants when young, often moving to hawthorn and other shrubs later.

CENTRE-BARRED SALLOW *Atethmia centrago* fw 15–20mm
The broad pinkish-brown central band and the dark band on the outer margin both fail to reach the front edge of the forewing and distinguish this moth from other sallows. August–September, mainly in and around woodland. FP: ash.

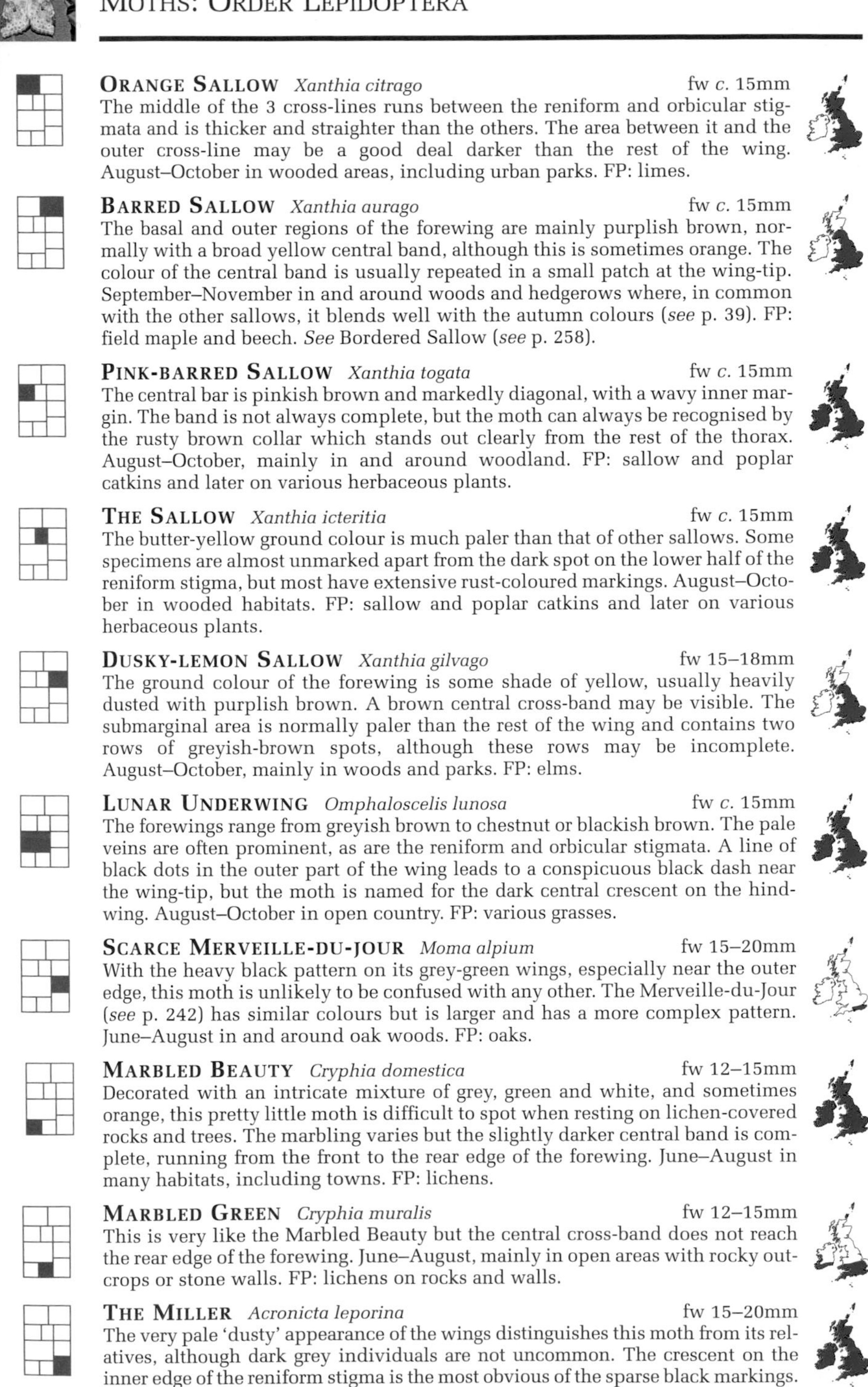

ORANGE SALLOW *Xanthia citrago* fw *c.* 15mm
The middle of the 3 cross-lines runs between the reniform and orbicular stigmata and is thicker and straighter than the others. The area between it and the outer cross-line may be a good deal darker than the rest of the wing. August–October in wooded areas, including urban parks. FP: limes.

BARRED SALLOW *Xanthia aurago* fw *c.* 15mm
The basal and outer regions of the forewing are mainly purplish brown, normally with a broad yellow central band, although this is sometimes orange. The colour of the central band is usually repeated in a small patch at the wing-tip. September–November in and around woods and hedgerows where, in common with the other sallows, it blends well with the autumn colours (*see* p. 39). FP: field maple and beech. *See* Bordered Sallow (*see* p. 258).

PINK-BARRED SALLOW *Xanthia togata* fw *c.* 15mm
The central bar is pinkish brown and markedly diagonal, with a wavy inner margin. The band is not always complete, but the moth can always be recognised by the rusty brown collar which stands out clearly from the rest of the thorax. August–October, mainly in and around woodland. FP: sallow and poplar catkins and later on various herbaceous plants.

THE SALLOW *Xanthia icteritia* fw *c.* 15mm
The butter-yellow ground colour is much paler than that of other sallows. Some specimens are almost unmarked apart from the dark spot on the lower half of the reniform stigma, but most have extensive rust-coloured markings. August–October in wooded habitats. FP: sallow and poplar catkins and later on various herbaceous plants.

DUSKY-LEMON SALLOW *Xanthia gilvago* fw 15–18mm
The ground colour of the forewing is some shade of yellow, usually heavily dusted with purplish brown. A brown central cross-band may be visible. The submarginal area is normally paler than the rest of the wing and contains two rows of greyish-brown spots, although these rows may be incomplete. August–October, mainly in woods and parks. FP: elms.

LUNAR UNDERWING *Omphaloscelis lunosa* fw *c.* 15mm
The forewings range from greyish brown to chestnut or blackish brown. The pale veins are often prominent, as are the reniform and orbicular stigmata. A line of black dots in the outer part of the wing leads to a conspicuous black dash near the wing-tip, but the moth is named for the dark central crescent on the hindwing. August–October in open country. FP: various grasses.

SCARCE MERVEILLE-DU-JOUR *Moma alpium* fw 15–20mm
With the heavy black pattern on its grey-green wings, especially near the outer edge, this moth is unlikely to be confused with any other. The Merveille-du-Jour (*see* p. 242) has similar colours but is larger and has a more complex pattern. June–August in and around oak woods. FP: oaks.

MARBLED BEAUTY *Cryphia domestica* fw 12–15mm
Decorated with an intricate mixture of grey, green and white, and sometimes orange, this pretty little moth is difficult to spot when resting on lichen-covered rocks and trees. The marbling varies but the slightly darker central band is complete, running from the front to the rear edge of the forewing. June–August in many habitats, including towns. FP: lichens.

MARBLED GREEN *Cryphia muralis* fw 12–15mm
This is very like the Marbled Beauty but the central cross-band does not reach the rear edge of the forewing. June–August, mainly in open areas with rocky outcrops or stone walls. FP: lichens on rocks and walls.

THE MILLER *Acronicta leporina* fw 15–20mm
The very pale 'dusty' appearance of the wings distinguishes this moth from its relatives, although dark grey individuals are not uncommon. The crescent on the inner edge of the reniform stigma is the most obvious of the sparse black markings. May–August in many habitats. FP: mainly birch and alder. Caterpillar p. 294.

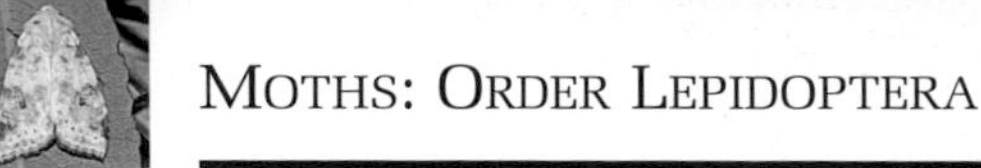

POPLAR GREY *Acronicta megacephala* fw 15–20mm
The forewings are light grey with dark grey or black markings and a heavy dusting of dark grey scales. The orbicular stigma is always paler than the surrounding parts of the wing and there is often an indistinct pale patch just beyond the reniform stigma. The hindwings are largely white. May–August in many habitats. FP: poplars and sometimes willows. Caterpillar p. 294.

THE SYCAMORE *Acronicta aceris* fw 18–25mm

This moth resembles the Poplar Grey but is usually paler and has a narrower outline at rest. There is usually a slim branching black streak near the wing-base, the orbicular stigma is not normally any paler than the rest of the wing, and there is no paler patch beyond the reniform stigma. June–August in woods and parkland; common in urban areas. FP: sycamore, maples, horse chestnut, and probably other deciduous trees. Caterpillar p. 294.

ALDER MOTH *Acronicta alni* fw 15–20mm

The sooty-black rear half and central band of the forewing unmistakable. The grey may be replaced by brown, but the dark pattern is still clear. May–June, mainly in damp woodland. FP: alder, birch and other deciduous trees. Caterpillar p. 294.

GREY DAGGER *Acronicta psi* fw 17–20mm

Named for the black daggers on the grey wings, this moth is impossible to separate with certainty from the **Dark Dagger** (*A. tridens*) without examining the genitalia, although the larvae (*see* p. 294) are very different. May–October almost everywhere. FP: many deciduous trees and shrubs. The Dark Dagger has similar habits, with a preference for rosaceous trees and shrubs.

LIGHT KNOTGRASS *Acronicta menyanthidis* fw 15–20mm

The forewings range from pale grey to slate-grey. The small, black-ringed orbicular stigma is always clearly visible but the reniform stigma may be obscured. A black streak pointing towards the rear corner of the wing is usually visible to the rear of the reniform stigma. May–August on heaths and moors. FP: mainly heather and bog myrtle. The Light Feathered Rustic (p. 223) holds its wings flat at rest.

KNOTGRASS *Acronicta rumicis* fw 15–20mm

Can be confused with the Poplar Grey, but can usually be recognised by a bright white spot close to the rear margin of each forewing. The outline is also more slender at rest, and the hindwings are greyish brown. May–September in all kinds of open habitats. FP: a wide range of woody and herbaceous plants. Caterpillar p. 296.

REED DAGGER *Simyra albovenosa* fw 15–20mm

This rather rare moth is easily mistaken for a wainscot (pp. 236–238 and p.258) but can be distinguished from all superficially similar species by its pure white hindwings. May–September in marshes and other damp habitats. FP: various marsh plants, but mainly common reed.

THE CORONET *Craniophora ligustri* fw 17–20mm

The pale circular area in the outer part of the forewing readily identifies this moth, although it is reduced to a faint white ring in some darker specimens. June–August in wooded areas. FP: mainly ash.

COPPER UNDERWING *Amphipyra pyramidea* fw 20–25mm

The size and wing pattern, with a prominent pale orbicular stigma, should distinguish this from all except the next species. The copper-coloured hindwings have no black borders and on the underside the copper is restricted to a marginal band. July–October in all kinds of wooded areas. FP: wide range of deciduous trees and shrubs. Caterpillar p. 296.

SVENSSON'S COPPER UNDERWING *Amphipyra berbera* fw 20–25mm

The main difference between this and the previous species lies on the underside of the hindwing, where the copper colour is more extensive. The abdomen also lacks the chequered fringe. Flight times and habits are the same for both species.

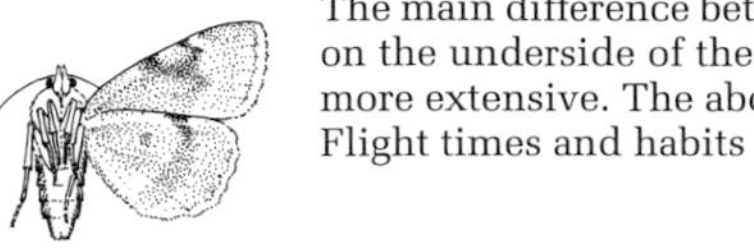

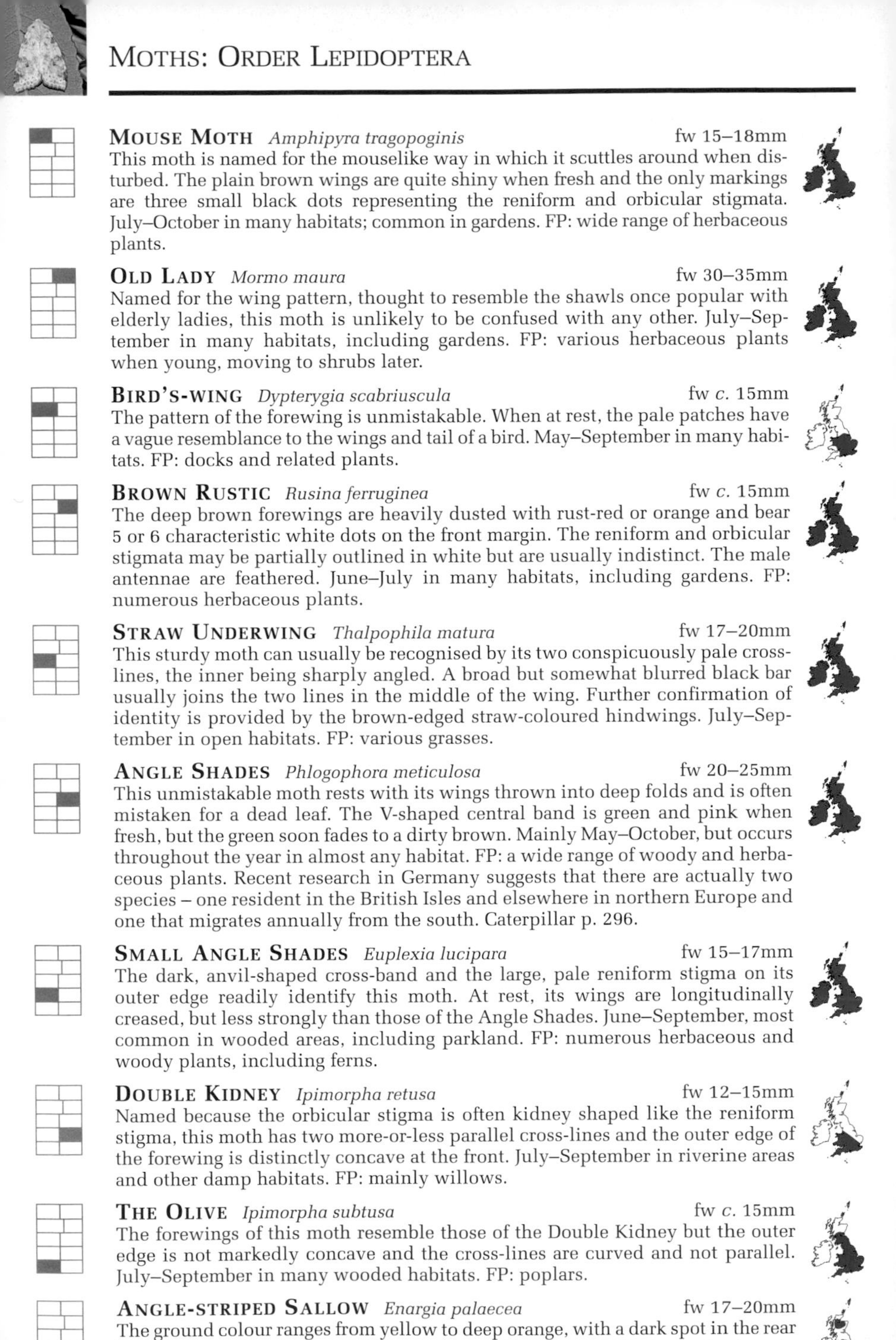

MOUSE MOTH *Amphipyra tragopoginis* fw 15–18mm
This moth is named for the mouselike way in which it scuttles around when disturbed. The plain brown wings are quite shiny when fresh and the only markings are three small black dots representing the reniform and orbicular stigmata. July–October in many habitats; common in gardens. FP: wide range of herbaceous plants.

OLD LADY *Mormo maura* fw 30–35mm
Named for the wing pattern, thought to resemble the shawls once popular with elderly ladies, this moth is unlikely to be confused with any other. July–September in many habitats, including gardens. FP: various herbaceous plants when young, moving to shrubs later.

BIRD'S-WING *Dypterygia scabriuscula* fw *c.* 15mm
The pattern of the forewing is unmistakable. When at rest, the pale patches have a vague resemblance to the wings and tail of a bird. May–September in many habitats. FP: docks and related plants.

BROWN RUSTIC *Rusina ferruginea* fw *c.* 15mm
The deep brown forewings are heavily dusted with rust-red or orange and bear 5 or 6 characteristic white dots on the front margin. The reniform and orbicular stigmata may be partially outlined in white but are usually indistinct. The male antennae are feathered. June–July in many habitats, including gardens. FP: numerous herbaceous plants.

STRAW UNDERWING *Thalpophila matura* fw 17–20mm
This sturdy moth can usually be recognised by its two conspicuously pale cross-lines, the inner being sharply angled. A broad but somewhat blurred black bar usually joins the two lines in the middle of the wing. Further confirmation of identity is provided by the brown-edged straw-coloured hindwings. July–September in open habitats. FP: various grasses.

ANGLE SHADES *Phlogophora meticulosa* fw 20–25mm
This unmistakable moth rests with its wings thrown into deep folds and is often mistaken for a dead leaf. The V-shaped central band is green and pink when fresh, but the green soon fades to a dirty brown. Mainly May–October, but occurs throughout the year in almost any habitat. FP: a wide range of woody and herbaceous plants. Recent research in Germany suggests that there are actually two species – one resident in the British Isles and elsewhere in northern Europe and one that migrates annually from the south. Caterpillar p. 296.

SMALL ANGLE SHADES *Euplexia lucipara* fw 15–17mm
The dark, anvil-shaped cross-band and the large, pale reniform stigma on its outer edge readily identify this moth. At rest, its wings are longitudinally creased, but less strongly than those of the Angle Shades. June–September, most common in wooded areas, including parkland. FP: numerous herbaceous and woody plants, including ferns.

DOUBLE KIDNEY *Ipimorpha retusa* fw 12–15mm
Named because the orbicular stigma is often kidney shaped like the reniform stigma, this moth has two more-or-less parallel cross-lines and the outer edge of the forewing is distinctly concave at the front. July–September in riverine areas and other damp habitats. FP: mainly willows.

THE OLIVE *Ipimorpha subtusa* fw *c.* 15mm
The forewings of this moth resemble those of the Double Kidney but the outer edge is not markedly concave and the cross-lines are curved and not parallel. July–September in many wooded habitats. FP: poplars.

ANGLE-STRIPED SALLOW *Enargia palaecea* fw 17–20mm
The ground colour ranges from yellow to deep orange, with a dark spot in the rear half of the reniform stigma. The basal cross-line, nearest the body, is strongly elbowed. July–September in birch woods and on heathland. FP: birches.

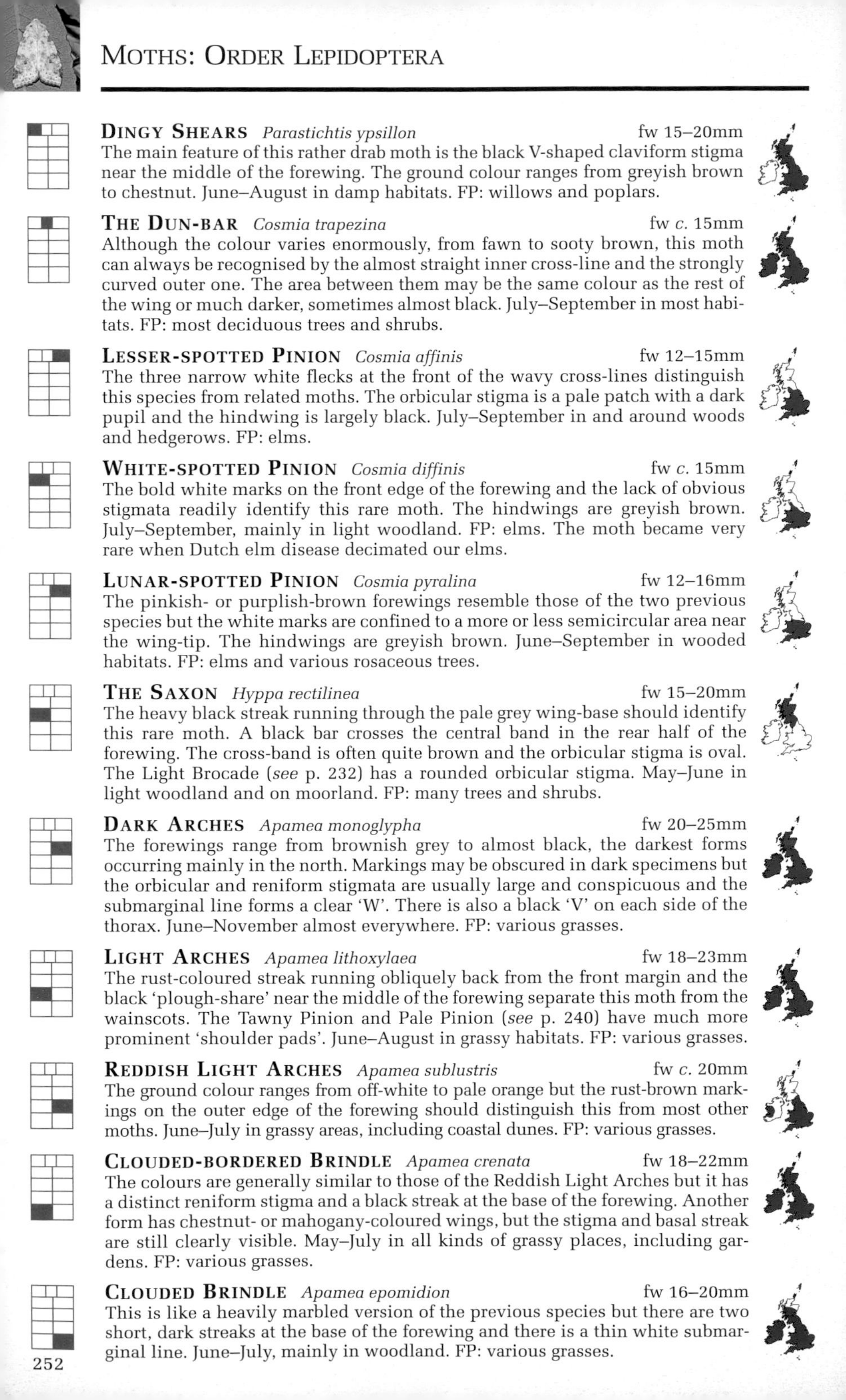

DINGY SHEARS *Parastichtis ypsillon* fw 15–20mm
The main feature of this rather drab moth is the black V-shaped claviform stigma near the middle of the forewing. The ground colour ranges from greyish brown to chestnut. June–August in damp habitats. FP: willows and poplars.

THE DUN-BAR *Cosmia trapezina* fw *c.* 15mm
Although the colour varies enormously, from fawn to sooty brown, this moth can always be recognised by the almost straight inner cross-line and the strongly curved outer one. The area between them may be the same colour as the rest of the wing or much darker, sometimes almost black. July–September in most habitats. FP: most deciduous trees and shrubs.

LESSER-SPOTTED PINION *Cosmia affinis* fw 12–15mm
The three narrow white flecks at the front of the wavy cross-lines distinguish this species from related moths. The orbicular stigma is a pale patch with a dark pupil and the hindwing is largely black. July–September in and around woods and hedgerows. FP: elms.

WHITE-SPOTTED PINION *Cosmia diffinis* fw *c.* 15mm
The bold white marks on the front edge of the forewing and the lack of obvious stigmata readily identify this rare moth. The hindwings are greyish brown. July–September, mainly in light woodland. FP: elms. The moth became very rare when Dutch elm disease decimated our elms.

LUNAR-SPOTTED PINION *Cosmia pyralina* fw 12–16mm
The pinkish- or purplish-brown forewings resemble those of the two previous species but the white marks are confined to a more or less semicircular area near the wing-tip. The hindwings are greyish brown. June–September in wooded habitats. FP: elms and various rosaceous trees.

THE SAXON *Hyppa rectilinea* fw 15–20mm
The heavy black streak running through the pale grey wing-base should identify this rare moth. A black bar crosses the central band in the rear half of the forewing. The cross-band is often quite brown and the orbicular stigma is oval. The Light Brocade (*see* p. 232) has a rounded orbicular stigma. May–June in light woodland and on moorland. FP: many trees and shrubs.

DARK ARCHES *Apamea monoglypha* fw 20–25mm
The forewings range from brownish grey to almost black, the darkest forms occurring mainly in the north. Markings may be obscured in dark specimens but the orbicular and reniform stigmata are usually large and conspicuous and the submarginal line forms a clear 'W'. There is also a black 'V' on each side of the thorax. June–November almost everywhere. FP: various grasses.

LIGHT ARCHES *Apamea lithoxylaea* fw 18–23mm
The rust-coloured streak running obliquely back from the front margin and the black 'plough-share' near the middle of the forewing separate this moth from the wainscots. The Tawny Pinion and Pale Pinion (*see* p. 240) have much more prominent 'shoulder pads'. June–August in grassy habitats. FP: various grasses.

REDDISH LIGHT ARCHES *Apamea sublustris* fw *c.* 20mm
The ground colour ranges from off-white to pale orange but the rust-brown markings on the outer edge of the forewing should distinguish this from most other moths. June–July in grassy areas, including coastal dunes. FP: various grasses.

CLOUDED-BORDERED BRINDLE *Apamea crenata* fw 18–22mm
The colours are generally similar to those of the Reddish Light Arches but it has a distinct reniform stigma and a black streak at the base of the forewing. Another form has chestnut- or mahogany-coloured wings, but the stigma and basal streak are still clearly visible. May–July in all kinds of grassy places, including gardens. FP: various grasses.

CLOUDED BRINDLE *Apamea epomidion* fw 16–20mm
This is like a heavily marbled version of the previous species but there are two short, dark streaks at the base of the forewing and there is a thin white submarginal line. June–July, mainly in woodland. FP: various grasses.

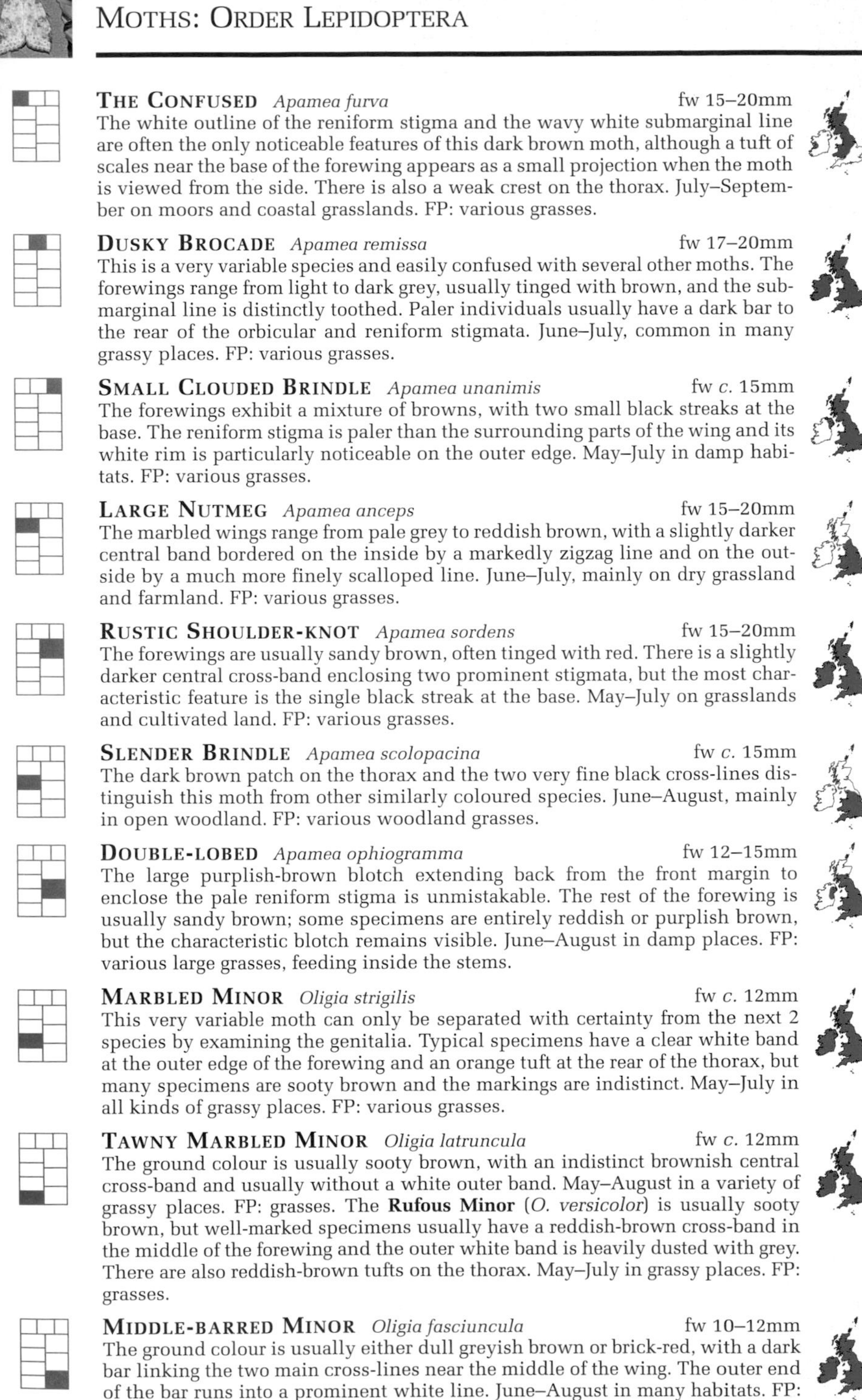

THE CONFUSED *Apamea furva* fw 15–20mm
The white outline of the reniform stigma and the wavy white submarginal line are often the only noticeable features of this dark brown moth, although a tuft of scales near the base of the forewing appears as a small projection when the moth is viewed from the side. There is also a weak crest on the thorax. July–September on moors and coastal grasslands. FP: various grasses.

DUSKY BROCADE *Apamea remissa* fw 17–20mm
This is a very variable species and easily confused with several other moths. The forewings range from light to dark grey, usually tinged with brown, and the submarginal line is distinctly toothed. Paler individuals usually have a dark bar to the rear of the orbicular and reniform stigmata. June–July, common in many grassy places. FP: various grasses.

SMALL CLOUDED BRINDLE *Apamea unanimis* fw *c.* 15mm
The forewings exhibit a mixture of browns, with two small black streaks at the base. The reniform stigma is paler than the surrounding parts of the wing and its white rim is particularly noticeable on the outer edge. May–July in damp habitats. FP: various grasses.

LARGE NUTMEG *Apamea anceps* fw 15–20mm
The marbled wings range from pale grey to reddish brown, with a slightly darker central band bordered on the inside by a markedly zigzag line and on the outside by a much more finely scalloped line. June–July, mainly on dry grassland and farmland. FP: various grasses.

RUSTIC SHOULDER-KNOT *Apamea sordens* fw 15–20mm
The forewings are usually sandy brown, often tinged with red. There is a slightly darker central cross-band enclosing two prominent stigmata, but the most characteristic feature is the single black streak at the base. May–July on grasslands and cultivated land. FP: various grasses.

SLENDER BRINDLE *Apamea scolopacina* fw *c.* 15mm
The dark brown patch on the thorax and the two very fine black cross-lines distinguish this moth from other similarly coloured species. June–August, mainly in open woodland. FP: various woodland grasses.

DOUBLE-LOBED *Apamea ophiogramma* fw 12–15mm
The large purplish-brown blotch extending back from the front margin to enclose the pale reniform stigma is unmistakable. The rest of the forewing is usually sandy brown; some specimens are entirely reddish or purplish brown, but the characteristic blotch remains visible. June–August in damp places. FP: various large grasses, feeding inside the stems.

MARBLED MINOR *Oligia strigilis* fw *c.* 12mm
This very variable moth can only be separated with certainty from the next 2 species by examining the genitalia. Typical specimens have a clear white band at the outer edge of the forewing and an orange tuft at the rear of the thorax, but many specimens are sooty brown and the markings are indistinct. May–July in all kinds of grassy places. FP: various grasses.

TAWNY MARBLED MINOR *Oligia latruncula* fw *c.* 12mm
The ground colour is usually sooty brown, with an indistinct brownish central cross-band and usually without a white outer band. May–August in a variety of grassy places. FP: grasses. The **Rufous Minor** (*O. versicolor*) is usually sooty brown, but well-marked specimens usually have a reddish-brown cross-band in the middle of the forewing and the outer white band is heavily dusted with grey. There are also reddish-brown tufts on the thorax. May–July in grassy places. FP: grasses.

MIDDLE-BARRED MINOR *Oligia fasciuncula* fw 10–12mm
The ground colour is usually either dull greyish brown or brick-red, with a dark bar linking the two main cross-lines near the middle of the wing. The outer end of the bar runs into a prominent white line. June–August in many habitats. FP: various grasses.

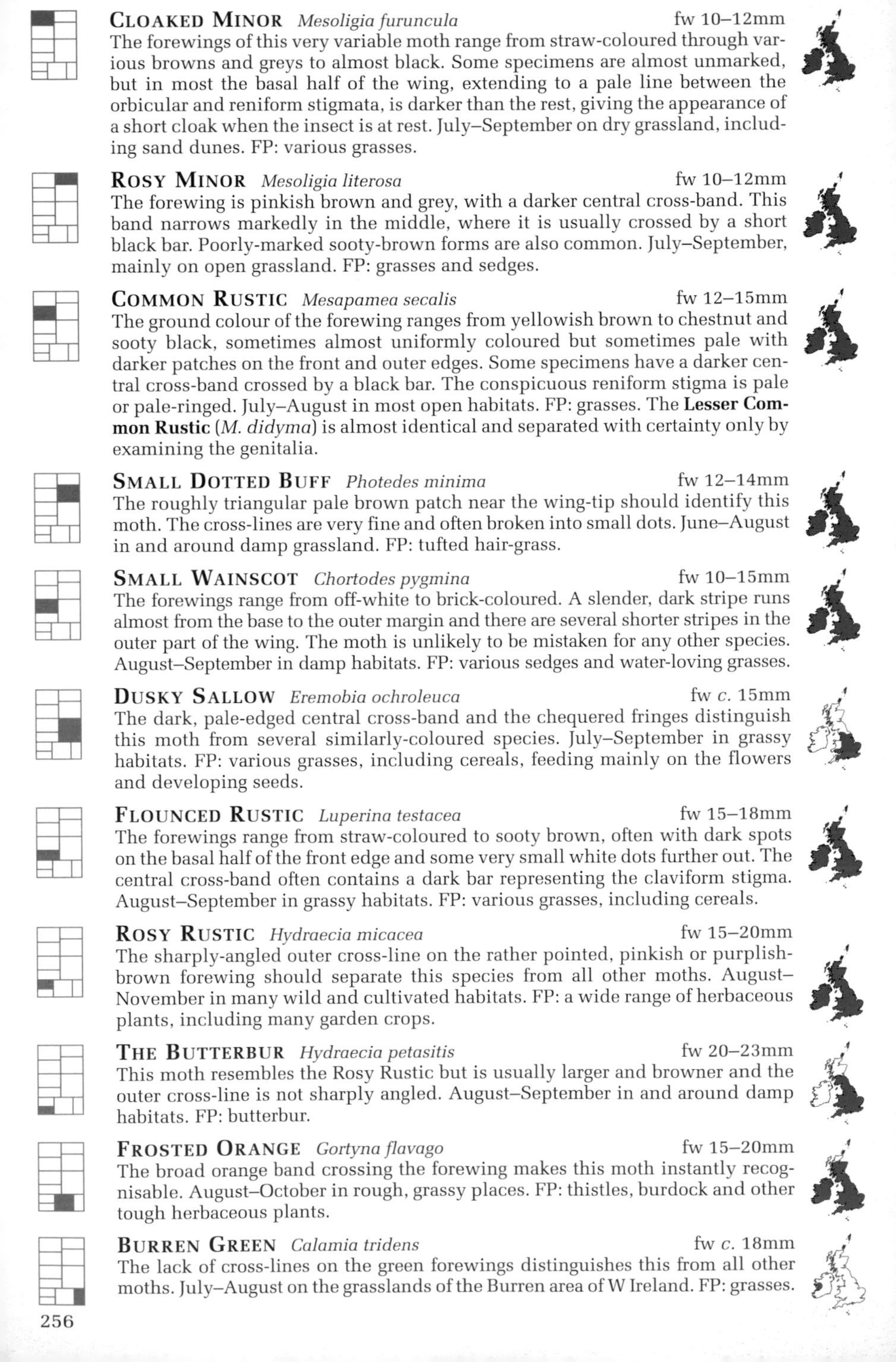

CLOAKED MINOR *Mesoligia furuncula* fw 10–12mm
The forewings of this very variable moth range from straw-coloured through various browns and greys to almost black. Some specimens are almost unmarked, but in most the basal half of the wing, extending to a pale line between the orbicular and reniform stigmata, is darker than the rest, giving the appearance of a short cloak when the insect is at rest. July–September on dry grassland, including sand dunes. FP: various grasses.

ROSY MINOR *Mesoligia literosa* fw 10–12mm
The forewing is pinkish brown and grey, with a darker central cross-band. This band narrows markedly in the middle, where it is usually crossed by a short black bar. Poorly-marked sooty-brown forms are also common. July–September, mainly on open grassland. FP: grasses and sedges.

COMMON RUSTIC *Mesapamea secalis* fw 12–15mm
The ground colour of the forewing ranges from yellowish brown to chestnut and sooty black, sometimes almost uniformly coloured but sometimes pale with darker patches on the front and outer edges. Some specimens have a darker central cross-band crossed by a black bar. The conspicuous reniform stigma is pale or pale-ringed. July–August in most open habitats. FP: grasses. The **Lesser Common Rustic** (*M. didyma*) is almost identical and separated with certainty only by examining the genitalia.

SMALL DOTTED BUFF *Photedes minima* fw 12–14mm
The roughly triangular pale brown patch near the wing-tip should identify this moth. The cross-lines are very fine and often broken into small dots. June–August in and around damp grassland. FP: tufted hair-grass.

SMALL WAINSCOT *Chortodes pygmina* fw 10–15mm
The forewings range from off-white to brick-coloured. A slender, dark stripe runs almost from the base to the outer margin and there are several shorter stripes in the outer part of the wing. The moth is unlikely to be mistaken for any other species. August–September in damp habitats. FP: various sedges and water-loving grasses.

DUSKY SALLOW *Eremobia ochroleuca* fw *c.* 15mm
The dark, pale-edged central cross-band and the chequered fringes distinguish this moth from several similarly-coloured species. July–September in grassy habitats. FP: various grasses, including cereals, feeding mainly on the flowers and developing seeds.

FLOUNCED RUSTIC *Luperina testacea* fw 15–18mm
The forewings range from straw-coloured to sooty brown, often with dark spots on the basal half of the front edge and some very small white dots further out. The central cross-band often contains a dark bar representing the claviform stigma. August–September in grassy habitats. FP: various grasses, including cereals.

ROSY RUSTIC *Hydraecia micacea* fw 15–20mm
The sharply-angled outer cross-line on the rather pointed, pinkish or purplish-brown forewing should separate this species from all other moths. August–November in many wild and cultivated habitats. FP: a wide range of herbaceous plants, including many garden crops.

THE BUTTERBUR *Hydraecia petasitis* fw 20–23mm
This moth resembles the Rosy Rustic but is usually larger and browner and the outer cross-line is not sharply angled. August–September in and around damp habitats. FP: butterbur.

FROSTED ORANGE *Gortyna flavago* fw 15–20mm
The broad orange band crossing the forewing makes this moth instantly recognisable. August–October in rough, grassy places. FP: thistles, burdock and other tough herbaceous plants.

BURREN GREEN *Calamia tridens* fw *c.* 18mm
The lack of cross-lines on the green forewings distinguishes this from all other moths. July–August on the grasslands of the Burren area of W Ireland. FP: grasses.

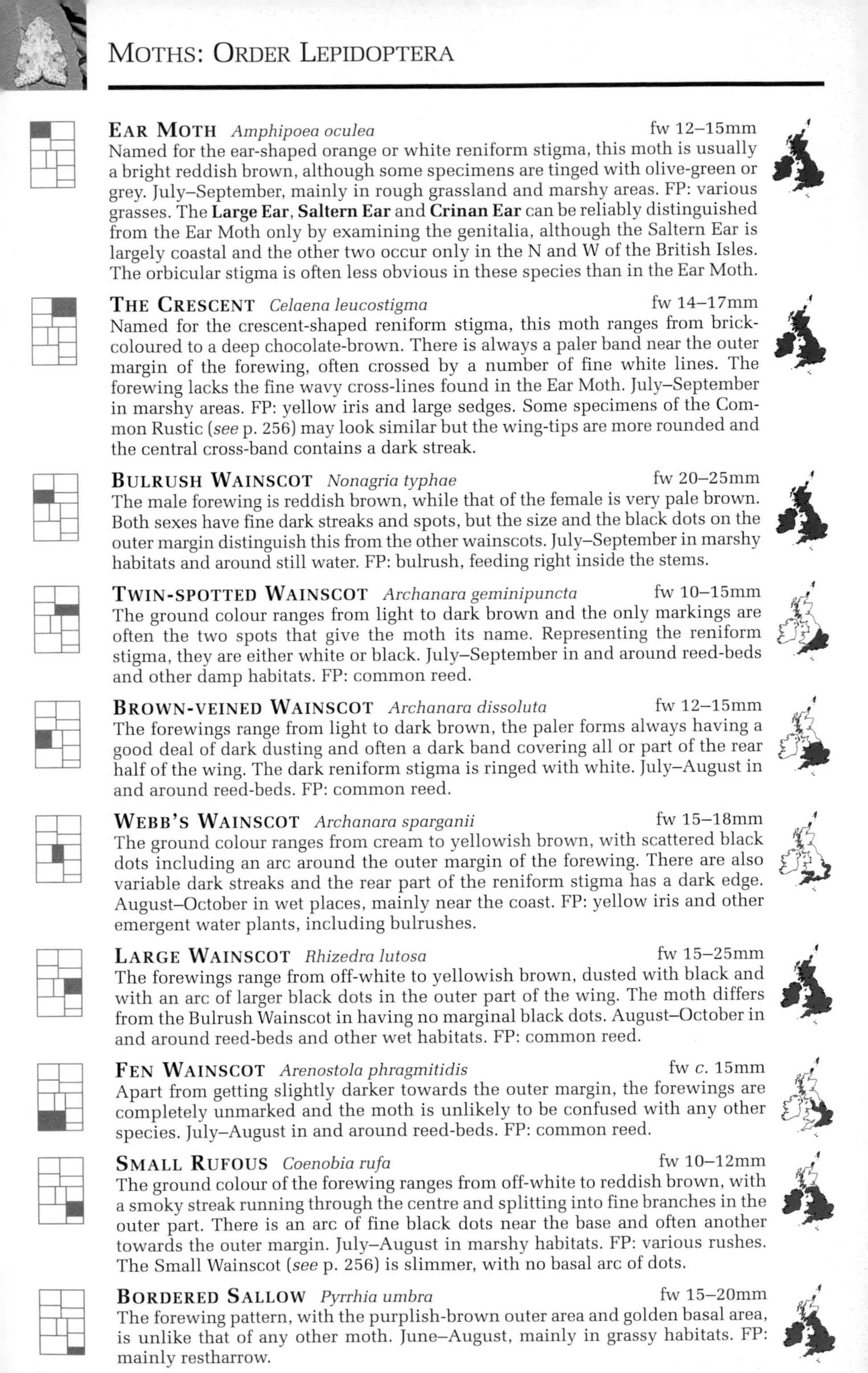

EAR MOTH *Amphipoea oculea* fw 12–15mm
Named for the ear-shaped orange or white reniform stigma, this moth is usually a bright reddish brown, although some specimens are tinged with olive-green or grey. July–September, mainly in rough grassland and marshy areas. FP: various grasses. The **Large Ear**, **Saltern Ear** and **Crinan Ear** can be reliably distinguished from the Ear Moth only by examining the genitalia, although the Saltern Ear is largely coastal and the other two occur only in the N and W of the British Isles. The orbicular stigma is often less obvious in these species than in the Ear Moth.

THE CRESCENT *Celaena leucostigma* fw 14–17mm
Named for the crescent-shaped reniform stigma, this moth ranges from brick-coloured to a deep chocolate-brown. There is always a paler band near the outer margin of the forewing, often crossed by a number of fine white lines. The forewing lacks the fine wavy cross-lines found in the Ear Moth. July–September in marshy areas. FP: yellow iris and large sedges. Some specimens of the Common Rustic (*see* p. 256) may look similar but the wing-tips are more rounded and the central cross-band contains a dark streak.

BULRUSH WAINSCOT *Nonagria typhae* fw 20–25mm
The male forewing is reddish brown, while that of the female is very pale brown. Both sexes have fine dark streaks and spots, but the size and the black dots on the outer margin distinguish this from the other wainscots. July–September in marshy habitats and around still water. FP: bulrush, feeding right inside the stems.

TWIN-SPOTTED WAINSCOT *Archanara geminipuncta* fw 10–15mm
The ground colour ranges from light to dark brown and the only markings are often the two spots that give the moth its name. Representing the reniform stigma, they are either white or black. July–September in and around reed-beds and other damp habitats. FP: common reed.

BROWN-VEINED WAINSCOT *Archanara dissoluta* fw 12–15mm
The forewings range from light to dark brown, the paler forms always having a good deal of dark dusting and often a dark band covering all or part of the rear half of the wing. The dark reniform stigma is ringed with white. July–August in and around reed-beds. FP: common reed.

WEBB'S WAINSCOT *Archanara sparganii* fw 15–18mm
The ground colour ranges from cream to yellowish brown, with scattered black dots including an arc around the outer margin of the forewing. There are also variable dark streaks and the rear part of the reniform stigma has a dark edge. August–October in wet places, mainly near the coast. FP: yellow iris and other emergent water plants, including bulrushes.

LARGE WAINSCOT *Rhizedra lutosa* fw 15–25mm
The forewings range from off-white to yellowish brown, dusted with black and with an arc of larger black dots in the outer part of the wing. The moth differs from the Bulrush Wainscot in having no marginal black dots. August–October in and around reed-beds and other wet habitats. FP: common reed.

FEN WAINSCOT *Arenostola phragmitidis* fw *c.* 15mm
Apart from getting slightly darker towards the outer margin, the forewings are completely unmarked and the moth is unlikely to be confused with any other species. July–August in and around reed-beds. FP: common reed.

SMALL RUFOUS *Coenobia rufa* fw 10–12mm
The ground colour of the forewing ranges from off-white to reddish brown, with a smoky streak running through the centre and splitting into fine branches in the outer part. There is an arc of fine black dots near the base and often another towards the outer margin. July–August in marshy habitats. FP: various rushes. The Small Wainscot (*see* p. 256) is slimmer, with no basal arc of dots.

BORDERED SALLOW *Pyrrhia umbra* fw 15–20mm
The forewing pattern, with the purplish-brown outer area and golden basal area, is unlike that of any other moth. June–August, mainly in grassy habitats. FP: mainly restharrow.

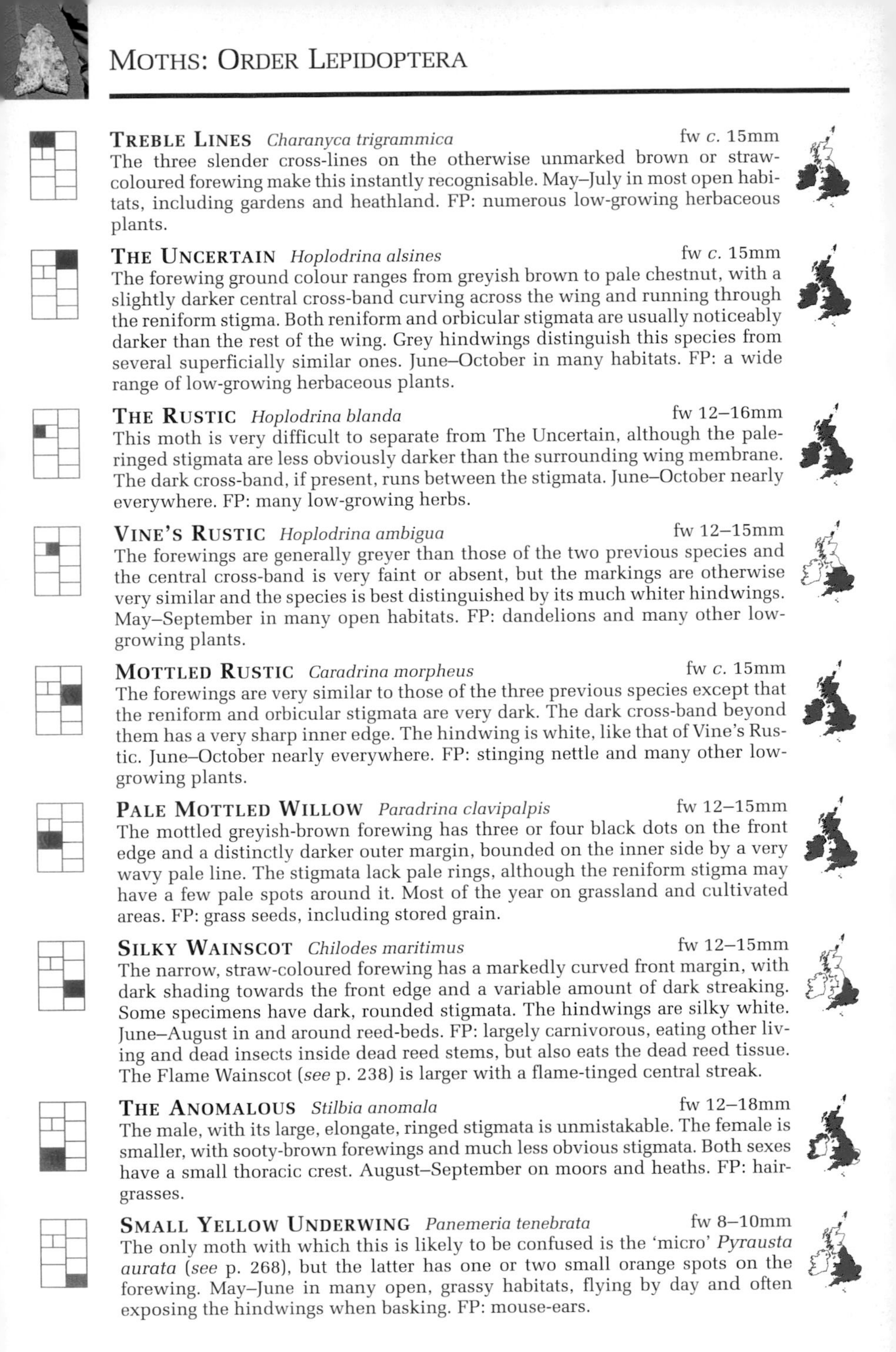

TREBLE LINES *Charanyca trigrammica* fw *c.* 15mm
The three slender cross-lines on the otherwise unmarked brown or straw-coloured forewing make this instantly recognisable. May–July in most open habitats, including gardens and heathland. FP: numerous low-growing herbaceous plants.

THE UNCERTAIN *Hoplodrina alsines* fw *c.* 15mm
The forewing ground colour ranges from greyish brown to pale chestnut, with a slightly darker central cross-band curving across the wing and running through the reniform stigma. Both reniform and orbicular stigmata are usually noticeably darker than the rest of the wing. Grey hindwings distinguish this species from several superficially similar ones. June–October in many habitats. FP: a wide range of low-growing herbaceous plants.

THE RUSTIC *Hoplodrina blanda* fw 12–16mm
This moth is very difficult to separate from The Uncertain, although the pale-ringed stigmata are less obviously darker than the surrounding wing membrane. The dark cross-band, if present, runs between the stigmata. June–October nearly everywhere. FP: many low-growing herbs.

VINE'S RUSTIC *Hoplodrina ambigua* fw 12–15mm
The forewings are generally greyer than those of the two previous species and the central cross-band is very faint or absent, but the markings are otherwise very similar and the species is best distinguished by its much whiter hindwings. May–September in many open habitats. FP: dandelions and many other low-growing plants.

MOTTLED RUSTIC *Caradrina morpheus* fw *c.* 15mm
The forewings are very similar to those of the three previous species except that the reniform and orbicular stigmata are very dark. The dark cross-band beyond them has a very sharp inner edge. The hindwing is white, like that of Vine's Rustic. June–October nearly everywhere. FP: stinging nettle and many other low-growing plants.

PALE MOTTLED WILLOW *Paradrina clavipalpis* fw 12–15mm
The mottled greyish-brown forewing has three or four black dots on the front edge and a distinctly darker outer margin, bounded on the inner side by a very wavy pale line. The stigmata lack pale rings, although the reniform stigma may have a few pale spots around it. Most of the year on grassland and cultivated areas. FP: grass seeds, including stored grain.

SILKY WAINSCOT *Chilodes maritimus* fw 12–15mm
The narrow, straw-coloured forewing has a markedly curved front margin, with dark shading towards the front edge and a variable amount of dark streaking. Some specimens have dark, rounded stigmata. The hindwings are silky white. June–August in and around reed-beds. FP: largely carnivorous, eating other living and dead insects inside dead reed stems, but also eats the dead reed tissue. The Flame Wainscot (*see* p. 238) is larger with a flame-tinged central streak.

THE ANOMALOUS *Stilbia anomala* fw 12–18mm
The male, with its large, elongate, ringed stigmata is unmistakable. The female is smaller, with sooty-brown forewings and much less obvious stigmata. Both sexes have a small thoracic crest. August–September on moors and heaths. FP: hair-grasses.

SMALL YELLOW UNDERWING *Panemeria tenebrata* fw 8–10mm
The only moth with which this is likely to be confused is the 'micro' *Pyrausta aurata* (*see* p. 268), but the latter has one or two small orange spots on the forewing. May–June in many open, grassy habitats, flying by day and often exposing the hindwings when basking. FP: mouse-ears.

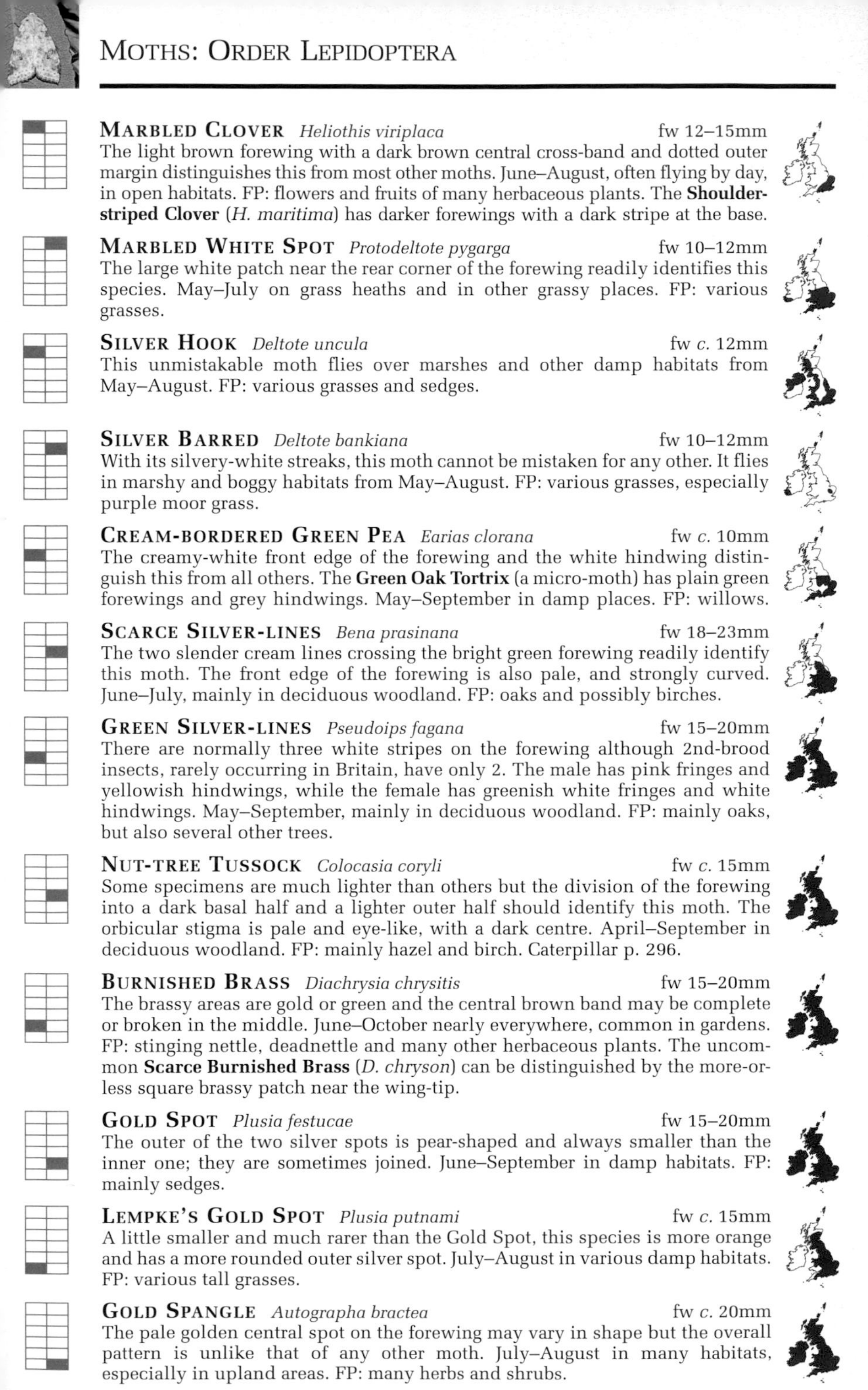

MARBLED CLOVER *Heliothis viriplaca* fw 12–15mm
The light brown forewing with a dark brown central cross-band and dotted outer margin distinguishes this from most other moths. June–August, often flying by day, in open habitats. FP: flowers and fruits of many herbaceous plants. The **Shoulder-striped Clover** (*H. maritima*) has darker forewings with a dark stripe at the base.

MARBLED WHITE SPOT *Protodeltote pygarga* fw 10–12mm
The large white patch near the rear corner of the forewing readily identifies this species. May–July on grass heaths and in other grassy places. FP: various grasses.

SILVER HOOK *Deltote uncula* fw *c.* 12mm
This unmistakable moth flies over marshes and other damp habitats from May–August. FP: various grasses and sedges.

SILVER BARRED *Deltote bankiana* fw 10–12mm
With its silvery-white streaks, this moth cannot be mistaken for any other. It flies in marshy and boggy habitats from May–August. FP: various grasses, especially purple moor grass.

CREAM-BORDERED GREEN PEA *Earias clorana* fw *c.* 10mm
The creamy-white front edge of the forewing and the white hindwing distinguish this from all others. The **Green Oak Tortrix** (a micro-moth) has plain green forewings and grey hindwings. May–September in damp places. FP: willows.

SCARCE SILVER-LINES *Bena prasinana* fw 18–23mm
The two slender cream lines crossing the bright green forewing readily identify this moth. The front edge of the forewing is also pale, and strongly curved. June–July, mainly in deciduous woodland. FP: oaks and possibly birches.

GREEN SILVER-LINES *Pseudoips fagana* fw 15–20mm
There are normally three white stripes on the forewing although 2nd-brood insects, rarely occurring in Britain, have only 2. The male has pink fringes and yellowish hindwings, while the female has greenish white fringes and white hindwings. May–September, mainly in deciduous woodland. FP: mainly oaks, but also several other trees.

NUT-TREE TUSSOCK *Colocasia coryli* fw *c.* 15mm
Some specimens are much lighter than others but the division of the forewing into a dark basal half and a lighter outer half should identify this moth. The orbicular stigma is pale and eye-like, with a dark centre. April–September in deciduous woodland. FP: mainly hazel and birch. Caterpillar p. 296.

BURNISHED BRASS *Diachrysia chrysitis* fw 15–20mm
The brassy areas are gold or green and the central brown band may be complete or broken in the middle. June–October nearly everywhere, common in gardens. FP: stinging nettle, deadnettle and many other herbaceous plants. The uncommon **Scarce Burnished Brass** (*D. chryson*) can be distinguished by the more-or-less square brassy patch near the wing-tip.

GOLD SPOT *Plusia festucae* fw 15–20mm
The outer of the two silver spots is pear-shaped and always smaller than the inner one; they are sometimes joined. June–September in damp habitats. FP: mainly sedges.

LEMPKE'S GOLD SPOT *Plusia putnami* fw *c.* 15mm
A little smaller and much rarer than the Gold Spot, this species is more orange and has a more rounded outer silver spot. July–August in various damp habitats. FP: various tall grasses.

GOLD SPANGLE *Autographa bractea* fw *c.* 20mm
The pale golden central spot on the forewing may vary in shape but the overall pattern is unlike that of any other moth. July–August in many habitats, especially in upland areas. FP: many herbs and shrubs.

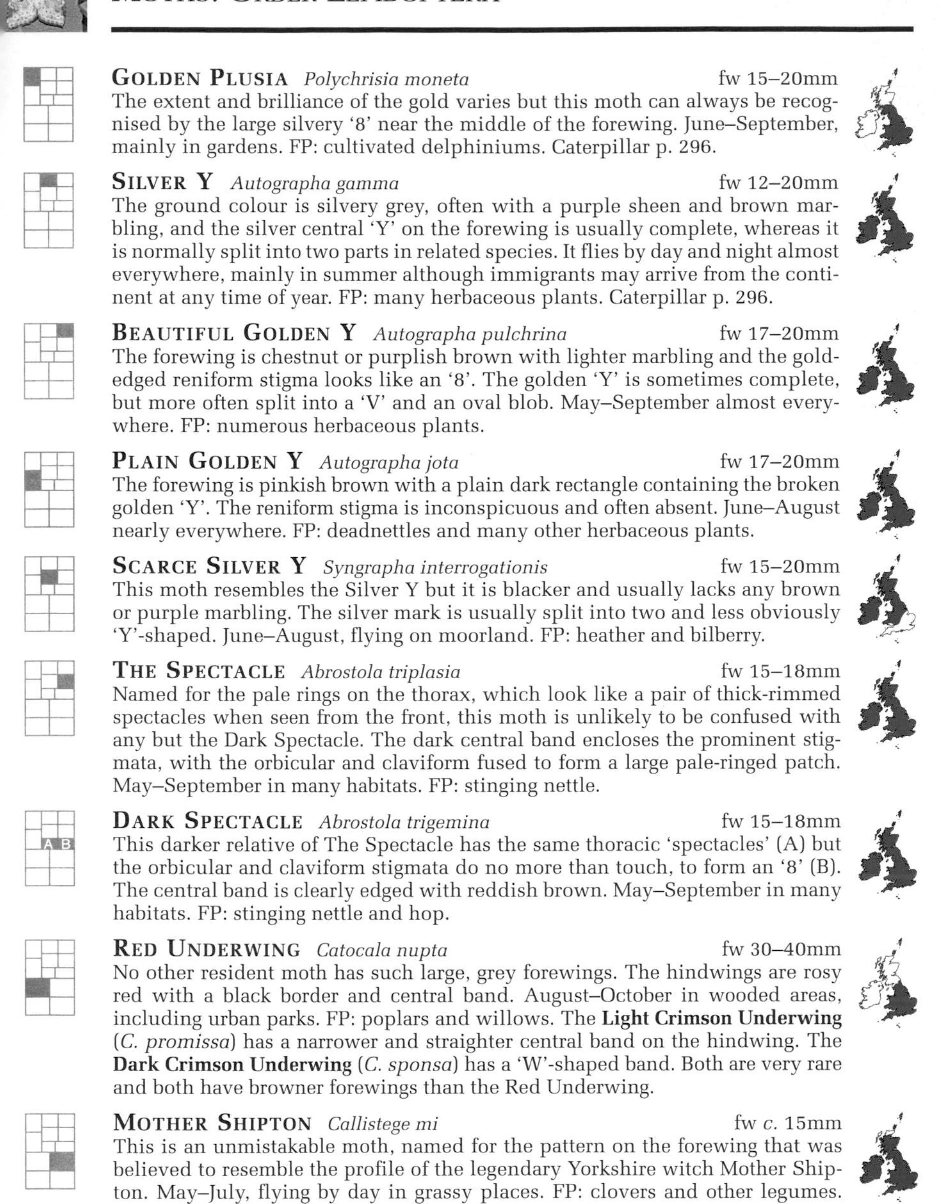

GOLDEN PLUSIA *Polychrisia moneta* fw 15–20mm
The extent and brilliance of the gold varies but this moth can always be recognised by the large silvery '8' near the middle of the forewing. June–September, mainly in gardens. FP: cultivated delphiniums. Caterpillar p. 296.

SILVER Y *Autographa gamma* fw 12–20mm
The ground colour is silvery grey, often with a purple sheen and brown marbling, and the silver central 'Y' on the forewing is usually complete, whereas it is normally split into two parts in related species. It flies by day and night almost everywhere, mainly in summer although immigrants may arrive from the continent at any time of year. FP: many herbaceous plants. Caterpillar p. 296.

BEAUTIFUL GOLDEN Y *Autographa pulchrina* fw 17–20mm
The forewing is chestnut or purplish brown with lighter marbling and the gold-edged reniform stigma looks like an '8'. The golden 'Y' is sometimes complete, but more often split into a 'V' and an oval blob. May–September almost everywhere. FP: numerous herbaceous plants.

PLAIN GOLDEN Y *Autographa jota* fw 17–20mm
The forewing is pinkish brown with a plain dark rectangle containing the broken golden 'Y'. The reniform stigma is inconspicuous and often absent. June–August nearly everywhere. FP: deadnettles and many other herbaceous plants.

SCARCE SILVER Y *Syngrapha interrogationis* fw 15–20mm
This moth resembles the Silver Y but it is blacker and usually lacks any brown or purple marbling. The silver mark is usually split into two and less obviously 'Y'-shaped. June–August, flying on moorland. FP: heather and bilberry.

THE SPECTACLE *Abrostola triplasia* fw 15–18mm
Named for the pale rings on the thorax, which look like a pair of thick-rimmed spectacles when seen from the front, this moth is unlikely to be confused with any but the Dark Spectacle. The dark central band encloses the prominent stigmata, with the orbicular and claviform fused to form a large pale-ringed patch. May–September in many habitats. FP: stinging nettle.

DARK SPECTACLE *Abrostola trigemina* fw 15–18mm
This darker relative of The Spectacle has the same thoracic 'spectacles' (A) but the orbicular and claviform stigmata do no more than touch, to form an '8' (B). The central band is clearly edged with reddish brown. May–September in many habitats. FP: stinging nettle and hop.

RED UNDERWING *Catocala nupta* fw 30–40mm
No other resident moth has such large, grey forewings. The hindwings are rosy red with a black border and central band. August–October in wooded areas, including urban parks. FP: poplars and willows. The **Light Crimson Underwing** (*C. promissa*) has a narrower and straighter central band on the hindwing. The **Dark Crimson Underwing** (*C. sponsa*) has a 'W'-shaped band. Both are very rare and both have browner forewings than the Red Underwing.

MOTHER SHIPTON *Callistege mi* fw *c.* 15mm
This is an unmistakable moth, named for the pattern on the forewing that was believed to resemble the profile of the legendary Yorkshire witch Mother Shipton. May–July, flying by day in grassy places. FP: clovers and other legumes. Caterpillar p. 296.

BURNET COMPANION *Euclidia glyphica* fw *c.* 15mm
The purplish-brown forewing, with chocolate-brown stripes and a darker, more-or-less triangular patch near the tip, is unlike that of any other moth. The hindwing is yellow and brown. May–July, flying by day in grassy places. FP: clovers and other low-growing legumes.

THE FOUR-SPOTTED *Tyta luctuosa* fw *c.* 12mm
This moth is easily identified by the large white spot on each black forewing. A white band on each hindwing gives the impression of two more spots when the wings are not fully closed. May–September in open grassland; uncommon. FP: field bindweed.

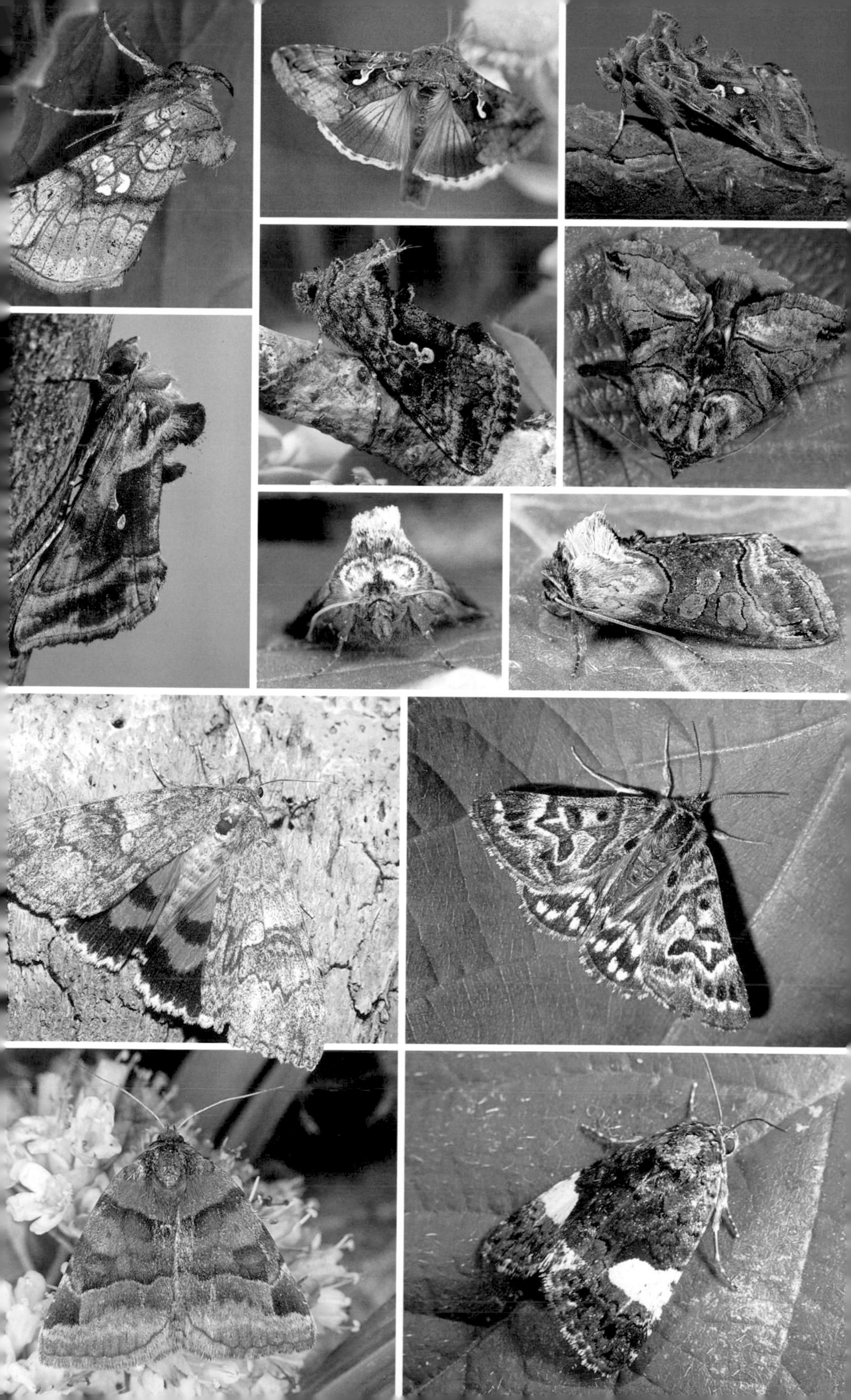

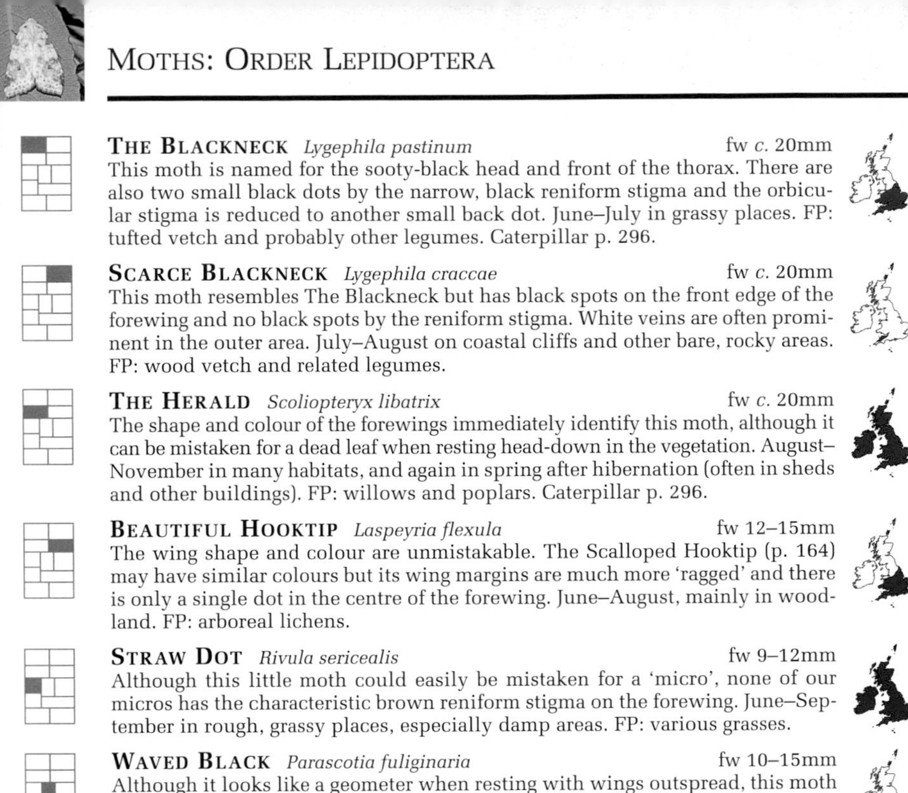

THE BLACKNECK *Lygephila pastinum* fw *c.* 20mm
This moth is named for the sooty-black head and front of the thorax. There are also two small black dots by the narrow, black reniform stigma and the orbicular stigma is reduced to another small back dot. June–July in grassy places. FP: tufted vetch and probably other legumes. Caterpillar p. 296.

SCARCE BLACKNECK *Lygephila craccae* fw *c.* 20mm
This moth resembles The Blackneck but has black spots on the front edge of the forewing and no black spots by the reniform stigma. White veins are often prominent in the outer area. July–August on coastal cliffs and other bare, rocky areas. FP: wood vetch and related legumes.

THE HERALD *Scoliopteryx libatrix* fw *c.* 20mm
The shape and colour of the forewings immediately identify this moth, although it can be mistaken for a dead leaf when resting head-down in the vegetation. August–November in many habitats, and again in spring after hibernation (often in sheds and other buildings). FP: willows and poplars. Caterpillar p. 296.

BEAUTIFUL HOOKTIP *Laspeyria flexula* fw 12–15mm
The wing shape and colour are unmistakable. The Scalloped Hooktip (p. 164) may have similar colours but its wing margins are much more 'ragged' and there is only a single dot in the centre of the forewing. June–August, mainly in woodland. FP: arboreal lichens.

STRAW DOT *Rivula sericealis* fw 9–12mm
Although this little moth could easily be mistaken for a 'micro', none of our micros has the characteristic brown reniform stigma on the forewing. June–September in rough, grassy places, especially damp areas. FP: various grasses.

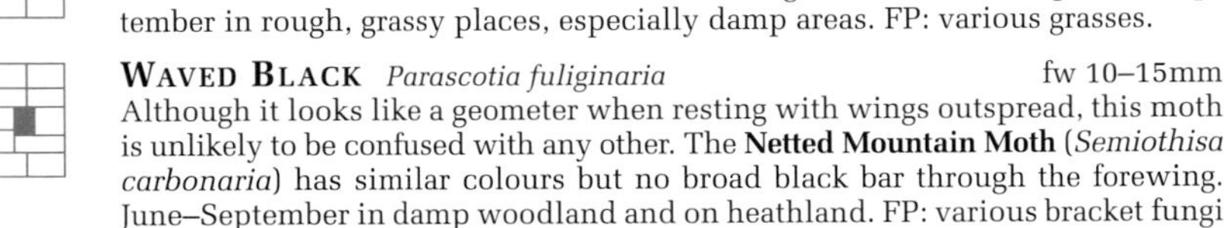

WAVED BLACK *Parascotia fuliginaria* fw 10–15mm
Although it looks like a geometer when resting with wings outspread, this moth is unlikely to be confused with any other. The **Netted Mountain Moth** (*Semiothisa carbonaria*) has similar colours but no broad black bar through the forewing. June–September in damp woodland and on heathland. FP: various bracket fungi and other wood-feeding fungi.

THE SNOUT *Hypena proboscidalis* fw 15–20mm
The snouts are named for the long, up-turned palps. The forewings are slightly hooked at the tip and range from pale to deep brown. The central cross-band is usually distinct and has a very angular inner margin. June–October nearly everywhere. FP: stinging nettle.

BEAUTIFUL SNOUT *Hypena crassalis* fw *c.* 15mm
The dark patch covering much of the forewing is unmistakable. The paler areas are greyish brown in the male and largely white in the female. May–August, mainly in open woodland and heathland. FP: bilberry

BUTTONED SNOUT *Hypena rostralis* fw 12–15mm
Named for the button-like tufts of scales on the narrow forewings, this moth has a very large 'snout'. The forewings may be plain brown but often have paler patches. August–October and again in spring after hibernation, mainly in and around hop fields, hedgerows and woodland edges; rare. FP: hop.

THE FAN-FOOT *Zanclognatha tarsipennalis* fw 12–15mm
The forewing is crossed by three fine lines, the central one being strongly curved and the outer one meeting the front margin before the tip. As in several other fan-foots, the male has a large tuft of hairs on the front leg. June–September, mainly in wooded areas. FP: fallen leaves.

SMALL FAN-FOOT *Herminia grisealis* fw *c.* 12mm
This moth differs from The Fan-foot in that the outer cross-line ends at the wing-tip. June–September in and around woods and hedgerows. FP: many deciduous trees and shrubs.

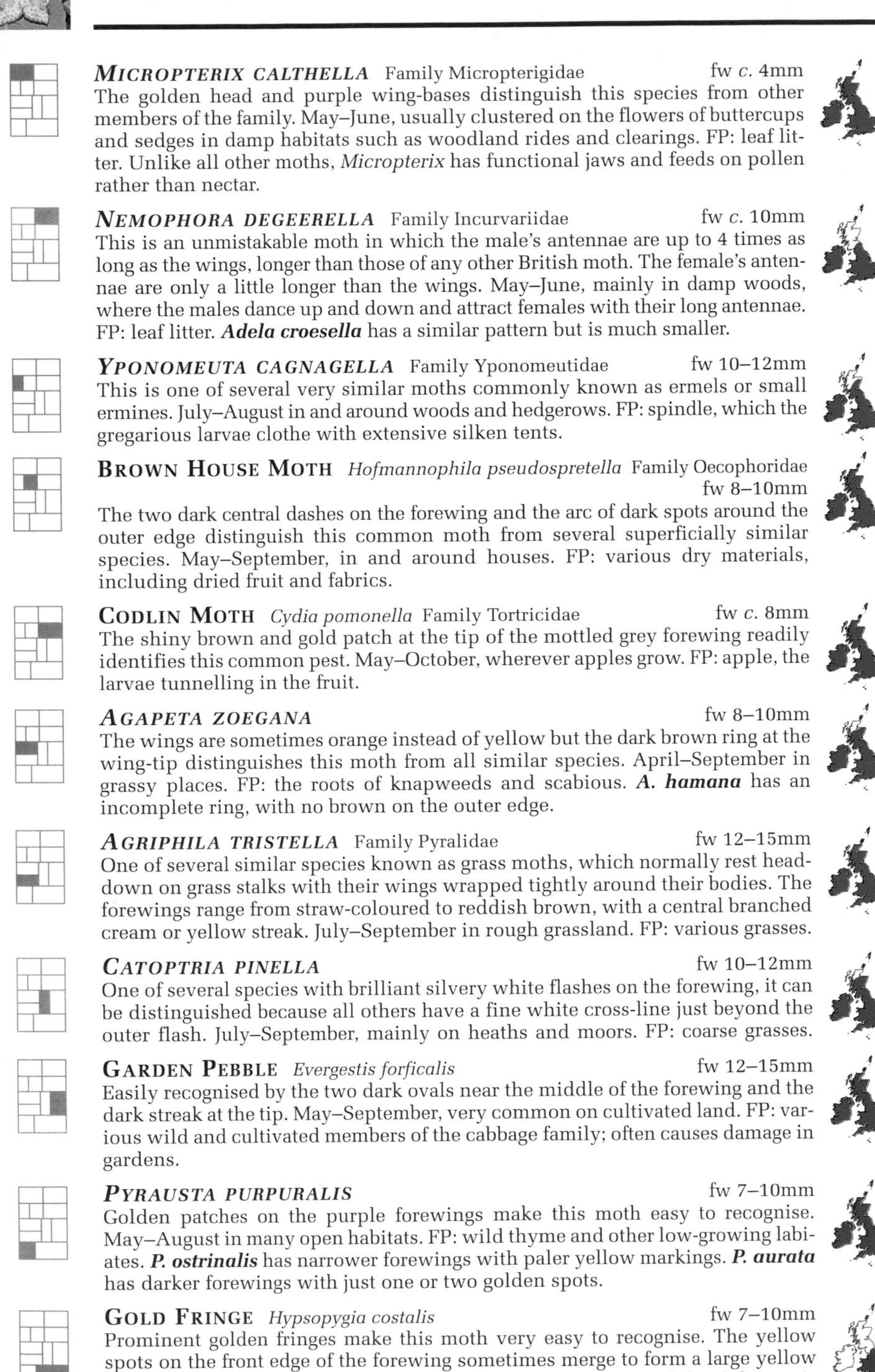

MICROPTERIX CALTHELLA Family Micropterigidae fw *c.* 4mm
The golden head and purple wing-bases distinguish this species from other members of the family. May–June, usually clustered on the flowers of buttercups and sedges in damp habitats such as woodland rides and clearings. FP: leaf litter. Unlike all other moths, *Micropterix* has functional jaws and feeds on pollen rather than nectar.

NEMOPHORA DEGEERELLA Family Incurvariidae fw *c.* 10mm
This is an unmistakable moth in which the male's antennae are up to 4 times as long as the wings, longer than those of any other British moth. The female's antennae are only a little longer than the wings. May–June, mainly in damp woods, where the males dance up and down and attract females with their long antennae. FP: leaf litter. ***Adela croesella*** has a similar pattern but is much smaller.

YPONOMEUTA CAGNAGELLA Family Yponomeutidae fw 10–12mm
This is one of several very similar moths commonly known as ermels or small ermines. July–August in and around woods and hedgerows. FP: spindle, which the gregarious larvae clothe with extensive silken tents.

BROWN HOUSE MOTH *Hofmannophila pseudospretella* Family Oecophoridae fw 8–10mm
The two dark central dashes on the forewing and the arc of dark spots around the outer edge distinguish this common moth from several superficially similar species. May–September, in and around houses. FP: various dry materials, including dried fruit and fabrics.

CODLIN MOTH *Cydia pomonella* Family Tortricidae fw *c.* 8mm
The shiny brown and gold patch at the tip of the mottled grey forewing readily identifies this common pest. May–October, wherever apples grow. FP: apple, the larvae tunnelling in the fruit.

AGAPETA ZOEGANA fw 8–10mm
The wings are sometimes orange instead of yellow but the dark brown ring at the wing-tip distinguishes this moth from all similar species. April–September in grassy places. FP: the roots of knapweeds and scabious. ***A. hamana*** has an incomplete ring, with no brown on the outer edge.

AGRIPHILA TRISTELLA Family Pyralidae fw 12–15mm
One of several similar species known as grass moths, which normally rest head-down on grass stalks with their wings wrapped tightly around their bodies. The forewings range from straw-coloured to reddish brown, with a central branched cream or yellow streak. July–September in rough grassland. FP: various grasses.

CATOPTRIA PINELLA fw 10–12mm
One of several species with brilliant silvery white flashes on the forewing, it can be distinguished because all others have a fine white cross-line just beyond the outer flash. July–September, mainly on heaths and moors. FP: coarse grasses.

GARDEN PEBBLE *Evergestis forficalis* fw 12–15mm
Easily recognised by the two dark ovals near the middle of the forewing and the dark streak at the tip. May–September, very common on cultivated land. FP: various wild and cultivated members of the cabbage family; often causes damage in gardens.

PYRAUSTA PURPURALIS fw 7–10mm
Golden patches on the purple forewings make this moth easy to recognise. May–August in many open habitats. FP: wild thyme and other low-growing labiates. ***P. ostrinalis*** has narrower forewings with paler yellow markings. ***P. aurata*** has darker forewings with just one or two golden spots.

GOLD FRINGE *Hypsopygia costalis* fw 7–10mm
Prominent golden fringes make this moth very easy to recognise. The yellow spots on the front edge of the forewing sometimes merge to form a large yellow patch. At rest, the wings are pressed flat against the surface and the abdomen is usually raised. July–August in a wide range of habitats. FP: dried plant matter, including hay and even straw thatch.

SCOPARIA PYRALELLA fw *c.* 8mm
One of several very similar species, distinguished from most by its grey-bordered white hindwings and the central pale brown stigmata on each forewing. June–July in rough, dry grassland, especially on chalk and limestone; usually rests on the ground and is very difficult to spot. FP: mainly decaying plant matter. ***S. ambigualis*** has darker hindwings.

BROWN CHINA-MARK MOTH *Elophila nymphaeata* fw 12–15mm
The brown markings range from tan to dark chocolate, but the moth is unlikely to be confused with any other. May–August, rarely far from still or slow-moving water. FP: various aquatic plants, the larva living under the water in a case made from silk and leaf fragments.

SMALL MAGPIE *Eurrhypara hortulata* fw *c.* 15mm
The dark markings range from dark grey to almost black, but the moth cannot be mistaken for any other species. The Magpie Moth (*see* p. 192) is much larger and has a yellow band across the forewing. June–July in rough vegetation, especially hedgerows and woodland margins, and also in gardens. FP: stinging nettles and, less often, mint and other labiates.

MOTHER-OF-PEARL *Pleuroptya ruralis* fw *c.* 15mm
This well-named and easily recognised moth has shiny wings, often with a pinkish tinge. The grey markings vary in intensity. July–August in many habitats, but especially common in hedgerows and on wasteland. FP: stinging nettle.

SYNAPHE PUNCTALIS fw 10–12mm
The conspicuous white dots along the front edge of the forewing and the way in which the wings overlap at rest immediately identify this moth. The male's forewings are reddish brown, while the smaller female has paler brown wings. June–August, mainly in sandy places; day- and night-flying. FP: mosses.

MEAL MOTH *Pyralis farinalis* fw 10–15mm
The central area of the forewing ranges from yellow to brick-red, but the two conspicuous white cross-lines readily identify this moth. The abdomen is commonly turned up at rest. June–August, usually in and around buildings where grain is stored. FP: stored grain and cereal products.

MYELOIS CRIBRELLA fw *c.* 15mm
This moth is superficially like the *Yponomeuta* species (*see* p. 268) but has no black spots on the thorax and, unlike *Yponomeuta*, has a pure white fringe on the hindwing. June–July in a variety of open habitats. FP: thistles, the larvae feeding inside the stems.

ONCOCERA SEMIRUBELLA fw *c.* 12mm
The pink and yellow forewings of this moth are very distinctive. There may also be a white or pale grey streak along the front edge. June–July, mainly on chalk and limestone grassland. FP: bird's-foot trefoil and white clover.

WHITE PLUME MOTH *Pterophorus pentadactyla* Family Pterophoridae fw *c.* 12mm
This is an unmistakable moth, whose pure white wings are split from about halfway into feathery plumes – two on the forewing and three on the hindwing. At rest, the wings are rolled around each other and look like small feathers. May–August in gardens and hedgerows. FP: hedge bindweed.

EMMELINA MONODACTYLA fw 12–15mm
The wings vary from off-white to light brown, usually dusted with darker scales. Only the outer third of each wing is split into plumes, but the wings are so tightly rolled at rest that they look solid. A dark central spot on the forewing is usually visible even when the wings are rolled. All year but dormant in winter; common in many habitats, especially hedgerows. FP: bindweeds.

MANY-PLUMED MOTH *Alucita hexadactyla* Family Alucitidae fw *c.* 7mm
Each wing is divided into 6 slender 'feathers', but the feathery nature is not obvious when the moth is at rest with its wings swept back like an arrowhead. All year in woods, hedgerows, and gardens; may come to lighted windows. FP: honeysuckle.

SWALLOWTAIL *Papilio machaon* Family Papilionidae 40mm
Newly-hatched caterpillars are black and white and resemble bird droppings. They get greener at each moult and then blend well with their food-plants. After the 3rd moult the insect acquires another means of defence: when alarmed it everts the orange osmeterium just behind its head and the scent from this may deter parasites. FP: milk parsley, May–September. Adult and map p. 280. (*See also* Emperor Moth p. 144.)

LARGE WHITE *Pieris brassicae* Family Pieridae 40mm
No other British caterpillar has the bold black and yellow warning coloration of this species. FP: mainly cultivated brassicas and nasturtiums, May–September; gregarious for much of its life and a serious garden pest. Adult and map p. 142.

SMALL WHITE *Pieris rapae* 25mm
This caterpillar can be distinguished from several superficially similar moth larvae by its coating of short, dark hairs. There is a fine yellow line along the back and each spiracle sits on a yellow patch. FP: mainly cultivated brassicas and nasturtiums, April–September. Adult and map p. 142 The Green-veined White (*P. napi*) is very similar but has no yellow line on the back.

ORANGE-TIP *Anthocharis cardamines* 30mm
The body is clearly divided into a greyish-green or bluish-green upperside and a dark green underside, with a broad white line between them. This pattern makes the caterpillar extremely difficult to see among the slender seed capsules of its food-plants. FP: garlic mustard, cuckoo flower and other crucifers, including garden honesty and sweet rocket, June–July. Adult and map p. 142.

BRIMSTONE *Gonepteryx rhamni* 35mm
This is similar to the Orange-tip caterpillar although usually less blue on top and the lower surface is a much brighter yellowish green. No other green caterpillar of this size is likely to be found on buckthorns, although some smaller moth larvae, mostly green with yellow lines, do occur. FP: buckthorn and alder buckthorn, May–July. Adult and map p. 144.

CLOUDED YELLOW *Colias croceus* 30–35mm
The body is *very finely* speckled with black, with a fine white hair springing from each spot. The pale yellow line running through the spiracles contains a slender red or orange dash on each segment and there is a black dot below each dash. FP: clover, lucerne and other legumes, June–October. Adult and map p. 144.

PALE CLOUDED YELLOW *Colias hyale* 30–35mm
This caterpillar is very difficult to distinguish from the Clouded Yellow but the orange dashes on the spiracular line tend to be longer and the black spots below them are smaller. The fine hairs coating the body are black but, as they all come from black spots, it is not easy to appreciate this difference without a lens. FP: clovers and lucerne, June–October, but very much rarer in the British Isles than the Clouded Yellow. Adult and map p. 144.

BERGER'S CLOUDED YELLOW *Colias alfacariensis* 30–35mm
Although the adults of this and the Pale Clouded Yellow are almost identical, the mature caterpillars are quite different. This one has 4 bright yellow lines running the length of the body, with a series of black patches on each segment. FP: horseshoe vetch, June–October, but very rare in the British Isles. Adult and map p. 144.

WOOD WHITE *Leptidea sinapis* 20mm
This slender, pale green caterpillar differs from that of the Small White in having a continuous yellow spiracular stripe and a darker line along its back. FP: meadow vetchling and other leguminous plants, June–July. Adult and map p. 142.

DUKE OF BURGUNDY *Hamearis lucina* Family Riodinidae 15mm
The mature caterpillar, pictured here, is clothed with white, brown and black hair and resembles a hairy slug. Younger larvae are much paler and have only sparse hair. FP: primrose and cowslip, June–August, feeding mainly at night and hiding under foliage by day. Adult and map p. 144.

FAMILY NYMPHALIDAE

Most of the caterpillars in this family can be recognised as such by their branched spines, many of which spring from fleshy tubercles on the top of the body. The head may also bear spines.

WHITE ADMIRAL *Limenitis camilla* 30mm
This unmistakable caterpillar has a very characteristic humped resting position. Most of the segments bear prominent spines arising from paired fleshy tubercles. FP: honeysuckle, all year but dormant from early autumn to April. Adult and map p. 150.

PURPLE EMPEROR *Apatura iris* 40–45mm
The slug-like shape and the two 'horns' at the front immediately identify this caterpillar. FP: sallows, August–June, but dormant on a twig throughout the winter, during which time it is dull brown. Adult and map p. 150.

RED ADMIRAL *Vanessa atalanta* 35mm
The body colour varies from dirty green to dark grey or almost black, with branched grey or black spines. Most segments have a yellow flash low down on each side. FP: stinging nettle, mainly May–October, spending most of its life concealed in folded leaves. Adult and map p. 150.

PAINTED LADY *Vanessa cardui* 30mm
This caterpillar varies from greyish green to black and is similar to that of the Red Admiral, but the spines are either yellow or black and yellow. The legs and the underside are brown. FP: mainly thistles, sometimes mallows, April–October, spending much of its life concealed in folded leaves. Adult and map p. 150.

SMALL TORTOISESHELL *Aglais urticae* 20–25mm
The body colour varies a good deal but is most commonly yellow with densely packed black dots. It may also be black with yellow lines on the back and sides. Young larvae are always predominantly yellowish. The spines are usually black with yellow bases. FP: stinging nettle, May–August, living gregariously until the final instar. Adult and map p. 150.

PEACOCK *Inachis io* 40–45mm
Newly-hatched larvae are greenish grey but mature caterpillars are velvety black with conspicuous white dots and long, branched black spines. The prolegs are brown. FP: stinging nettle, May–July, living gregariously until nearly fully grown. Adult and map p. 150.

COMMA *Polygonia c-album* 35mm
This caterpillar is easily recognised by the large white patch on the rear half of the body. The rest of the body is black with bright orange markings and it is quite easy to mistake the whole insect for a bird dropping. FP: mainly hop and stinging nettle, sometimes elm and sallow, May–August. Adult and map p. 150.

DARK GREEN FRITILLARY *Argynnis aglaja* 40mm
The black legs and the conspicuous red patches on the sides immediately distinguish this rather glossy caterpillar from that of the Peacock. FP: violets, March–May; newly-hatched larvae hibernate immediately after eating their eggshells and move to the food-plant in spring. Adult and map p. 152.

SILVER-WASHED FRITILLARY *Argynnis paphia* 40mm
This caterpillar is dark brown or black on top, with two yellow or orange lines running from front to rear. The branched spines are brown with black tips, although the front pair are completely black and point forward like horns. The sides of the body are mottled brown. FP: violets, March–May. Adult and map p. 152.

FAMILY SATYRIDAE

The caterpillars of this family all feed on grasses and the eleven British species are all green or brown and clothed with short hair. They taper noticeably towards the rear and longitudinal stripes help to camouflage them in the grass, but the most characteristic features are the two short prongs at the rear. These will always distinguish satyrid caterpillars from those of the white butterflies (*see* p. 272).

MARBLED WHITE *Melanargia galathea* 25–30mm
The head is reddish brown and the body is either pale green or light brown with two darker lines running along the top. A yellowish line runs just above the spiracles and the terminal prongs are pink. FP: mainly red fescue grass. February–June. Adult and map p. 154.

SPECKLED WOOD *Pararge aegeria* 25–30mm
The bright, yellowish green body has a pale-edged dark stripe along the top and pale lines along the sides. The terminal prongs are white. FP: a wide range of grasses, all year, although dormant in the coldest months. Adult and map p. 154.

MEADOW BROWN *Maniola jurtina* 25mm
Yellowish green above and darker below, this caterpillar has relatively long white hair. There is a pale yellow line just below the spiracles on each side and the terminal prongs are white. FP: mainly fine grasses, July–May, active in all but the coldest weather. Adult and map p. 154.

GRAYLING *Hipparchia semele* 30mm
The three broad, dark brown lines running the length of the body are enough to distinguish this from our other brown satyrid caterpillars. The Gatekeeper (*Pyronia tithonus*) also has three bands, but those on the sides are below the level of the spiracles and the body is usually tinged with green. FP: various grasses, August–June, active in all but the coldest weather. Adult and map p. 154.

RINGLET *Aphantopus hyperantus* 20–22mm
Somewhat more bristly than other satyrid larvae, this species has a dark line along the top and another, often broken, below the spiracles on each side. There is also a white line below the spiracles and it is edged with pinkish brown. FP: assorted grasses, August–June, most often seen in spring. Adult and map p. 154.

LARGE HEATH *Coenonympha tullia* 25mm
There is a pale-edged, dark green line along the middle of the back, but the most striking feature is the bright white band running along each side of the body. The tail prongs are tipped with pink. FP: various sedges and grasses on bogs and moorland, August–May, but dormant in winter. Adult and map p. 154.

FAMILY LYCAENIDAE

The caterpillars in this family are short and squat. With the legs rarely visible and the head pulled into the body when resting, they could easily be mistaken for slugs were it not for their hairs.

HOLLY BLUE *Celastrina argiolus* *c.* 12mm
The body ranges from bright green to almost yellow and is sometimes tinged with pink. There is a pale line below the spiracles and many individuals also have a deep pink line along the back, sometimes surrounded by pale triangles. FP: holly, dogwood, and many other shrubs in the spring; ivy in late summer. The caterpillars feed on the flower buds and then on the flowers and developing fruits. Adult and map p. 146.

COMMON BLUE *Polyommatus icarus* *c.* 12mm
The main feature of this common caterpillar is the pale edged, dark green line along the top of the body. There is also a pale stripe below the spiracles on each side, although this is often indistinct. FP: bird's-foot trefoil, clovers and related plants. Adult and map p. 148.

CHALKHILL BLUE *Lysandra coridon* *c.* 15mm
The two bright yellow lines on the top and two more low down on each side distinguish this from most other lycaenids. FP: horseshoe vetch, April–early June. Adult and map p. 148. Larvae of the **Adonis Blue** (*L. bellargus*) are a little darker and can be found in late summer as well as in spring.

SMALL COPPER *Lycaena phlaeas* 15mm
The pink line along the top and another low down on each side immediately identify this caterpillar. FP: sorrels and docks, all year, but dormant in winter. Adult and map p. 146.

GREEN HAIRSTREAK *Callophrys rubi* 15mm
The yellow and dark green triangles along the sides of the body, together with the yellow or cream line running through the spiracles, make this caterpillar quite easy to recognise. FP: gorse, broom and other low-growing shrubs, including rockroses, from May–July. Adult and map p. 146.

BROWN HAIRSTREAK *Thecla betulae* 15–18mm
There are two pale yellow ridges close together along the back, diverging on the thorax to form a triangle. The yellow diagonal stripes are usually prominent, as is the line running below the spiracles. FP: blackthorn, from April–June. Adult and map p. 146.

PURPLE HAIRSTREAK *Quercusia quercus* 15mm
This reddish-brown caterpillar, marked with dark brown diagonal stripes, is unlikely to be mistaken for any other species, although it is not often seen because it feeds at night and hides in the leaf bases by day. FP: oak, March–June. Adult and map p. 146.

WHITE-LETTER HAIRSTREAK *Satyrium w-album* 15mm
Dark green triangles on the back and lighter green diagonal stripes on the sides distinguish this from our other hairstreak caterpillars. The red patches are more extensive in some individuals, but completely absent in others. FP: elms, March–May, but inside the flower-buds at first. Adult and map p. 146.

BLACK HAIRSTREAK *Satyrium pruni* 15mm
The two rows of small purple-tipped humps on the back immediately identify this caterpillar. There may also be a row of oblique yellow stripes on each side, although these are often very faint. FP: blackthorn from March–May. Adult and map p. 146.

FAMILY HESPERIIDAE

This family contains the skipper butterflies, whose caterpillars have relatively large heads and small but obvious necks. Clothed with short, fine hairs, they generally hide in folded leaves by day and emerge to feed at night. Unlike most butterfly caterpillars, they pupate in flimsy silken cocoons.

LARGE SKIPPER *Ochlodes venatus* 25–30mm
The body is bluish green, with a darker line along the back and an inconspicuous yellow line along the spiracles. FP: mainly cock's-foot and false brome grasses from September–May, but dormant between grass blades in winter; often common in sunny woodland rides and roadside verges. Adult and map p. 140.

SMALL SKIPPER *Thymelicus sylvestris* *c.* 20mm
The body is pale green, with yellow inter-segmental membranes and a conspicuous cream stripe below the spiracles on each side. There are narrower pale stripes higher up and a dark green line along the middle of the back. FP: various grasses with relatively wide blades, August–June, but dormant in a leaf sheath in winter. Adult and map p. 140. The caterpillar of the Essex Skipper is bluer, while that of the Lulworth Skipper has a brown head. All three species nibble conspicuous V-shaped notches in the grass blades.

GRIZZLED SKIPPER *Pyrgus malvae* *c.* 20mm
The body is essentially pale green, but the upper half is densely speckled with brown and contrasts strongly with the lower half. Some individuals bear noticeable dark green or brown stripes on the back. FP: wild strawberry, cinquefoils, and some other rosaceous plants; usually May–July, but occasionally a 2nd brood in the autumn. Adult and map p. 140.

DINGY SKIPPER *Erynnis tages* 15–18mm
This caterpillar is usually plain green, although some individuals have a brown tinge on the back. The head is purplish black, with or without red patches. FP: mainly bird's-foot trefoil, May–August. Mature caterpillars pass the winter in leafy shelters and pupate in the spring without further feeding. Adult and map p. 140.

COMMON SWIFT *Hepialus lupulinus* Family Hepialidae 35mm
This is the commonest of the caterpillars likely to be dug up in the garden. The head and 'collar' are shiny brown and the body bears numerous indistinct yellow spots. FP: the roots of many wild and cultivated herbaceous plants. Adult and map p. 156. The Ghost Swift caterpillar has more obvious brown spots.

SIX-SPOT BURNET *Zygaena filipendulae* Family Zygaenidae 25mm
The ground colour is yellow, often with a greenish tinge, and there are black spots on the sides as well as on the top. FP: bird's-foot trefoil, July–May, but dormant in winter. Adult and map p. 160. The Five-spot Burnet and Narrow-bordered Five-spot Burnet have a greenish-white ground colour.

B
A

EMPEROR MOTH *Saturnia pavonia* Family Saturniidae 60mm
Newly-hatched larvae (A) are velvety black, gradually acquiring yellow spots on the sides and then becoming green with black rings (B). Each ring bears pink or yellow pimples bearing black bristles. FP: heather, blackthorn, bramble and many other shrubs, May–August. Adult and map p. 220. Swallowtail butterfly caterpillars (*see* p. 272) are superficially similar but lack the bristles.

KENTISH GLORY *Endromis versicolora* Family Endromidae *c.* 60mm
The body tapers strongly towards the front and there is a small horn on the rear end, but the most obvious feature is the pale stripe above the thoracic legs. FP: birch and sometimes alder, from May–July, the caterpillars being gregarious when young. Adult and map p. 220.

OAK EGGAR *Lasiocampa quercus* Family Lasiocampidae 75mm
Patches of rich brown hair on the back alternate with the very black intersegmental membranes, giving the caterpillar a strongly banded appearance. There is a white line just above the spiracles. FP: bramble, hawthorn, heather and many other trees and shrubs, hatching in August but soon becoming dormant and seen mainly April–July. Adult and map p. 162. The caterpillar of the Fox Moth is similar but lacks the pale markings on the sides.

SMALL EGGAR *Eriogaster lanestris* 45mm
Patches of chestnut-brown hairs, edged with cream, readily identify this caterpillar. Chestnut prolegs contrast strongly with the velvety black sides. FP: mainly blackthorn and hawthorn, April–July. The caterpillars live communally in large silken tents, on which they bask in warm weather. Adult and map p. 162.

DECEMBER MOTH *Poecilocampa populi* 50mm
The body is clothed with fine grey hair and heavily speckled with dark grey or black. There may also be some light brown mottling on the back and there is a conspicuous reddish-brown patch just behind the head. FP: a wide range of deciduous trees and shrubs, April–June. Adult and map p. 162.

THE LACKEY *Malacosoma neustria* 45mm
The red, white and blue lines make this caterpillar easy to recognise. FP: hawthorn, blackthorn and many other deciduous trees and shrubs, April–June, living communally until the final instar. Adult and map p. 162.

THE DRINKER *Euthrix potatoria* 75mm
Named for its supposed habit of drinking dew, this caterpillar is easily identified by its coloration and the hornlike tuft of hair at each end. FP: assorted coarse grasses, feeding in late summer and then hibernating before completing growth in spring. Adult and map p. 162.

BARRED HOOKTIP *Watsonalla cultraria* Family Drepanidae *c.* 20mm
In common with all hooktip larvae, this one lacks proper claspers and rests with both ends raised clear of the surface. It is easily recognised by its pinkish-brown colour and the pale saddle-shaped patch on its back. There is also a double 'wart' on the top of the 3rd segment. FP: beech, June–September. Adult and map p. 164.

OAK HOOKTIP *Watsonalla binaria* 25mm
This caterpillar resembles that of the Barred Hooktip in pattern, but a network of fine black and brown lines gives it a much darker appearance. June–October in and around oakwoods. FP: oak. Adult and map p. 164.

FAMILY GEOMETRIDAE

The caterpillars of this large family, with over 300 British species, are known as loopers. They move by stretching forward to grip with their front legs and then bringing the rear end forward, causing the body to form a loop. There are normally just two pairs of prolegs, including the claspers right at the back. The caterpillars have little or no hair and many are incredibly twig-like (*see* p. 42).

OBLIQUE CARPET *Orthonama vittata* 15–20mm
Seen here in the characteristic resting position common to several geometrid larvae, this caterpillar is normally yellowish green with reddish-brown patches on the back. There are also several darker lines on the back. FP: bedstraws, throughout the year, although dormant in winter. Adult and map p. 172.

GARDEN CARPET *Xanthoroe fluctuata* 20–25mm
The upper surface is brown, green, or grey with a variable series of pale blotches, but there is always a clear division between this and the much paler underside. FP: wild and cultivated crucifers, including garden cabbages and wallflowers, April–September, active mainly at night. Adult and map p. 174.

DARK UMBER *Philereme transversata* 20–25mm
The purple patch around the last three spiracles is the most obvious and constant feature of this caterpillar, although there may be other purple patches as well. The body is usually green with a white stripe on each side, but some individuals are grey with purple spots. FP: buckthorn, April–June. Adult and map p. 182.

NOVEMBER MOTH *Epirrita dilutata* 25mm
The body is basically green with yellowish inter-segmental rings and a rust-red line along the back, but the back may also be marked with rust-red or purplish blotches. The underside is greenish white and there is a white line along each side. FP: many deciduous trees and shrubs, April–July. Adult and map p. 184.

LIME-SPECK PUG *Eupithecia centaureata* 15–20mm
The ground colour of this slender caterpillar ranges from yellow, through green, to brown or grey. The chestnut spotting varies in extent and is sometimes absent. FP: the flowers of a wide range of plants, including ragwort and hemp agrimony, on which the larvae are usually very well camouflaged; June–September. Adult and map p. 186.

GOLDEN-ROD PUG *Eupithecia virgaureata* 15–20mm
The ground colour varies a great deal but the more or less triangular dark marks are usually visible on the sides, together with the characteristically wavy white line. FP: mainly golden-rod and ragwort, feeding on the flowers June–October. Adult and map p. 188.

SLOE PUG *Pasiphila chloerata* *c.* 15mm
A little plumper than most pug moth larvae, this species can be recognised by its pale body with a broad pinkish stripe along the back. FP: the flowers of blackthorn in March–April. Adult and map p. 190.

THE MAGPIE *Abraxas grossulariata* 30mm
The warning colours of this caterpillar make it unlikely that it will be confused with any other species. FP: a wide range of trees and shrubs, but especially damaging to currants and gooseberries; most of the year, but dormant in winter. Adult and map p. 192.

SCORCHED CARPET *Ligdia adustata* 20mm
This bright green caterpillar has a rust-coloured head and legs and several brownish patches elsewhere on its body, and is a wonderful match for the young twigs among which it lives. FP: spindle, June–September. Adult and map p. 192.

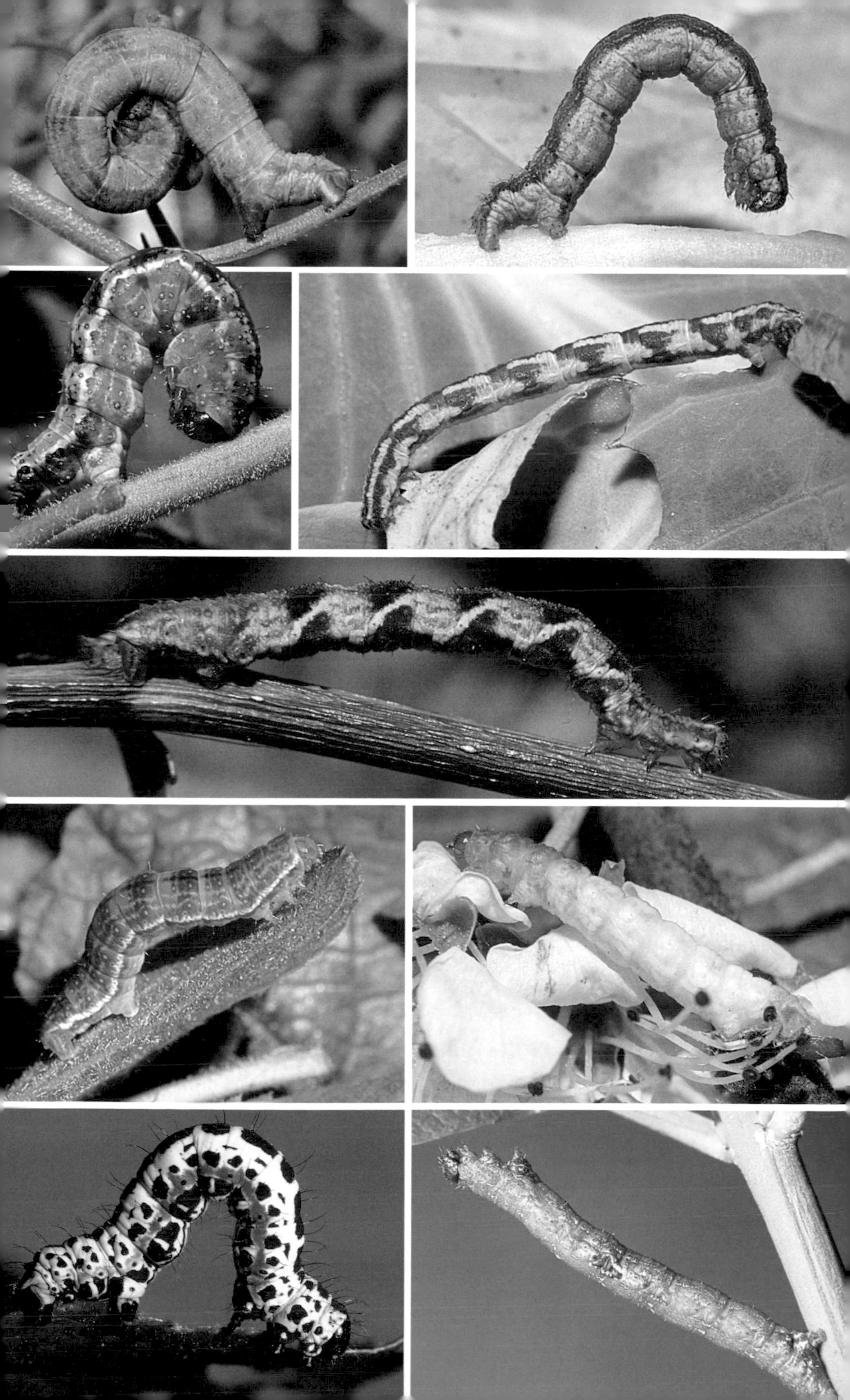

BRIMSTONE MOTH *Opisthograptis luteolata* 35mm
The body is generally some shade of brown, heavily mottled with green, which resembles the algae found on most twigs. Little humps on the rear half of the body resemble dormant buds, as do two pairs of vestigial prolegs. FP: hawthorn, blackthorn and other rosaceous trees, April–October. Adult and map p. 194.

LILAC BEAUTY *Apeira syringaria* 30mm
Hooked outgrowths on the 8th segment give this caterpillar a vague likeness to a coat hanger and make it easy to recognise. It rests in the looped position and is easily mistaken for a dead leaf. FP: honeysuckle, privet and lilac, mainly May–June after hibernating when very small. Adult and map p. 196.

EARLY THORN *Selenia dentaria* 40mm
Ranging from orange-yellow to chestnut-brown, this caterpillar is one of the best of the 'living twigs'. The 3rd pair of thoracic legs are swollen and bud-like and the 7th and 8th segments also bear bud-like swellings. FP: a wide range of deciduous trees and shrubs, especially blackthorn, May–October. Adult and map p. 196.

LUNAR THORN *Selenia lunularia* 40mm
This caterpillar is even more twig-like than that of the Early Thorn, with a surface texture just like bark, and bumps and bulges all over it. The 3rd pair of thoracic legs are particularly swollen. FP: many deciduous trees and shrubs, especially *Prunus* species, July–October. Adult and map p. 196.

SCALLOPED OAK *Crocallis elinguaria* 45mm
Ranging from pale brown to purplish, this caterpillar usually has a dark streak on each side that looks very much like a split in the bark of a twig. There may also be a row of dark diamonds on the back. FP: a wide range of deciduous trees and shrubs, especially blackthorn, March–May. Adult and map p. 194.

SWALLOWTAILED MOTH *Ourapteryx sambucaria* 50mm
This very slender brown or dirty green caterpillar is best identified by its very flat head. FP: hawthorn and other shrubs, including ivy, all year, but dormant in winter; often freely exposed but well-camouflaged on bare twigs. Adult and map p. 198.

PALE BRINDLED BEAUTY *Phigalia pilosaria* 40mm
This is a very warty caterpillar, with one or more stout bristles arising from each wart. The body is dark brown, often with a reddish or purplish tinge, and each spiracle is surrounded by a small rust-coloured patch. FP: a wide range of deciduous trees and shrubs, April–June. Adult and map p. 198.

PEPPERED MOTH *Biston betularia* 60mm
This caterpillar is dark green or purplish brown, with bark-like sculpturing and a deeply notched head. FP: numerous trees and shrubs, including garden raspberries, June–September. Adult and map p. 198.

OAK BEAUTY *Biston strataria* 60mm
This caterpillar resembles that of the Peppered Moth but, with a very rough surface and bumps on both the back and the belly, it blends in with somewhat older twigs. FP: mainly oak, but also many other trees, May–August. Adult and map p. 198.

DOTTED BORDER *Agriopis marginaria* 30mm
The body ranges from dirty yellow, through various greens, to dark brown or almost black. Dark 'X'-shaped marks may be visible on the back. The head and claspers are brown. FP: numerous deciduous trees and shrubs, April–June. Adult and map p. 198.

MOTTLED UMBER *Erannis defoliaria* 30mm
The body is yellowish-brown or rust-coloured above and yellow underneath, with a black line just above the spiracles. Each of the latter may sit on a rust-coloured patch. FP: almost any deciduous tree or shrub, March–June; sometimes a pest in orchards. Adult and map p. 200.

WAVED UMBER *Menophra abruptaria* 35–40mm
This remarkably twig-like caterpillar is normally greyish brown but is sometimes heavily dusted with green. There are usually a number of dark pink or black blotches on the back. FP: mainly privet and lilac, but also redcurrant, from May–August. Adult and map p. 200.

GREAT OAK BEAUTY *Hypomecis roboraria* 45–50mm
The reddish-brown body has a hump on segments 5 and 11 and there is a similar swelling on the underside of segment 6, giving the caterpillar amazing similarity to the knobbly twigs among which it feeds and rests. FP: mainly oak, August–May but dormant in winter. Adult and map p. 200.

BRUSSELS LACE *Cleorodes lichenaria* 25mm
This caterpillar has almost exactly the same colour and warty texture as the lichens among which it lives, and it is extremely difficult to spot. FP: various branched arboreal lichens, August–May, although dormant in winter and most likely to be noticed when approaching full size in spring. Adult and map p. 200.

BORDERED WHITE *Bupalus piniaria* 30mm
This is one of several quite similar conifer-eating geometers, but it differs from most in having a green head and no brown markings. There are 1 or more yellow lines on the sides and a single, wider white line along the back. FP: mainly pine and spruce, but other conifers are eaten, July–September. Adult and map p. 202.

FAMILY SPHINGIDAE

The hawkmoth caterpillars nearly all have a curved horn at the rear. They are virtually hairless, although the skin is often rather rough and warty, and many of the species are decorated with slanting stripes. These break up the body outlines so well that the caterpillars are often quite hard to spot even from a short distance.

PRIVET HAWKMOTH *Sphinx ligustri* 100mm
With its bright green body and purple and white stripes, this caterpillar, often found in gardens, cannot be mistaken for any other. At rest, it adopts the 'sphinx' position that is responsible for its scientific name and that of the whole family. FP: mainly privet, but also lilac and ash, June–August. Adult and map p. 206.

DEATH'S-HEAD HAWKMOTH *Acherontia atropos* 130mm
This is the largest caterpillar likely to be found in the British Isles. It is normally yellow with blue or purplish stripes. Some individuals are brown with a largely white thorax, but all can be recognised by the rough, curled tail-horn. FP: mainly potatoes, but also privet and nightshades, August–October. Adult and map p. 206.

PINE HAWKMOTH *Hyloicus pinastri* 80mm
Young larvae are green with pale stripes that blend with the pine needles, but older individuals, too large to hide among the needles, develop darker stripes and a rusty band along the back. They then blend well with the twigs. FP: pine and spruce, July–September. Adult and map p. 206.

LIME HAWKMOTH *Mimas tiliae* 65mm
This caterpillar is bright green with yellow stripes and dense yellow spots. The horn is blue on the top and largely red beneath. FP: mainly lime, but also elm, birch and alder, July–September. Adult and map p. 206.

POPLAR HAWKMOTH *Laothoe populi* 65mm
This caterpillar often has a bluish tinge. The horn is often tipped with red and there may be a line of small red blotches just above the spiracles. FP: poplars and willows, July–September. Adult and map p. 208.

EYED HAWKMOTH *Smerinthus ocellata* 75mm
The body often has a strong blue tinge. The spots and stripes are white or pale green and the horn is generally bluish. Unlike most caterpillars, it normally rests upside-down and its ventral surface is darker than the dorsal side. FP: mainly willows and poplars, but also apple, June–September. Adult and map p. 286.

NARROW-BORDERED BEE HAWKMOTH *Hemaris tityus* 40mm
This caterpillar has a very obvious yellow line above the spiracles, with a row of red or rust-coloured spots above it, although these spots may be quite small. FP: scabious, July–August. Adult and map p. 208. The caterpillar of the Broad-bordered Bee Hawkmoth (*see* p. 208) is a little larger, usually densely speckled with white and often with a bluish tinge. The underside is rusty brown and the spiracles sit on a yellow line or a series of yellow blotches. FP: mainly honeysuckle, July–September.

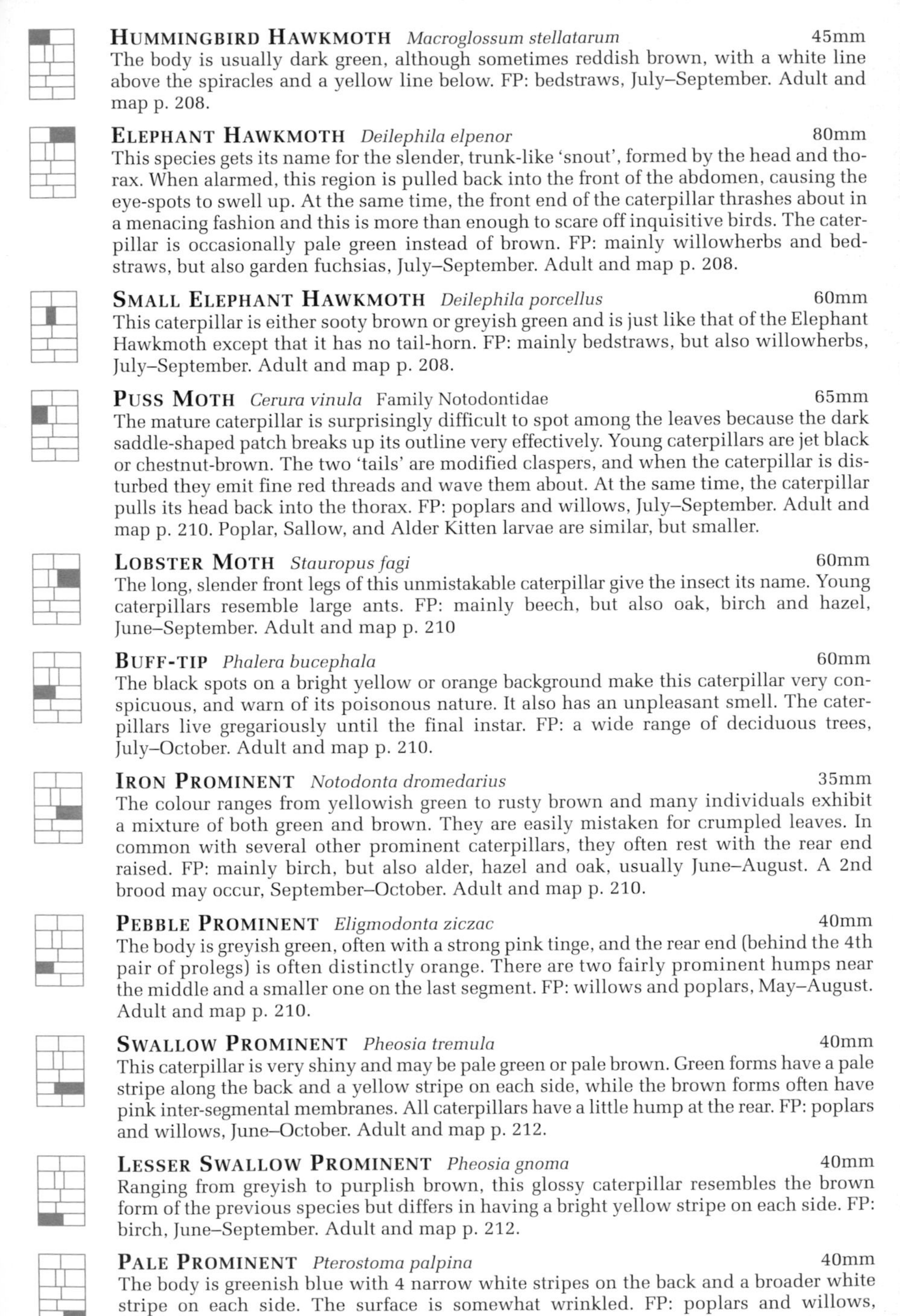

Hummingbird Hawkmoth *Macroglossum stellatarum* 45mm
The body is usually dark green, although sometimes reddish brown, with a white line above the spiracles and a yellow line below. FP: bedstraws, July–September. Adult and map p. 208.

Elephant Hawkmoth *Deilephila elpenor* 80mm
This species gets its name for the slender, trunk-like 'snout', formed by the head and thorax. When alarmed, this region is pulled back into the front of the abdomen, causing the eye-spots to swell up. At the same time, the front end of the caterpillar thrashes about in a menacing fashion and this is more than enough to scare off inquisitive birds. The caterpillar is occasionally pale green instead of brown. FP: mainly willowherbs and bedstraws, but also garden fuchsias, July–September. Adult and map p. 208.

Small Elephant Hawkmoth *Deilephila porcellus* 60mm
This caterpillar is either sooty brown or greyish green and is just like that of the Elephant Hawkmoth except that it has no tail-horn. FP: mainly bedstraws, but also willowherbs, July–September. Adult and map p. 208.

Puss Moth *Cerura vinula* Family Notodontidae 65mm
The mature caterpillar is surprisingly difficult to spot among the leaves because the dark saddle-shaped patch breaks up its outline very effectively. Young caterpillars are jet black or chestnut-brown. The two 'tails' are modified claspers, and when the caterpillar is disturbed they emit fine red threads and wave them about. At the same time, the caterpillar pulls its head back into the thorax. FP: poplars and willows, July–September. Adult and map p. 210. Poplar, Sallow, and Alder Kitten larvae are similar, but smaller.

Lobster Moth *Stauropus fagi* 60mm
The long, slender front legs of this unmistakable caterpillar give the insect its name. Young caterpillars resemble large ants. FP: mainly beech, but also oak, birch and hazel, June–September. Adult and map p. 210

Buff-tip *Phalera bucephala* 60mm
The black spots on a bright yellow or orange background make this caterpillar very conspicuous, and warn of its poisonous nature. It also has an unpleasant smell. The caterpillars live gregariously until the final instar. FP: a wide range of deciduous trees, July–October. Adult and map p. 210.

Iron Prominent *Notodonta dromedarius* 35mm
The colour ranges from yellowish green to rusty brown and many individuals exhibit a mixture of both green and brown. They are easily mistaken for crumpled leaves. In common with several other prominent caterpillars, they often rest with the rear end raised. FP: mainly birch, but also alder, hazel and oak, usually June–August. A 2nd brood may occur, September–October. Adult and map p. 210.

Pebble Prominent *Eligmodonta ziczac* 40mm
The body is greyish green, often with a strong pink tinge, and the rear end (behind the 4th pair of prolegs) is often distinctly orange. There are two fairly prominent humps near the middle and a smaller one on the last segment. FP: willows and poplars, May–August. Adult and map p. 210.

Swallow Prominent *Pheosia tremula* 40mm
This caterpillar is very shiny and may be pale green or pale brown. Green forms have a pale stripe along the back and a yellow stripe on each side, while the brown forms often have pink inter-segmental membranes. All caterpillars have a little hump at the rear. FP: poplars and willows, June–October. Adult and map p. 212.

Lesser Swallow Prominent *Pheosia gnoma* 40mm
Ranging from greyish to purplish brown, this glossy caterpillar resembles the brown form of the previous species but differs in having a bright yellow stripe on each side. FP: birch, June–September. Adult and map p. 212.

Pale Prominent *Pterostoma palpina* 40mm
The body is greenish blue with 4 narrow white stripes on the back and a broader white stripe on each side. The surface is somewhat wrinkled. FP: poplars and willows, June–September. Adult and map p. 210.

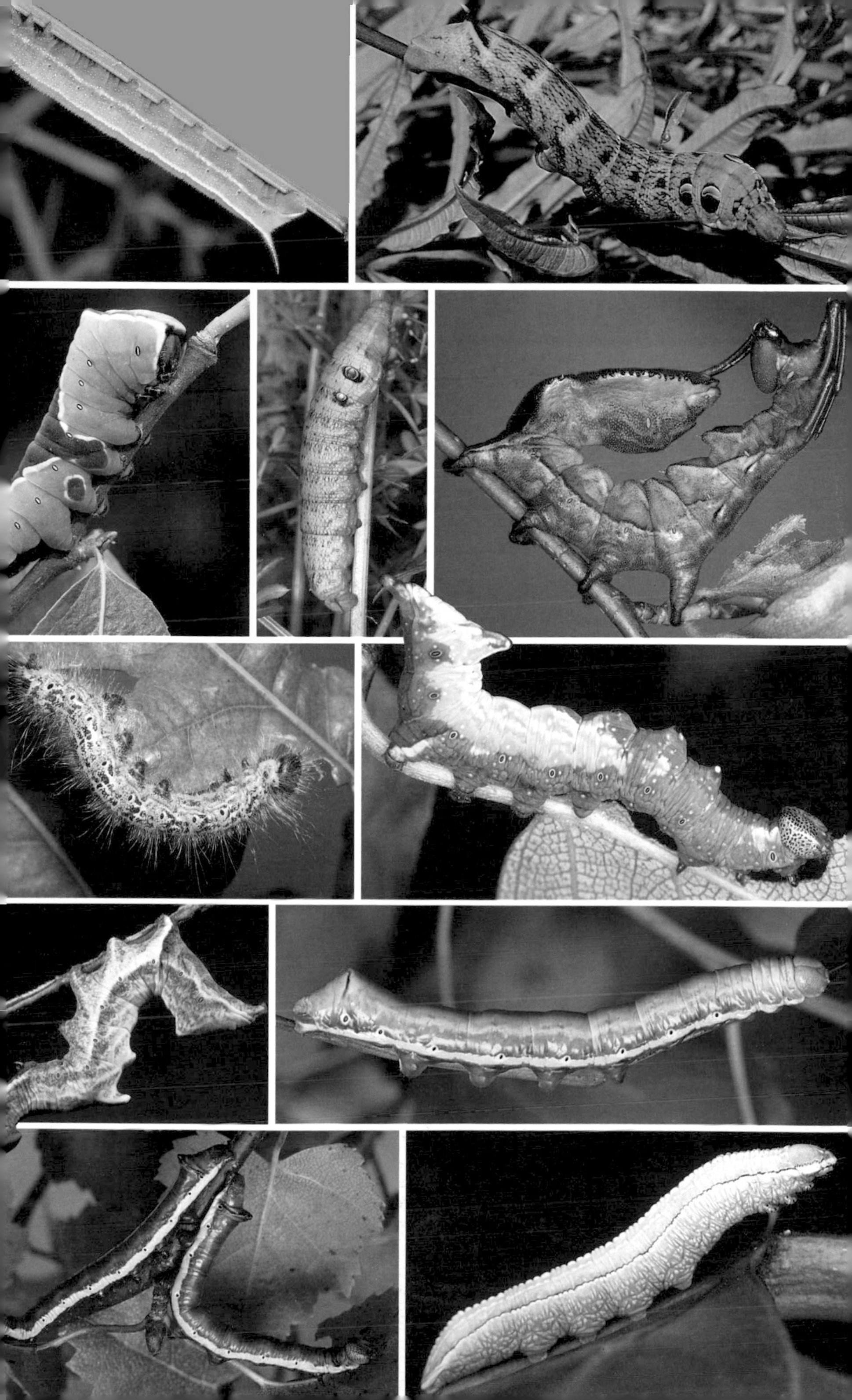

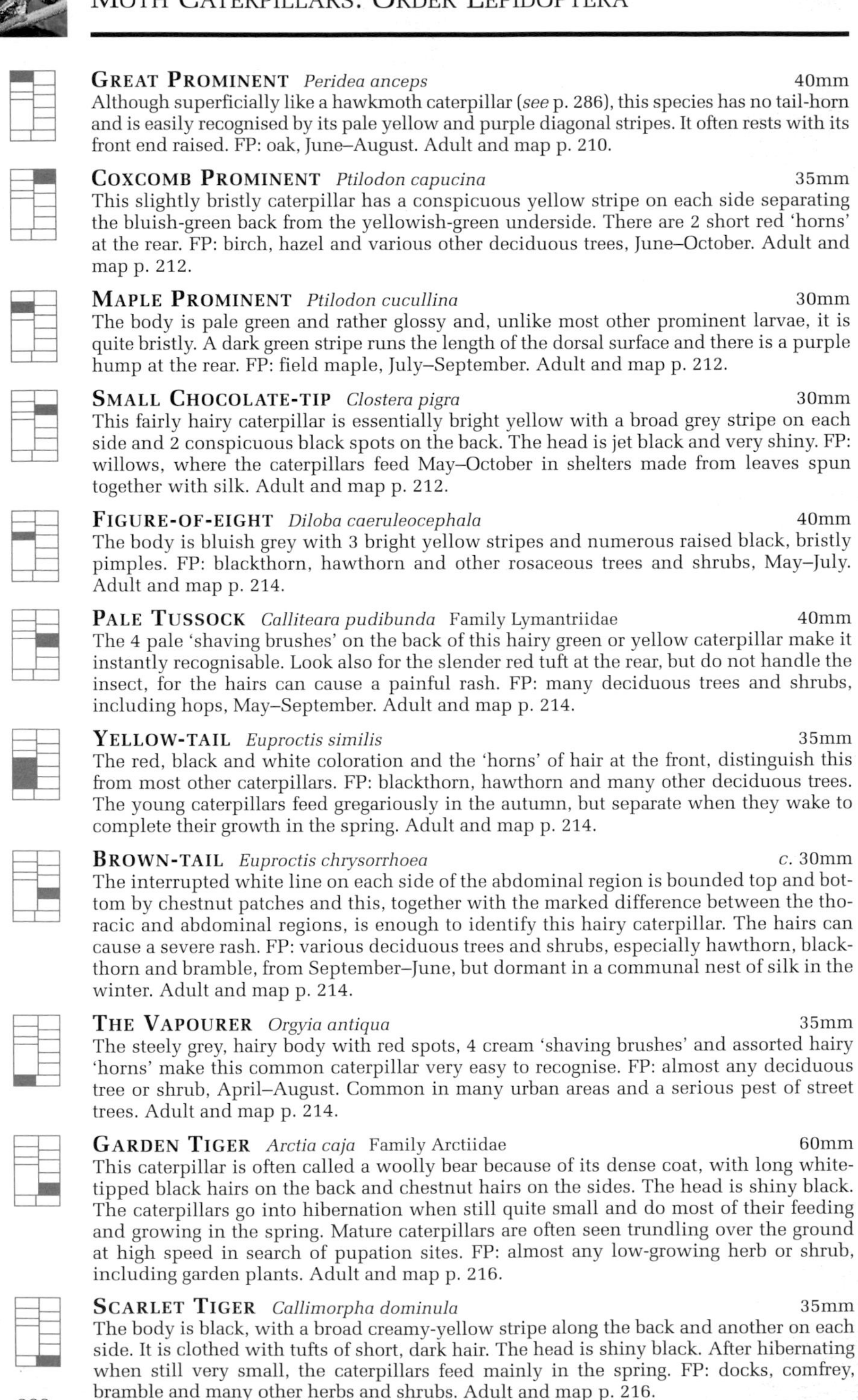

GREAT PROMINENT *Peridea anceps* 40mm
Although superficially like a hawkmoth caterpillar (*see* p. 286), this species has no tail-horn and is easily recognised by its pale yellow and purple diagonal stripes. It often rests with its front end raised. FP: oak, June–August. Adult and map p. 210.

COXCOMB PROMINENT *Ptilodon capucina* 35mm
This slightly bristly caterpillar has a conspicuous yellow stripe on each side separating the bluish-green back from the yellowish-green underside. There are 2 short red 'horns' at the rear. FP: birch, hazel and various other deciduous trees, June–October. Adult and map p. 212.

MAPLE PROMINENT *Ptilodon cucullina* 30mm
The body is pale green and rather glossy and, unlike most other prominent larvae, it is quite bristly. A dark green stripe runs the length of the dorsal surface and there is a purple hump at the rear. FP: field maple, July–September. Adult and map p. 212.

SMALL CHOCOLATE-TIP *Clostera pigra* 30mm
This fairly hairy caterpillar is essentially bright yellow with a broad grey stripe on each side and 2 conspicuous black spots on the back. The head is jet black and very shiny. FP: willows, where the caterpillars feed May–October in shelters made from leaves spun together with silk. Adult and map p. 212.

FIGURE-OF-EIGHT *Diloba caeruleocephala* 40mm
The body is bluish grey with 3 bright yellow stripes and numerous raised black, bristly pimples. FP: blackthorn, hawthorn and other rosaceous trees and shrubs, May–July. Adult and map p. 214.

PALE TUSSOCK *Calliteara pudibunda* Family Lymantriidae 40mm
The 4 pale 'shaving brushes' on the back of this hairy green or yellow caterpillar make it instantly recognisable. Look also for the slender red tuft at the rear, but do not handle the insect, for the hairs can cause a painful rash. FP: many deciduous trees and shrubs, including hops, May–September. Adult and map p. 214.

YELLOW-TAIL *Euproctis similis* 35mm
The red, black and white coloration and the 'horns' of hair at the front, distinguish this from most other caterpillars. FP: blackthorn, hawthorn and many other deciduous trees. The young caterpillars feed gregariously in the autumn, but separate when they wake to complete their growth in the spring. Adult and map p. 214.

BROWN-TAIL *Euproctis chrysorrhoea* *c.* 30mm
The interrupted white line on each side of the abdominal region is bounded top and bottom by chestnut patches and this, together with the marked difference between the thoracic and abdominal regions, is enough to identify this hairy caterpillar. The hairs can cause a severe rash. FP: various deciduous trees and shrubs, especially hawthorn, blackthorn and bramble, from September–June, but dormant in a communal nest of silk in the winter. Adult and map p. 214.

THE VAPOURER *Orgyia antiqua* 35mm
The steely grey, hairy body with red spots, 4 cream 'shaving brushes' and assorted hairy 'horns' make this common caterpillar very easy to recognise. FP: almost any deciduous tree or shrub, April–August. Common in many urban areas and a serious pest of street trees. Adult and map p. 214.

GARDEN TIGER *Arctia caja* Family Arctiidae 60mm
This caterpillar is often called a woolly bear because of its dense coat, with long white-tipped black hairs on the back and chestnut hairs on the sides. The head is shiny black. The caterpillars go into hibernation when still quite small and do most of their feeding and growing in the spring. Mature caterpillars are often seen trundling over the ground at high speed in search of pupation sites. FP: almost any low-growing herb or shrub, including garden plants. Adult and map p. 216.

SCARLET TIGER *Callimorpha dominula* 35mm
The body is black, with a broad creamy-yellow stripe along the back and another on each side. It is clothed with tufts of short, dark hair. The head is shiny black. After hibernating when still very small, the caterpillars feed mainly in the spring. FP: docks, comfrey, bramble and many other herbs and shrubs. Adult and map p. 216.

WHITE ERMINE *Spilosoma lubricipeda* 40mm
A conspicuous red or orange stripe running the length of the back distinguishes this fast-moving caterpillar from other similar species. The head and body hairs are black and the body is covered with distinct tufts of black hair. FP: a wide range of low-growing herbs, July–September. Adult and map p. 218.

BUFF ERMINE *Spilosoma luteum* 45mm
The head and hair tufts are dark brown and the dorsal stripe is yellowish and much less distinct than in the previous species. There is also a pale stripe on each side. FP: a wide range of low-growing herbaceous plants and shrubs, July–September. Adult and map p. 218. Muslin Ermine has yellowish-brown hairs and no pale stripe.

THE CINNABAR *Tyria jacobaeae* 25mm
This unmistakable caterpillar is a classic example of warning coloration (*see* p. 47). FP: ragwort June–August, feeding gregariously and freely exposed on the leaves and flower clusters. Adult and map p. 216.

FAMILY NOCTUIDAE

This family has about 400 species in the British Isles. Although some caterpillars, notably those of the *Acronicta* species, are extremely hairy, most of the larvae in this family have no more than a very short coat or a few scattered bristles. Most species feed on low-growing herbaceous plants and several of them are garden pests.

LARGE YELLOW UNDERWING *Noctua pronuba* 50mm
Usually some shade of brown, this rather flabby caterpillar can be recognised by the 2 rows of short, dark streaks on the back which are usually edged with light brown. Another row of dark streaks encloses the spiracles on each side. FP: a wide range of wild and cultivated herbaceous plants, August–May, although dormant in the coldest weather. Adult and map p. 224.

BEAUTIFUL YELLOW UNDERWING *Anarta myrtilli* 30mm
The green body, mottled with white and yellow, makes this caterpillar difficult to find among the heather shoots but, once found, it is easily recognised. FP: bell heather and ling, May–October. Adult and map p. 230.

CABBAGE MOTH *Mamestra brassicae* 45mm
The back ranges from greyish green to brown, but the underside is usually green, the 2 regions being separated by a pale line. FP: mainly brassicas, often being found right inside cabbages, June–September. Adult and map p. 230.

DOT MOTH *Melanchra persicariae* 40mm
The body colour ranges from dull green to purplish brown, but the caterpillar can usually be recognised by the rear-pointing 'V'-shaped marks on the back. A slender white line runs along the back and there is a slight hump at the rear. FP: a wide range of trees, shrubs and herbaceous plants, June–September. Adult and map p. 230.

BROOM MOTH *Melanchra pisi* 45mm
The colour ranges from dark green to purplish brown, but the caterpillar is easily recognised by its 3 bold yellow stripes. FP: broom, bracken, bramble, and many other woody and herbaceous plants, July–October. Adult and map p. 230.

BLOSSOM UNDERWING *Orthosia miniosa* 35mm
The ground colour is bluish grey to brown, but the caterpillar can always be recognised by its bold yellow and black stripes. FP: mainly oak, but also hawthorn and other rosaceous shrubs, April–June. Adult and map p. 234.

CLOUDED DRAB *Orthosia incerta* 35–40mm
The body varies from yellowish green, through bluish green, to olive or greyish green. A broad white line runs along the middle of the back and another, with a narrow black line above it, runs through the spiracles. FP: many deciduous trees and shrubs, from April–July. Adult and map p. 234.

HEBREW CHARACTER *Orthosia gothica* 40mm
The body is bright green and heavily speckled with pale yellow. It resembles the Clouded Drab but the spiracular stripe is much bolder. FP: a wide range of woody and herbaceous plants, April–August. Adult and map p. 236.

MULLEIN MOTH *Shargacucullia verbasci* 50mm
The ground colour ranges from white to a pale bluish-green. The black and yellow spots are shared with related species, but only this species has the narrow black streaks on the sides. FP: mainly mulleins, stripping the leaves and often turning the flower spikes into blackened stumps, June–August. Adult and map p. 238.

THE SPRAWLER *Asteroscopus sphinx* 50mm
The pale green body often has a bluish tinge. Three thin white lines run along the back and a broader, yellowish line runs through the spiracles. The most characteristic feature is the way in which the caterpillar throws its head and thorax back when disturbed. FP: wide range of woody plants, April–July. Adult and map p. 240.

BLAIR'S SHOULDER-KNOT *Lithophane leautieri* 40mm
There is a white line along the middle of the back and another on each side just below the spiracles. The latter are surrounded by purple blotches, and above them is a row of angular white blotches. This combination of green, white and purple blends well with the foliage and provides excellent camouflage. FP: Lawson's cypress and related trees, March–July. Adult and map p. 240.

SWORD-GRASS *Xylena exsoleta* 65mm
This colourful caterpillar is unlikely to be confused with any other, although the double row of black spots may be missing from the back. FP: numerous woody and herbaceous plants, May–August. Adult and map p. 240.

GREEN-BRINDLED CRESCENT *Allophyes oxyacanthae* 40–45mm
The body colour ranges from purplish brown to dark grey or dirty green and is broken up by an intricate pattern of lines that make the caterpillar very difficult to spot among the twigs of its food-plants. A small hump at the rear bears four tiny points, but the most diagnostic feature is the pale chevron on the 4th segment. FP: mainly hawthorn and blackthorn, feeding mostly by night. April–June. Adult and map p. 242.

POPLAR GREY *Acronicta megacephala* 35mm
This caterpillar can usually be recognised by the black dorsal band enclosing reddish spots, together with the large and rather conspicuous pale patch on the 10th segment. FP: poplars and sallows, July–September. Adult and map p. 248.

THE SYCAMORE *Acronicta aceris* 40mm
With its dense tufts of orange hair and the line of black-edged white diamonds along its back, this is an unmistakable caterpillar. FP: sycamore, maple and various other trees, July–September; one of the few caterpillars to feed on horse chestnut in Britain. Adult and map p. 248.

THE MILLER *Acronicta leporina* 40mm
This caterpillar begins life with a pale green and black pattern and blends beautifully with the partly chewed leaves. It later becomes green, and at the 3rd moult it develops long silvery or yellow hairs. It is then easily mistaken for a feather. FP: birch, alder and several other trees, July–October. Adult and map p. 246.

ALDER MOTH *Acronicta alni* 35mm
The young caterpillar is black and grey and looks like a bird dropping. The warning yellow patches and the strange paddle-shaped bristles do not develop until the final instar. FP: birch, alder and many other deciduous trees, June–August. Adult and map p. 248.

GREY DAGGER *Acronicta psi* 40mm
This caterpillar is easily identified by the broad yellow band on its back and the red spots on its sides, together with the prominent soft spike just behind the head. FP: a wide range of deciduous trees and shrubs, July–November. Adult and map p. 248.

DARK DAGGER *Acronicta tridens* 40mm
Although the adults of this and the Grey Dagger are visually indistinguishable, the larvae are quite different. The Dark Dagger larva has a white dorsal stripe with orange flashes and the hump near the front is short and broad. FP: mainly hawthorn and blackthorn and other rosaceous trees, June–September. Common in S Britain, the moth becomes local and rare in the north. The larva of the Yellow-tail (p. 290) has similar colours but it has a red line along the spiracles and it lacks the long hairs.

KNOTGRASS *Acronicta rumicis* 40mm
The alternating white and brick-coloured triangular patches below the spiracles distinguish this from all other caterpillars. There is also a line of white blotches on each side above the spiracles, and a line of red spots along the back. FP: docks and many other low-growing herbs and shrubs, May–October. Adult and map p. 248.

COPPER UNDERWING *Amphipyra pyramidea* 45mm
Bright green, often strongly tinged with blue, this caterpillar has a thin white line along the middle of the back leading to a prominent triangular, yellow-tipped hump. A broad white line runs through the spiracles and, like the other white markings, it becomes yellow towards the front. FP: mainly oak, but also other deciduous trees, April–June. Adult and map p. 248. **Svensson's Copper Underwing** (*A. berbera*) is similar but there are more obvious white lines on the back and sides and the hump has a red tip.

ANGLE SHADES *Phlogophora meticulosa* 40mm
The colour ranges from bright green to olive-green or pinkish brown, with a broken white line along the back and a pale line below the spiracles. There are oblique dark stripes above the spiracles, but the caterpillar differs from that of the Dot Moth larva (p. 292) in having no such stripes below the spiracles. The Dot Moth larva is also more angular. FP: a very wide range of woody and herbaceous plants, all year. Adult and map p. 250.

NUT-TREE TUSSOCK *Colocasia coryli* 35mm
The body is usually brick-coloured, with fine white hair and a line of dark rectangles along the back. There are 2 forward-pointing tufts of dark hair just behind the head. FP: mainly hazel and birch, but other deciduous trees may be used, May–October. Adult and map p. 262.

GOLDEN PLUSIA *Polychrisia moneta* 35mm
Tapering very strongly towards the front, this bright green caterpillar has a darker line along the back and a white line just above the spiracles. There are also narrow, yellowish rings between the segments. There are only 3 pairs of prolegs, including the claspers. FP: mainly garden delphiniums, often nibbling the bases of the veins and causing the leaves to collapse to form a shelter; mainly June–September. Adult and map p. 264.

SILVER Y *Autographa gamma* 25mm
The body ranges from yellowish green to a rather dirty greyish-green, with a slender, pale line running close to the spiracles. There is usually a black stripe on each side of the head. There are only 3 pairs of prolegs, including the claspers. FP: many wild and cultivated herbaceous plants, May–October. Adult and map p. 264.

THE BLACKNECK *Lygephila pastinum* 45–50mm
With its heavily-spotted, blue-grey and orange upper surface and brown underside, this caterpillar is unlikely to be mistaken for any other. The 1st pair of prolegs are noticeably shorter than the rest. FP: mainly tufted vetch, usually feeding by night in April–May, having passed the winter as a very small larva. Adult and map p. 266.

RED UNDERWING *Catocala nupta* 60–70mm
Tapering towards each end, this greyish-brown caterpillar clings tightly to the twigs of its food-plant and a fringe of hairlike outgrowths obliterates any shadow beneath its body, making it very difficult to spot. Bud-like 'warts' heighten its similarity to a twig. FP: mainly willows and poplars from Aprial–July. Adult and map p. 234.

MOTHER SHIPTON *Callistege mi* *c.* 40mm
The body ranges from off-white to pale yellow, with a number of brownish lines of variable thickness running along its length. There is also a pale yellow band below the spiracles. There are just three pairs of prolegs. FP: clovers and other legumes, June to September. Adult and map p. 264.

THE HERALD *Scoliopteryx libatrix* *c.* 50mm
This rich velvety-green caterpillar is not easy to spot because the lines on the back and sides effectively break up its outline among the leaves. FP: poplars and sallows, June–September. Adult and map p. 266.

TRUE FLIES: ORDER DIPTERA

CRANE-FLIES: FAMILY TIPULIDAE

This family of slim-bodied, long-legged flies can be recognised by the V-shaped groove on the thorax. The largest ones, commonly known as daddy-long-legs, usually rest with their wings extended, but most fold their wings flat over the body. Their legs often break off when handled. Adult crane-flies rarely feed, although some may lap nectar from flowers. The larvae are mostly subterranean or aquatic scavengers and some cause serious damage to farm crops and garden plants. Although some crane-flies are very large, most of the 300 or so species in the British Isles are mosquito-sized.

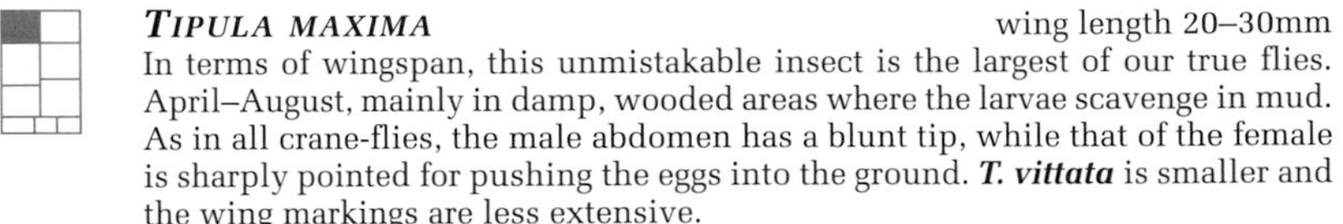

TIPULA MAXIMA wing length 20–30mm
In terms of wingspan, this unmistakable insect is the largest of our true flies. April–August, mainly in damp, wooded areas where the larvae scavenge in mud. As in all crane-flies, the male abdomen has a blunt tip, while that of the female is sharply pointed for pushing the eggs into the ground. ***T. vittata*** is smaller and the wing markings are less extensive.

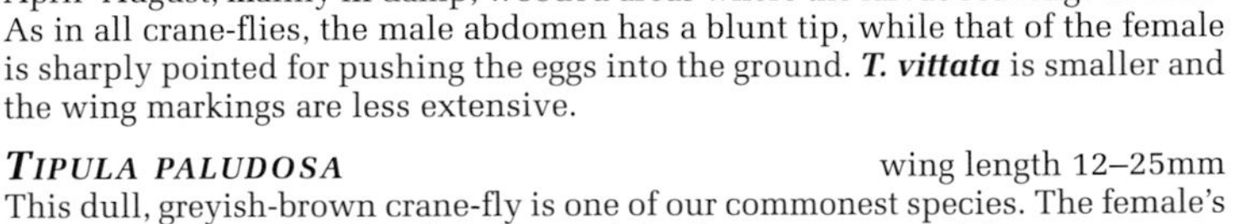

TIPULA PALUDOSA wing length 12–25mm
This dull, greyish-brown crane-fly is one of our commonest species. The female's wings are shorter than the abdomen and the first 2 segments of the antennae are brick-coloured. May–October, but most common in the autumn, when it is abundant in fields, parks and gardens and often enters houses. It is harmless, but its larvae are the infamous leatherjackets that damage lawns and crops by chewing the roots.

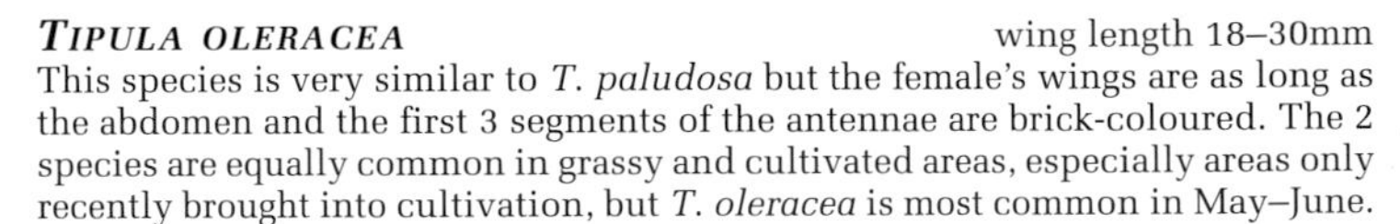

TIPULA OLERACEA wing length 18–30mm
This species is very similar to *T. paludosa* but the female's wings are as long as the abdomen and the first 3 segments of the antennae are brick-coloured. The 2 species are equally common in grassy and cultivated areas, especially areas only recently brought into cultivation, but *T. oleracea* is most common in May–June.

SPOTTED CRANE-FLY *Nephrotoma appendiculata* 15–20mm
The wings of this abundant garden pest are laid flat over the body at rest, but the yellow and black abdomen usually remains visible through the shiny wings. May–August in many well-vegetated habitats as well as gardens. ***N. quadrifaria*** has a dark smudge on the wing.

NEPHROTOMA CROCATA 15–20mm
The black and yellow banding on the abdomen is very conspicuous. There is a dark smudge on the wing just behind the pterostigma and the wing-tips are also lightly clouded. May–August in many habitats but especially in damp areas.

CTENOPHORA ORNATA 20–25mm
The black and yellow abdomen narrows markedly at the front and there is a brown smudge near the wing-tip. Males are easily identified by their large, feathery antennae; female antennae are toothed. May–July in and around woodlands, but not common. The larvae live in decaying timber.

ST MARK'S-FLY *Bibio marci* Family Bibionidae 10–15mm
Named for its appearance on or around St Mark's day (April 25th), this jet-black fly is most often seen drifting over the vegetation at about head height, with its long legs dangling limply beneath it. The front leg has a stout spine at the tip of the tibia. Females (A) have smoky wings and very much smaller heads and eyes than the males (B). Abundant April–June in rough grassland, hedgerows and woodland margins. The larvae live in the soil and in rotting vegetation. The **Fever-fly** (*Dilophus febrilis*) is smaller and can be distinguished by the circle of spines at the tip of the front tibia. It swarms over low-growing vegetation, mainly in spring and late summer.

BIBIO ANGLICUS 5–10mm
The shiny black male looks like a small St Mark's-fly, but the female's body is largely brick-coloured with a black head. Both sexes have black hair on the abdomen. The front edge of the wing is brown. April–May, often abundant on the flower heads of alexanders and related plants. ***B. hortulanus*** is similar but the abdominal hairs are pale in both sexes. It flies a little later in the year.

MOSQUITOES: FAMILY CULICIDAE

Mosquitoes are often lumped together with a lot of other small flies as gnats and midges, although these insects belong to several quite different families. The mosquitoes differ from the others in having scales on the veins and rear margins of the wings. The latter are laid flat over the body at rest. Female mosquitoes are bloodsuckers and each is equipped with a long, forward-pointing proboscis with which she pierces her victims. The males, recognisable by the bushy antennae with which they pick up the whining sounds of the females, also have long proboscі, but they feed only on nectar. The males have long palps as well and these are often ornately hairy. The insects all have aquatic larvae and several species breed in gardens ponds and water butts. There are about 30 species in the British Isles belonging to two distinct sub-families. The anophelines, which can transmit malaria, rest with their bodies at an angle to the surface, and the females have long, plain palps. The culicine mosquitoes, which include all but 5 of the British species, rest with their bodies more or less parallel to the surface. Female culicines have very short palps.

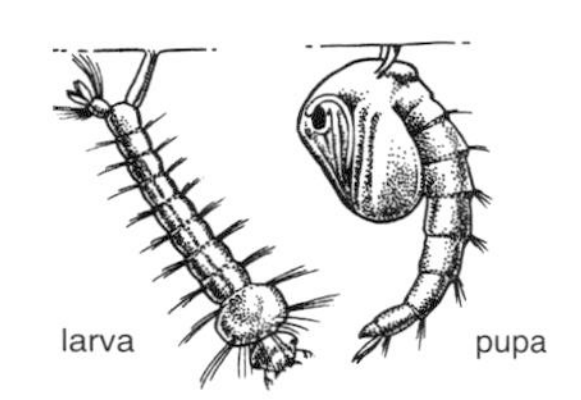

CULEX PIPIENS — *c.* 6mm

This is our commonest mosquito, distinguished from most other brownish species by the white band at the front of each abdominal segment. In common with most of our mosquitoes, it rests with its body more or less parallel to the surface. All year, almost everywhere, often hibernating in houses, although it rarely bites people.

CULISETA ANNULATA — 8–10mm

This is our largest mosquito, recognisable by its size and by its spotted wings and boldly banded legs and abdomen. All year in many habitats, including houses, but usually dormant in winter. It bites readily and causes severe reactions in some people.

AEDES CANTANS — 6–8mm

The abdomen is blackish brown, with a pale band at the front of each segment. The wing scales are a mixture of light and dark and the male has very ornate palps with swollen tips. The pale rings on the hind tarsi occupy about one third of each segment. March–October, mainly on marshes and in damp woodland where it bites readily. There are several very similar species.

CHIRONOMUS LURIDUS Family Chironomidae — *c.* 8mm

Although often confused with mosquitoes, this insect and its numerous relatives are easily distinguished by their short wings, which are held roofwise or almost vertically along the sides of the body at rest. The males have very bushy antennae but no long palps. Neither sex has a proboscis and the insects do not bite. Not all species have green bodies. April–October wherever there is standing water. Chironomids, often known as non-biting midges, breed in water and their red larvae are known as bloodworms.

WINTER GNAT *Trichocera annulata* — *c.* 8mm

This insect has a V-shaped groove on the thorax and resembles a small crane-fly, but its legs do not break off readily and the hindmost vein bends sharply back to the wing margin. The banded abdomen and the unspotted wings distinguish this from several related species. Abundant nearly everywhere throughout the year, but especially common in late autumn and winter when males gather in large swarms and dance up and down to attract females.

JAAPIELLA VERONICAE Family Cecidomyiidae — *c.* 2mm

This is one of a huge number of minute flies called gall midges, most often seen when they swarm to lighted windows on summer nights. The wings are hairy but have very few veins. The larvae cause the formation of hairy grey galls in the terminal buds of germander speedwell. May–August. The very similar ***Taxomyia taxi*** causes galls in terminal buds of yew trees.

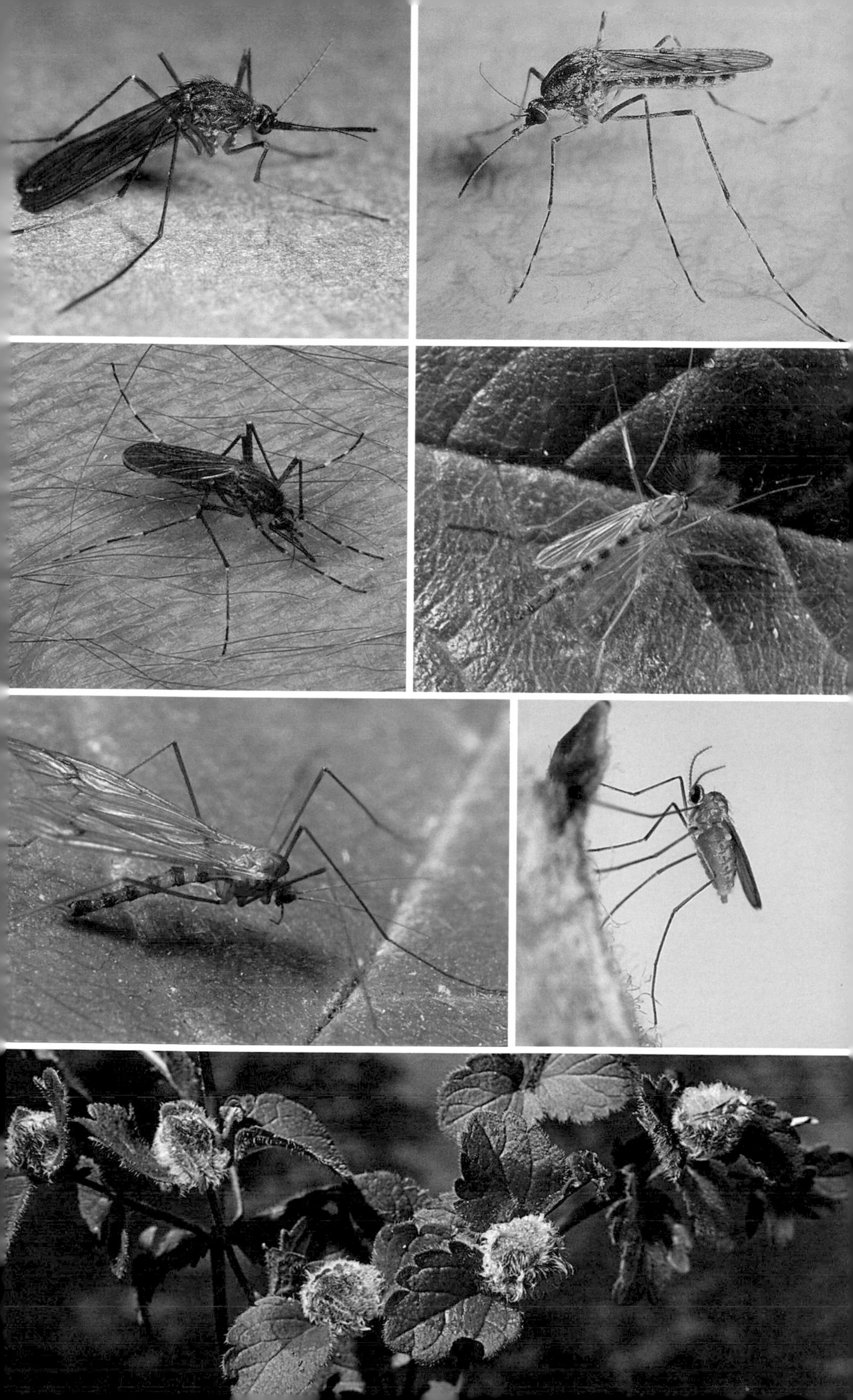

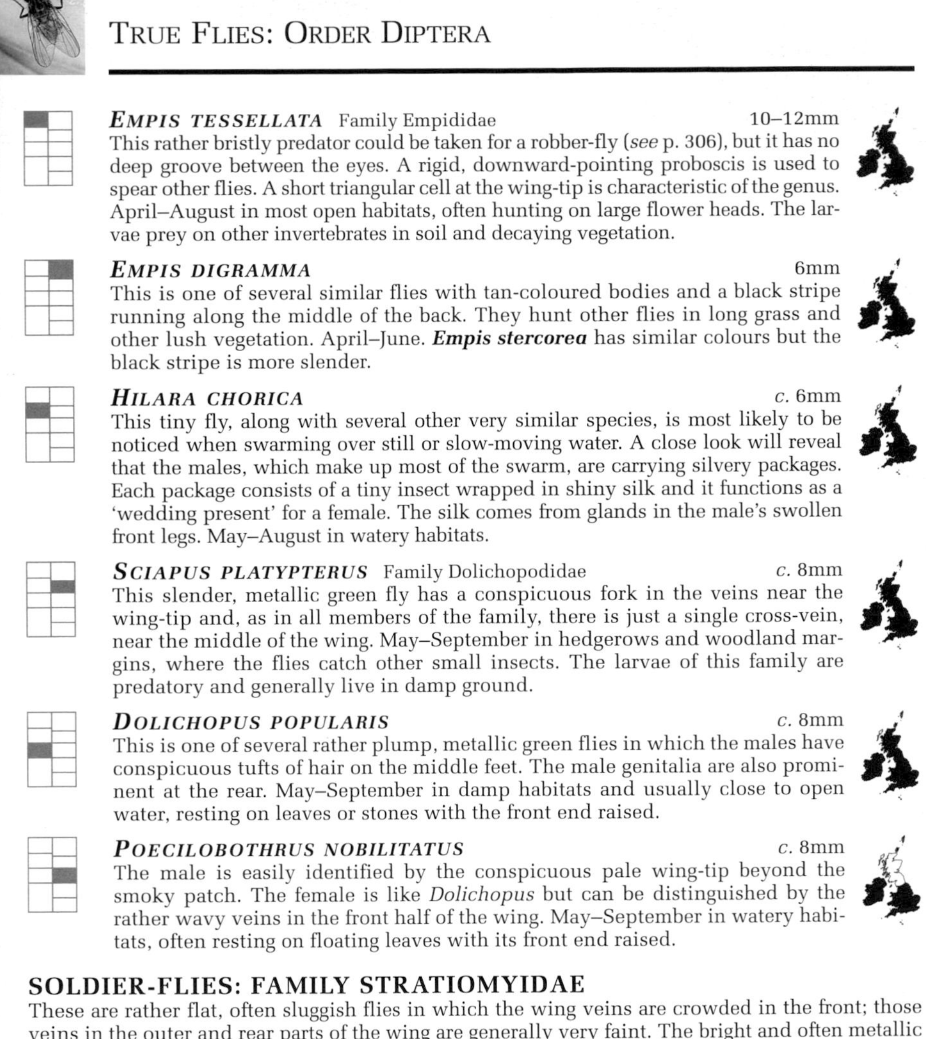

EMPIS TESSELLATA Family Empididae 10–12mm
This rather bristly predator could be taken for a robber-fly (*see* p. 306), but it has no deep groove between the eyes. A rigid, downward-pointing proboscis is used to spear other flies. A short triangular cell at the wing-tip is characteristic of the genus. April–August in most open habitats, often hunting on large flower heads. The larvae prey on other invertebrates in soil and decaying vegetation.

EMPIS DIGRAMMA 6mm
This is one of several similar flies with tan-coloured bodies and a black stripe running along the middle of the back. They hunt other flies in long grass and other lush vegetation. April–June. ***Empis stercorea*** has similar colours but the black stripe is more slender.

HILARA CHORICA *c.* 6mm
This tiny fly, along with several other very similar species, is most likely to be noticed when swarming over still or slow-moving water. A close look will reveal that the males, which make up most of the swarm, are carrying silvery packages. Each package consists of a tiny insect wrapped in shiny silk and it functions as a 'wedding present' for a female. The silk comes from glands in the male's swollen front legs. May–August in watery habitats.

SCIAPUS PLATYPTERUS Family Dolichopodidae *c.* 8mm
This slender, metallic green fly has a conspicuous fork in the veins near the wing-tip and, as in all members of the family, there is just a single cross-vein, near the middle of the wing. May–September in hedgerows and woodland margins, where the flies catch other small insects. The larvae of this family are predatory and generally live in damp ground.

DOLICHOPUS POPULARIS *c.* 8mm
This is one of several rather plump, metallic green flies in which the males have conspicuous tufts of hair on the middle feet. The male genitalia are also prominent at the rear. May–September in damp habitats and usually close to open water, resting on leaves or stones with the front end raised.

POECILOBOTHRUS NOBILITATUS *c.* 8mm
The male is easily identified by the conspicuous pale wing-tip beyond the smoky patch. The female is like *Dolichopus* but can be distinguished by the rather wavy veins in the front half of the wing. May–September in watery habitats, often resting on floating leaves with its front end raised.

SOLDIER-FLIES: FAMILY STRATIOMYIDAE

These are rather flat, often sluggish flies in which the wing veins are crowded in the front; those veins in the outer and rear parts of the wing are generally very faint. The bright and often metallic colours of many species are thought to be responsible for the name, for they were once thought to resemble the colours of military uniforms. The insects are most frequently seen basking on vegetation. The larvae are scavengers or carnivores living mainly in damp ground or in dung and rotting vegetation, including garden compost heaps. There are about 50 species in the British Isles.

STRATIOMYS SINGULARIOR *c.* 15mm
The narrow yellow spots on the sides of the abdomen give this fly a resemblance to various hover-flies (*see* p. 308), but the venation is quite different, as are the long, elbowed antennae. July and August, in and around saltmarshes and other brackish water. The larvae are aquatic. ***S. potamida*** and ***S. chamaeleon*** have much larger yellow spots at the front of the abdomen.

CHLOROMYIA FORMOSA 8–10mm
Although this fly is quite hairy, the metallic colours are always visible. The female is entirely green with blue and purple iridescence, but the male has a golden abdomen. May–August in damp, well-vegetated places, including gardens.

OPLODONTHA VIRIDULA 6–8mm
The abdominal colour ranges from cream to green or orange, but the insect can always be recognised by the bold black stripe along the top. June–August in watery habitats, where the fly commonly rests on reeds. The larvae are aquatic.

SNIPE-FLY *Rhagio scolopaceus* Family Rhagionidae 10–12mm
This fly and its relatives can be recognised by the way in which the 2nd long vein curves forward to enclose the lightly pigmented pterostigma. May–August, mainly in and around woods and hedgerows. It commonly rests in a head-down position and has been dubbed the 'down-looker-fly'. It darts from its perch to snatch other insects in mid-air. The soil-dwelling larvae are also carnivorous.

HORSE-FLIES: FAMILY TABANIDAE

These fast-flying, stoutly-built flies are sometimes called stouts. They have horn-like antennae and large eyes that commonly exhibit brilliantly iridescent colours in life. At rest, the wings are swept back and held slightly apart, making it easy to see the characteristic forking of the veins across the wing-tips. The females have blade-like mouthparts and are avid blood-feeders but the males, which often have different patterns from the females, feed on nectar. The larvae live as scavengers or predators in damp soil or decaying vegetation. There are about 28 species in the British Isles.

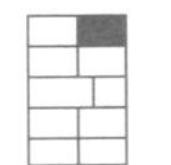

TABANUS BOVINUS *c.* 20mm
The abdomen is largely brick-coloured, with pale triangles in the middle that reach nearly to the front of each segment. The eyes are unbanded. June–August, mainly in and around grazing pastures and rarely far from water.

TABANUS SUDETICUS 20–25mm
The abdomen of this very large horse-fly is largely black and the pale triangles on the upper surface hardly reach the middle of each segment. July–August in grasslands and open woodland, especially in upland areas.

TABANUS BROMIUS 12–15mm
The abdomen is usually dark with 3 rows of pale triangles, although the markings vary a good deal and in some individuals the pale areas are more extensive than the dark ones. Each eye usually has a single reddish band in life. June–August in damp pastures.

CLEG-FLY *Haematopota pluvialis* *c.* 10mm
This is one of several similar horse-flies with mottled wings that are folded roof-wise over the body at rest. The more or less rectangular cell in the centre of the wing contains 4 pale streaks, whereas ***H. crassicornis***, the only other common species, has a pale 'V' or 'Y' in the cell. Unlike *Tabanus* species, *Haematopota* flies silently and the first indication we get of its presence is a sharp jab from its mouthparts. May–September, especially in damp, shady places.

CHRYSOPS RELICTUS 10–12mm
Long antennae, banded wings, and brilliant green and purple eyes characterise the 4 species of *Chrysops* found in the British Isles. They all have a yellow midriff as well. *C. relictus* has brownish middle tibiae and 2 black lobes on the 2nd abdominal segment, although the pattern differs slightly between the sexes. May–September, rarely far from water and most common on bogs and moorland.

CHRYSOPS CAECUTIENS 10–12mm
This species differs from *C. relictus* in having black middle tibiae and an inverted 'V' on the 2nd abdominal segment, although this mark is often indistinct in the male. May–September in many habitats but rarely far from water. ***C. viduatus*** has brownish middle tibiae and a rectangular black spot on the 2nd abdominal segment.

THEREVA NOBILITATA Family Therevidae 12–15mm
This is one of several similar species that are easily confused with robber-flies (*see* p. 306), although they are much more bristly about the abdomen and they lack the robber-flies' characteristic facial groove. The eyes of the males touch. Much of the thorax in *T. nobilitata* has yellow 'fur', as does the female's abdomen. May–August in many habitats.

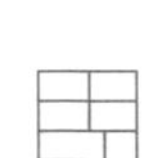

THEREVA ANNULATA 10–12mm
The silvery blue hair all over the thorax and abdomen makes this species easy to recognise. May–August, mainly on coastal dunes.

ROBBER-FLIES: FAMILY ASILIDAE

These predatory flies are equipped with a sturdy proboscis that is used to pierce other insects and suck them dry. Most species sit on a perch and dart out to catch their prey on the wing with the aid of their bristly legs. A bristly mound or 'beard' on the face keeps the struggling prey away from the robber-fly's eyes. There is a deep groove between the eyes in both sexes, so even the males' eyes are not in contact. The wings are folded flat over the body at rest. Robber-fly larvae generally live in the soil and are omnivorous, although animal matter probably makes up the bulk of their food. There are 27 species on the British list.

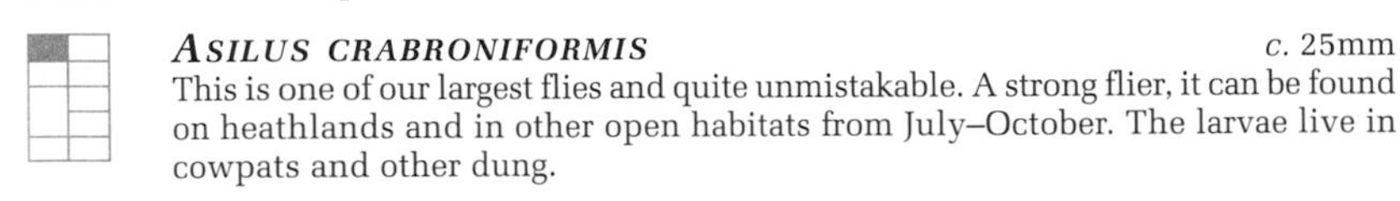

ASILUS CRABRONIFORMIS — *c.* 25mm

This is one of our largest flies and quite unmistakable. A strong flier, it can be found on heathlands and in other open habitats from July–October. The larvae live in cowpats and other dung.

LAPHRIA FLAVA — *c.* 25mm

This large, furry species is not likely to be confused with any other fly, although it could be mistaken for a bumble-bee. June–September in pinewoods, where the larvae live in decaying logs and stumps.

DYSMACHUS TRIGONUS — *c.* 15mm

This species is unusually bristly. Its whole body is clothed with stiff bristles and there are conspicuous white tufts around the head and at the rear of the strongly-domed thorax. May–August, mainly in sandy habitats.

NEOITAMUS CYANURUS — 10–15mm

The dark abdomen has a pale band at the front of each segment and the male has a silvery blue patch at the rear. The female has a long telescopic ovipositor. June–August, in and around oak woods, where weevils are among their commonest victims.

DIOCTRIA RUFIPES — 15–18mm

The yellow front and middle femora and the black hind femora distinguish this from other *Dioctria* species, all of which are rather bare and rather different from most other robber-flies. The white 'beard' is conspicuous, although not particularly dense. April–July in a variety of grassy places.

LEPTOGASTER CYLINDRICA — 12–15mm

This extremely slender species is even less like a robber-fly than *Dioctria*, and can easily be mistaken for a mosquito or a small crane-fly. June–August in rough grassland, flying slowly between the grass stems and plucking aphids from them. ***L. guttiventris*** has reddish hind femora instead of yellow. Both species darken with age.

BEE-FLY *Bombylius major* Family Bombyliidae — 10–12mm

The furry brown body and the long proboscis, together with the dark brown front edges of the wings make this fly very easy to recognise. Although the rigid proboscis alarms many people, the fly is quite harmless. March–June in many habitats, including gardens. It hovers with a high-pitched whine and feeds at a wide range of low-growing flowers. Although appearing to hover while feeding, it usually clings to the flowers with its spindly legs. The larvae live as parasitoids in the nests of mining bees. The rare ***B. discolor*** has dark spots on its wings as well as the brown front margin.

HEATH BEE-FLY *Bombylius minor* — 7–9mm

This rare fly has only a very pale brown smudge at the base of each wing. July–August on heathland. ***B. canescens*** is much more widely distributed and can be distinguished by the tufts of black bristles just behind the eyes.

THYRIDANTHRAX FENESTRATUS — 10–12mm

Easily identified by its mottled wings, this fly darts and hovers like *Bombylius* but has only a short proboscis. June–September, mainly in sandy habitats. The larvae live in the nests of sand wasps (*see* p. 338), feeding either on the sand wasp grubs or on the caterpillars stored in the nests.

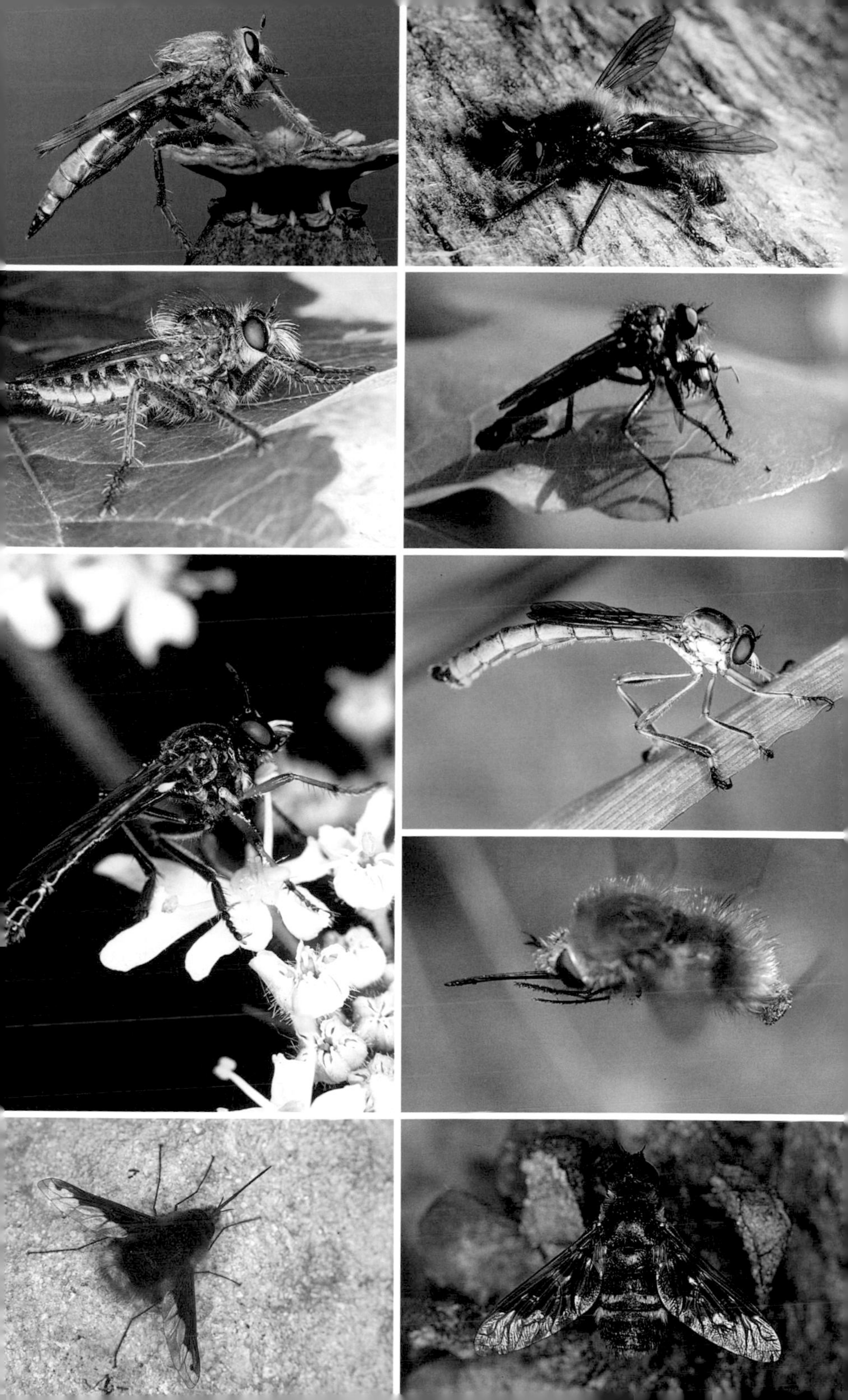

HOVER-FLIES: FAMILY SYRPHIDAE

The hover-flies include some of our most colourful and spectacular flies, although not all of them are brightly coloured. There are about 250 species in the British Isles and, as well as their remarkable hovering ability, most of them display a swift, darting flight. The insects can be recognised by the false margin, formed by veins running close to and more or less parallel to the hind edge of the wing, and also by the false vein running through the middle of the wing. Unconnected to any of the true veins, this is a thickened strip of the membrane giving extra support to the wing. Many hover-flies are excellent mimics of bees and wasps (*see* p. 50), although easily distinguished by the much shorter, drooping antennae. The adults feed mainly on nectar. The larvae are extremely varied in both appearance and behaviour and include carnivores, vegetarians and scavengers. Unless otherwise stated, the larvae of all species described here feed on aphids.

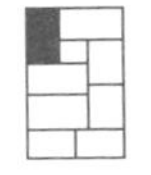

SYRPHUS RIBESII — *c.* 15mm

The first 2 long veins reach the wing margin and the squamae bear long hairs, although a lens is necessary to see them. April–November nearly everywhere. This is one of the commoner of many wasp-like hover-flies. ***S. vitripennis*** is almost identical but the female has largely black hind femora instead of the yellow of *S. ribesii.*

SCAEVA PYRASTRI — *c.* 15mm

The crescent-shaped white or cream abdominal spots, with the inner arm reaching further forwards than the outer one, distinguish this from other hover-flies. June–October in flowery places nearly everywhere; common in gardens. ***S. selenitica*** has a similar pattern but the outer arms of the crescents taper strongly and both arms reach forward to the same extent.

EPISYRPHUS BALTEATUS — 10–15mm

This slender species is easily identified by the abdominal pattern, with an additional narrow black band on the 3rd and 4th segments. March–November almost everywhere; abundant in gardens. The species sometimes migrates in swarms from continental Europe.

EUPEODES NITENS — *c.* 10mm

The yellow bands are often much narrower than pictured here and are not always broken in the middle. The abdomen is completely edged with black. May–September, mainly in ancient woodland.

XANTHOGRAMMA PEDISSEQUUM — 10–12mm

The yellow triangles at the front of the abdomen, together with the dark smudge on the wing, the yellow stripes on the sides of the thorax, and the yellow scutellum, separate this from our other black and yellow hover-flies. May–September in grassy places; uncommon.

LEUCOZONA LUCORUM — 12–15mm

The white or cream belt and the brown scutellum should separate this species from any other hover-fly, but *see Volucella pellucens* on p. 312. A dark band runs across the front half of the wing and the first 2 long veins both reach the margin. May–August, especially in woodland rides and hedgerows.

LEUCOZONA GLAUCIA — 12–15mm

The blue abdominal bands and yellowish scutellum make the female of this species unmistakable. The male has white patches on the abdomen, but is still easily recognised by the yellowish scutellum and the black pterostigma. June–September, mainly in woodland rides and margins.

MELANOSTOMA SCALARE — 8–10mm

The male has a very slender, parallel-sided abdomen with 3 pairs of yellow spots. The female abdomen gradually widens towards the tip and the 4 posterior spots are more-or-less triangular. April–November nearly everywhere.

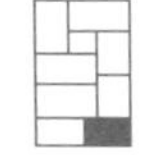

SPHAEROPHORIA SCRIPTA — *c.* 12mm

There is always a yellow stripe on each side of the thorax but the abdominal markings vary a great deal. The male is easily identified by the long abdomen reaching way beyond the folded wings. Abundant in all kinds of grassland, including urban wasteland, May–October. The larvae feed mainly on aphids.

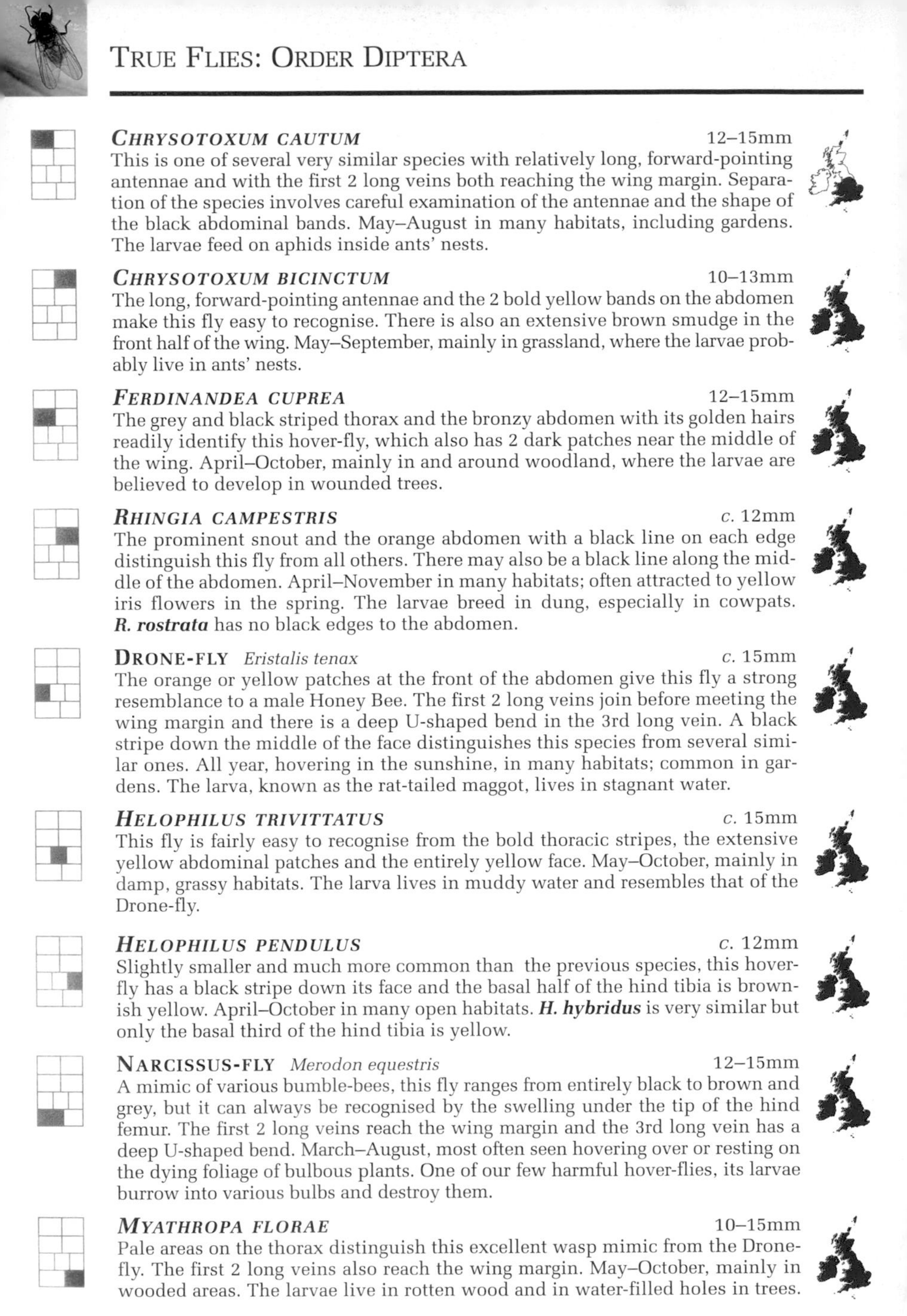

CHRYSOTOXUM CAUTUM 12–15mm
This is one of several very similar species with relatively long, forward-pointing antennae and with the first 2 long veins both reaching the wing margin. Separation of the species involves careful examination of the antennae and the shape of the black abdominal bands. May–August in many habitats, including gardens. The larvae feed on aphids inside ants' nests.

CHRYSOTOXUM BICINCTUM 10–13mm
The long, forward-pointing antennae and the 2 bold yellow bands on the abdomen make this fly easy to recognise. There is also an extensive brown smudge in the front half of the wing. May–September, mainly in grassland, where the larvae probably live in ants' nests.

FERDINANDEA CUPREA 12–15mm
The grey and black striped thorax and the bronzy abdomen with its golden hairs readily identify this hover-fly, which also has 2 dark patches near the middle of the wing. April–October, mainly in and around woodland, where the larvae are believed to develop in wounded trees.

RHINGIA CAMPESTRIS *c.* 12mm
The prominent snout and the orange abdomen with a black line on each edge distinguish this fly from all others. There may also be a black line along the middle of the abdomen. April–November in many habitats; often attracted to yellow iris flowers in the spring. The larvae breed in dung, especially in cowpats. ***R. rostrata*** has no black edges to the abdomen.

DRONE-FLY *Eristalis tenax* *c.* 15mm
The orange or yellow patches at the front of the abdomen give this fly a strong resemblance to a male Honey Bee. The first 2 long veins join before meeting the wing margin and there is a deep U-shaped bend in the 3rd long vein. A black stripe down the middle of the face distinguishes this species from several similar ones. All year, hovering in the sunshine, in many habitats; common in gardens. The larva, known as the rat-tailed maggot, lives in stagnant water.

HELOPHILUS TRIVITTATUS *c.* 15mm
This fly is fairly easy to recognise from the bold thoracic stripes, the extensive yellow abdominal patches and the entirely yellow face. May–October, mainly in damp, grassy habitats. The larva lives in muddy water and resembles that of the Drone-fly.

HELOPHILUS PENDULUS *c.* 12mm
Slightly smaller and much more common than the previous species, this hover-fly has a black stripe down its face and the basal half of the hind tibia is brownish yellow. April–October in many open habitats. ***H. hybridus*** is very similar but only the basal third of the hind tibia is yellow.

NARCISSUS-FLY *Merodon equestris* 12–15mm
A mimic of various bumble-bees, this fly ranges from entirely black to brown and grey, but it can always be recognised by the swelling under the tip of the hind femur. The first 2 long veins reach the wing margin and the 3rd long vein has a deep U-shaped bend. March–August, most often seen hovering over or resting on the dying foliage of bulbous plants. One of our few harmful hover-flies, its larvae burrow into various bulbs and destroy them.

MYATHROPA FLORAE 10–15mm
Pale areas on the thorax distinguish this excellent wasp mimic from the Drone-fly. The first 2 long veins also reach the wing margin. May–October, mainly in wooded areas. The larvae live in rotten wood and in water-filled holes in trees.

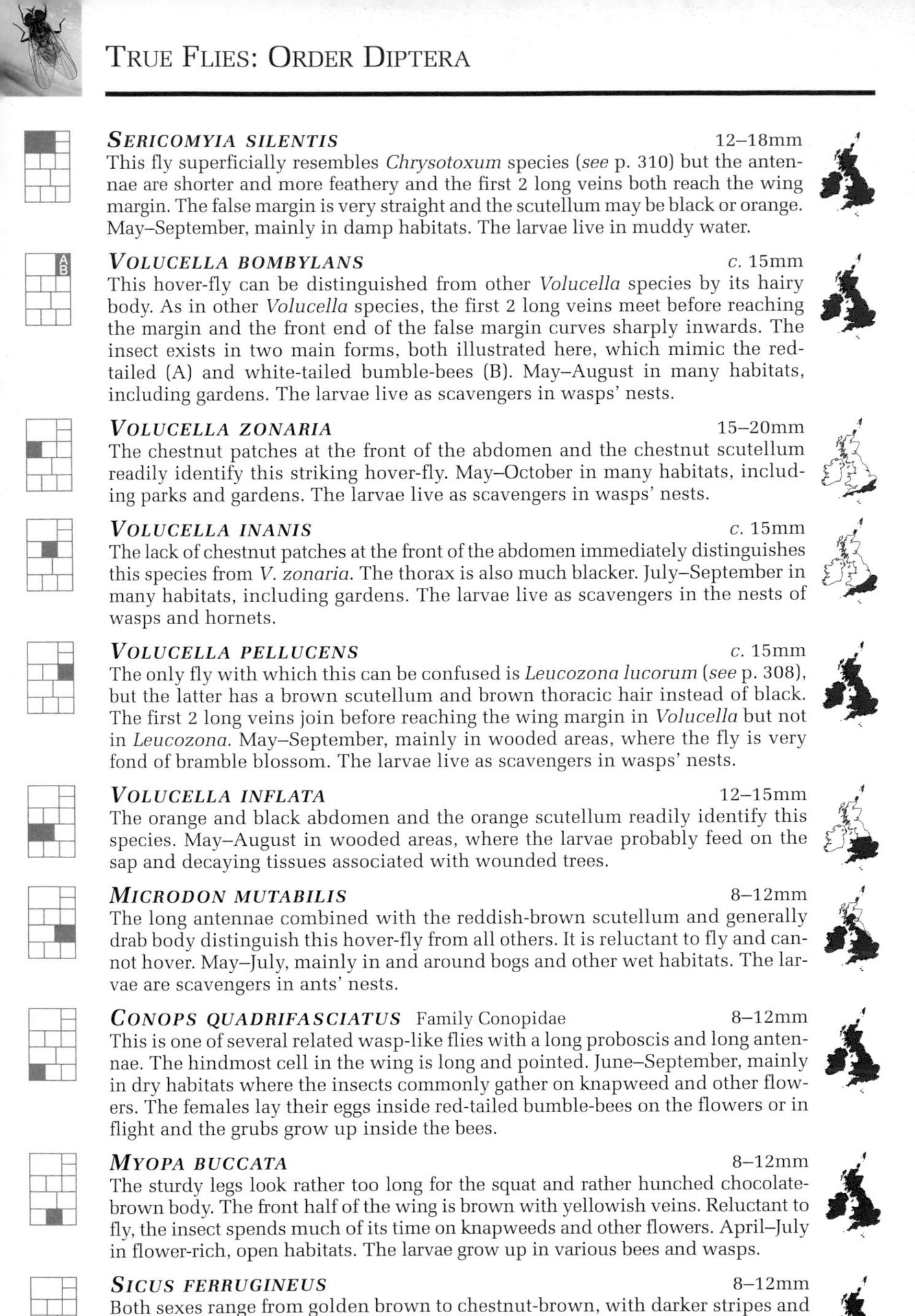

SERICOMYIA SILENTIS 12–18mm

This fly superficially resembles *Chrysotoxum* species (*see* p. 310) but the antennae are shorter and more feathery and the first 2 long veins both reach the wing margin. The false margin is very straight and the scutellum may be black or orange. May–September, mainly in damp habitats. The larvae live in muddy water.

VOLUCELLA BOMBYLANS *c.* 15mm

This hover-fly can be distinguished from other *Volucella* species by its hairy body. As in other *Volucella* species, the first 2 long veins meet before reaching the margin and the front end of the false margin curves sharply inwards. The insect exists in two main forms, both illustrated here, which mimic the red-tailed (A) and white-tailed bumble-bees (B). May–August in many habitats, including gardens. The larvae live as scavengers in wasps' nests.

VOLUCELLA ZONARIA 15–20mm

The chestnut patches at the front of the abdomen and the chestnut scutellum readily identify this striking hover-fly. May–October in many habitats, including parks and gardens. The larvae live as scavengers in wasps' nests.

VOLUCELLA INANIS *c.* 15mm

The lack of chestnut patches at the front of the abdomen immediately distinguishes this species from *V. zonaria*. The thorax is also much blacker. July–September in many habitats, including gardens. The larvae live as scavengers in the nests of wasps and hornets.

VOLUCELLA PELLUCENS *c.* 15mm

The only fly with which this can be confused is *Leucozona lucorum* (*see* p. 308), but the latter has a brown scutellum and brown thoracic hair instead of black. The first 2 long veins join before reaching the wing margin in *Volucella* but not in *Leucozona*. May–September, mainly in wooded areas, where the fly is very fond of bramble blossom. The larvae live as scavengers in wasps' nests.

VOLUCELLA INFLATA 12–15mm

The orange and black abdomen and the orange scutellum readily identify this species. May–August in wooded areas, where the larvae probably feed on the sap and decaying tissues associated with wounded trees.

MICRODON MUTABILIS 8–12mm

The long antennae combined with the reddish-brown scutellum and generally drab body distinguish this hover-fly from all others. It is reluctant to fly and cannot hover. May–July, mainly in and around bogs and other wet habitats. The larvae are scavengers in ants' nests.

CONOPS QUADRIFASCIATUS Family Conopidae 8–12mm

This is one of several related wasp-like flies with a long proboscis and long antennae. The hindmost cell in the wing is long and pointed. June–September, mainly in dry habitats where the insects commonly gather on knapweed and other flowers. The females lay their eggs inside red-tailed bumble-bees on the flowers or in flight and the grubs grow up inside the bees.

MYOPA BUCCATA 8–12mm

The sturdy legs look rather too long for the squat and rather hunched chocolate-brown body. The front half of the wing is brown with yellowish veins. Reluctant to fly, the insect spends much of its time on knapweeds and other flowers. April–July in flower-rich, open habitats. The larvae grow up in various bees and wasps.

SICUS FERRUGINEUS 8–12mm

Both sexes range from golden brown to chestnut-brown, with darker stripes and spots on the thorax and a broad, pale face between and below the eyes. Females are longer and slimmer than males. May–September in rough grassland, especially that of woodland rides and margins. Various bumble-bees serve as hosts for the larvae.

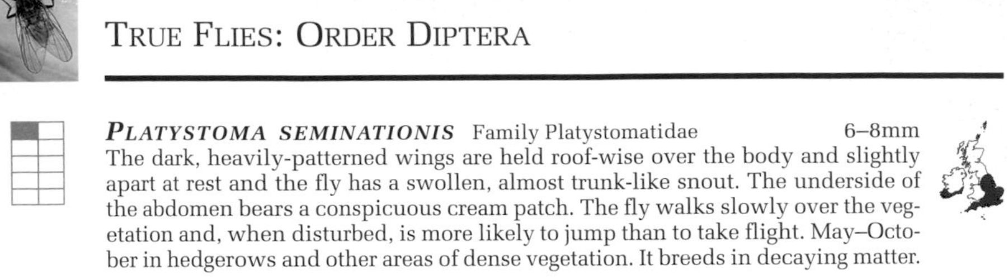

PLATYSTOMA SEMINATIONIS Family Platystomatidae 6–8mm

The dark, heavily-patterned wings are held roof-wise over the body and slightly apart at rest and the fly has a swollen, almost trunk-like snout. The underside of the abdomen bears a conspicuous cream patch. The fly walks slowly over the vegetation and, when disturbed, is more likely to jump than to take flight. May–October in hedgerows and other areas of dense vegetation. It breeds in decaying matter.

UROPHORA CARDUI Family Tephritidae 6–8mm

This is one of several very similar 'picture-winged' flies, although in most species the dark wing bands are not continuous. The bands are stronger in males than in females. May–August in open, grassy places. The larvae cause hard, egg-shaped galls in the stems of creeping thistle. Most related species cause inconspicuous galls in the flower heads of thistles and knapweeds.

CELERY-FLY *Euleia heraclei* *c.* 7mm

The wing markings vary from reddish brown to almost black and the body ranges from orange to black. April–November in many habitats, including gardens. The larvae live as leaf miners in the leaves of celery and many other wild and cultivated relatives.

MICROPEZA CORRIGIOLATA Family Micropezidae *c.* 10mm

This slender, large-headed fly walks slowly over the vegetation on its spindly legs and well deserves its common name of Stilt-legged Fly. The front coxae are yellow but the mid and hind coxae are brown. May–September in shady places where it breeds in decaying vegetation, including garden compost heaps. ***M. lateralis*** has all coxae yellow.

COREMACERA MARGINATA Family Sciomyzidae 8–10mm

With its long antennae and its heavily dappled wings laid flat over its body at rest, this insect is not likely to be mistaken for any other fly although, at first glance, it does not really look like a fly. June–October in hedgerows and damp grassland, where the grubs feed on small slugs and snails.

SEPSIS FULGENS Family Sepsidae *c.* 5mm

Although the individual flies are very small, they are sometimes very conspicuous because they swarm in their thousands on small areas of vegetation, waving their wings up and down as they scuttle about. Such swarms are most common in late summer and autumn. All year, although dormant in winter. The insects breed in dung of various kinds. There are many similar species, although not all of them swarm.

FAMILY TACHINIDAE

These flies are all are parasitoids, with larvae that grow up inside other insects. Some can be mistaken for blow-flies, but the tachinids are generally much more bristly and they have very large thoracic squamae (the membranous flaps at the sides of the thorax). There is also a post-scutellum bulging conspicuously from beneath the scutellum. Tachinids differ from house-flies and their relatives (*see* p. 316) in that the 4th long vein bends sharply forward towards the wing-tip. There are about 240 species in the British Isles.

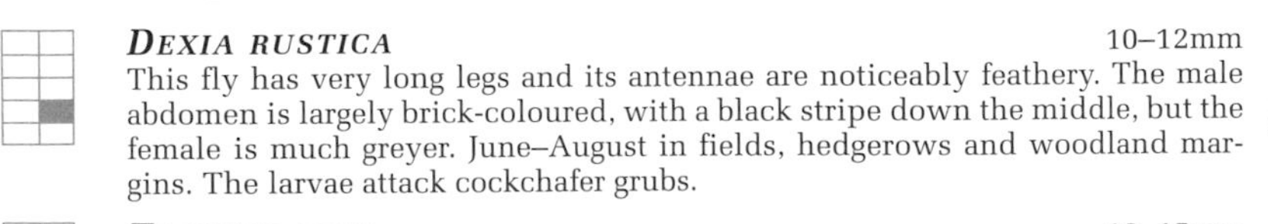

DEXIA RUSTICA 10–12mm

This fly has very long legs and its antennae are noticeably feathery. The male abdomen is largely brick-coloured, with a black stripe down the middle, but the female is much greyer. June–August in fields, hedgerows and woodland margins. The larvae attack cockchafer grubs.

TACHINA FERA 12–15mm

This large and handsome fly differs from most other species in its reddish legs and the yellow hair on its face. The wing bases are yellow and the abdomen has brick-coloured sides. April–September, mainly in damp habitats and often basking on waterside vegetation. It parasitises various caterpillars.

GONIA DIVISA *c.* 10mm

This species resembles a small *Tachina fera* but the legs are black and the wing bases are not yellow. March–June in grassy places, including woodland rides. It parasitises the caterpillars of various noctuid moths.

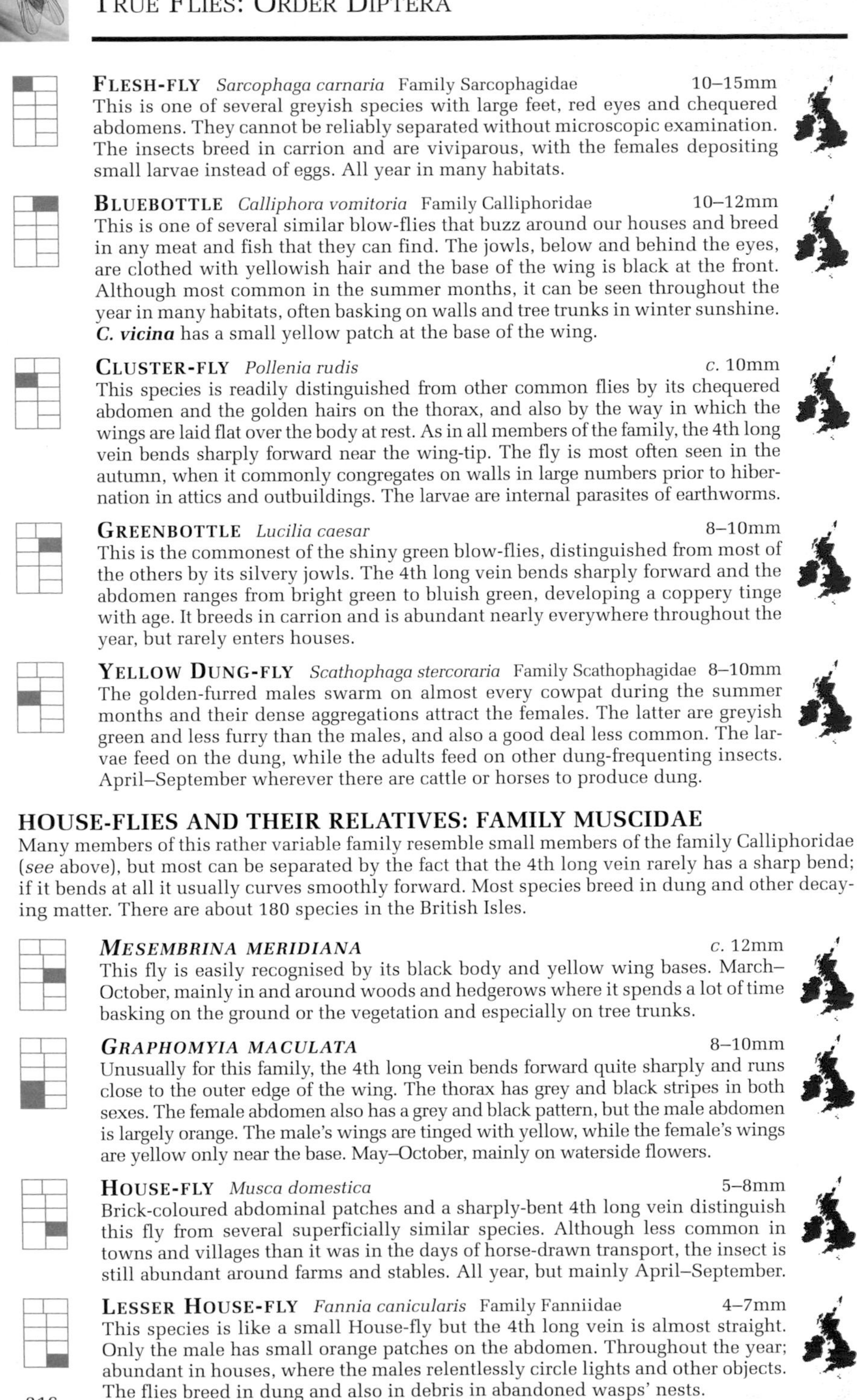

FLESH-FLY *Sarcophaga carnaria* Family Sarcophagidae 10–15mm
This is one of several greyish species with large feet, red eyes and chequered abdomens. They cannot be reliably separated without microscopic examination. The insects breed in carrion and are viviparous, with the females depositing small larvae instead of eggs. All year in many habitats.

BLUEBOTTLE *Calliphora vomitoria* Family Calliphoridae 10–12mm
This is one of several similar blow-flies that buzz around our houses and breed in any meat and fish that they can find. The jowls, below and behind the eyes, are clothed with yellowish hair and the base of the wing is black at the front. Although most common in the summer months, it can be seen throughout the year in many habitats, often basking on walls and tree trunks in winter sunshine. ***C. vicina*** has a small yellow patch at the base of the wing.

CLUSTER-FLY *Pollenia rudis* *c.* 10mm
This species is readily distinguished from other common flies by its chequered abdomen and the golden hairs on the thorax, and also by the way in which the wings are laid flat over the body at rest. As in all members of the family, the 4th long vein bends sharply forward near the wing-tip. The fly is most often seen in the autumn, when it commonly congregates on walls in large numbers prior to hibernation in attics and outbuildings. The larvae are internal parasites of earthworms.

GREENBOTTLE *Lucilia caesar* 8–10mm
This is the commonest of the shiny green blow-flies, distinguished from most of the others by its silvery jowls. The 4th long vein bends sharply forward and the abdomen ranges from bright green to bluish green, developing a coppery tinge with age. It breeds in carrion and is abundant nearly everywhere throughout the year, but rarely enters houses.

YELLOW DUNG-FLY *Scathophaga stercoraria* Family Scathophagidae 8–10mm
The golden-furred males swarm on almost every cowpat during the summer months and their dense aggregations attract the females. The latter are greyish green and less furry than the males, and also a good deal less common. The larvae feed on the dung, while the adults feed on other dung-frequenting insects. April–September wherever there are cattle or horses to produce dung.

HOUSE-FLIES AND THEIR RELATIVES: FAMILY MUSCIDAE

Many members of this rather variable family resemble small members of the family Calliphoridae (*see* above), but most can be separated by the fact that the 4th long vein rarely has a sharp bend; if it bends at all it usually curves smoothly forward. Most species breed in dung and other decaying matter. There are about 180 species in the British Isles.

MESEMBRINA MERIDIANA *c.* 12mm
This fly is easily recognised by its black body and yellow wing bases. March–October, mainly in and around woods and hedgerows where it spends a lot of time basking on the ground or the vegetation and especially on tree trunks.

GRAPHOMYIA MACULATA 8–10mm
Unusually for this family, the 4th long vein bends forward quite sharply and runs close to the outer edge of the wing. The thorax has grey and black stripes in both sexes. The female abdomen also has a grey and black pattern, but the male abdomen is largely orange. The male's wings are tinged with yellow, while the female's wings are yellow only near the base. May–October, mainly on waterside flowers.

HOUSE-FLY *Musca domestica* 5–8mm
Brick-coloured abdominal patches and a sharply-bent 4th long vein distinguish this fly from several superficially similar species. Although less common in towns and villages than it was in the days of horse-drawn transport, the insect is still abundant around farms and stables. All year, but mainly April–September.

LESSER HOUSE-FLY *Fannia canicularis* Family Fanniidae 4–7mm
This species is like a small House-fly but the 4th long vein is almost straight. Only the male has small orange patches on the abdomen. Throughout the year; abundant in houses, where the males relentlessly circle lights and other objects. The flies breed in dung and also in debris in abandoned wasps' nests.

The figures given at the beginning of each description refer to the number of spurs on the 1st, 2nd and 3rd tibiae. These are important in the classification and identification of the insects and care must be taken not to overlook broken ones.

Phryganea grandis Family Phryganeidae fw 20–30mm
2–4–4. The largest British caddis fly, this species has a long, narrow discal cell near the middle of the forewing. Only the female has the black stripe through the centre. May–August, breeding in still and slow-moving water with plenty of submerged plants. The larval case is made from plant fragments arranged in a neat spiral. ***P. striata*** has the black stripe broken into three dashes.

Odontocerum albicorne Family Odontoceridae fw 12–18mm
2–4–4. The wings of the freshly-emerged insect are silvery grey, but they become yellowish brown with age and the best way to identify the species is to look at the antennae. The basal segment is very thick and a lens will reveal that the rest are distinctly toothed. The male has a tuft of hair on the inner edge of the hindwing. June–October, breeding in stony streams. The larval case is a curved tube made of sand grains.

Limnephilus lunatus Family Limnephilidae fw *c.* 15mm
1–3–4. This is one of several similar species with clear patches on abruptly truncated forewings, but it can be distinguished from most by the pale crescent on the outer edge. The forewings are narrow and, as in all members of this large family, they are not very hairy. The hindwings are broad and transparent. May–November, breeding in still and slow-moving water. The larval case is made with debris of all kinds.

Limnephilus centralis fw *c.* 10mm
1–3–4. This species resembles *L. lunatus* but there is usually a fairly clear division between the pale and unmarked front half of the forewing and the darker, mottled rear half. June–September, breeding mainly in the shallow waters of marshy areas. The larval case is made with sand grains.

Glyphotaelius pellucidus fw 15–20mm
1–3–4. The forewings are usually brown with a variable pattern of light and dark patches. Some specimens are plain yellowish-brown, but the species can always be recognised by the strongly-notched outer margin. The abdomen is usually some shade of green. May–October, usually in still water. The larval case is made from irregular pieces of dead leaves that completely conceal the insect from above.

Anabolia nervosa fw 11–16mm
1–3–4. The forewing is plain brown, although the shade varies, and there are two pale patches in the rear half. June–October, breeding mainly in stony streams. The larval case is made from sand grains and has a number of small twigs attached to it as a defence against fish.

Halesus radiatus fw *c.* 20 mm
1–3–3. The forewing has a pattern of greyish streaks and blotches, the most noticeable of which are the radiating finger-like streaks near the tip. August–November, breeding in running water. The larval case is made of plant debris, with 2 or 3 slender twigs running along its length. *H. digitatus* is a little larger, with a straighter outer margin and a less distinct pattern.

Philopotamus montanus Family Philopotamidae fw 7–12mm
2–4–4. No other British caddis fly has mottled brown and yellow wings. April–August, breeding mainly in swift, upland streams. The larvae are net-spinners and make no cases.

Mystacides longicornis Family Leptoceridae fw 8–10mm
0–2–2, the two-spurred hind tibiae being diagnostic of the whole family. The species can usually be recognised by its long, pale antennae and the dark blotches on its yellowish-brown forewings. The long, hairy palps, another family feature, could even be mistaken for an extra pair of legs. May–September, breeding in still waters. The almost straight larval case is made of sand grains.

THE SAWFLIES: SUBORDER SYMPHYTA

The insects in this suborder have no obvious 'waist' between the thorax and the abdomen. They get their name because most females have a saw-like ovipositor, although in some species it works more like a drill. The eggs are nearly always laid inside plant tissues and the larvae are all vegetarians. Most adults feed on pollen and nectar, although some are partly carnivorous. Many are regular visitors to the flower heads of hogweed and related plants. Most species fly by day and fold their wings flat over the body at rest. Many sawfly larvae resemble the caterpillars of butterflies and moths, but they have more stumpy legs on the abdomen. Some species induce galls, especially on willows. There are over 400 British species.

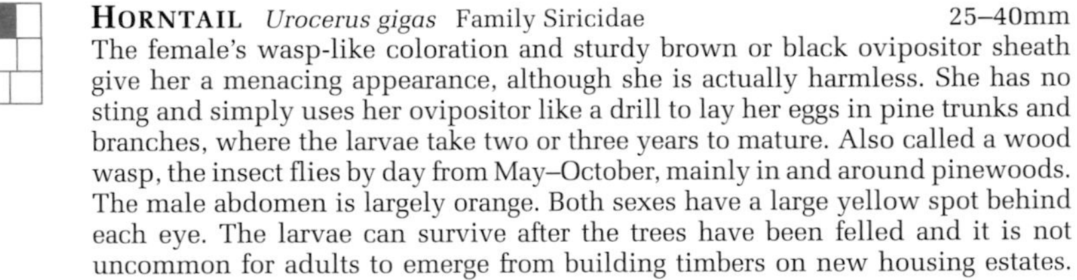

HORNTAIL *Urocerus gigas* Family Siricidae 25–40mm

The female's wasp-like coloration and sturdy brown or black ovipositor sheath give her a menacing appearance, although she is actually harmless. She has no sting and simply uses her ovipositor like a drill to lay her eggs in pine trunks and branches, where the larvae take two or three years to mature. Also called a wood wasp, the insect flies by day from May–October, mainly in and around pinewoods. The male abdomen is largely orange. Both sexes have a large yellow spot behind each eye. The larvae can survive after the trees have been felled and it is not uncommon for adults to emerge from building timbers on new housing estates.

SIREX JUVENCUS 15–30mm

The female, illustrated here, is black with a strong blue or purple sheen. The male resembles the male Horntail except that it has no pale patches behind the eyes. The antennae are usually reddish brown at the base although northern specimens may have entirely black antennae. The feet are yellowish. May–October, mainly in and around pinewoods and plantations, where the larvae spend two or three years tunnelling in the tree trunks. The larvae can survive in felled and sawn timber and adults often emerge from pallets and other timbers far from their original homes. ***Sirex noctilio*** is very similar and more common. Its antennae are entirely black and the last segment of each foot is brown.

XIPHYDRIA PROLONGATA Family Xiphidryiidae 6–18mm

The narrow but conspicuous neck and the broad red band on the abdomen should separate this insect from all other sawflies. The abdomen also has white spots on the sides and the first antennal segment is long and curved. June–August, mainly in damp areas. The larvae tunnel in willows and sallows.

XIPHYDRIA CAMELUS 6–18mm

This species resembles *Xiphydria prolongata* in shape but lacks the red abdominal band. It is more widely distributed and its larvae tunnel in the trunks and branches of birch and alder. May–August.

ARGE USTULATA Family Argidae 7–10mm

This is one of several similar metallic blue or green sawflies, but it can be distinguished from most by its yellow pseudostigmata. As in the whole family, each antenna consist of just three segments, but the first two are so small that it looks as if there is just a single segment. May–July in light woodland and scrubby habitats. FP: willows, birch and hawthorn.

ARGE PAGANA 7–10mm

The thorax and legs may be completely black or marked with yellow, but they always contrast strongly with the bright yellow abdomen. The wings are smoky with a black vein on the front edge. May–September, mainly in wooded areas. FP: roses.

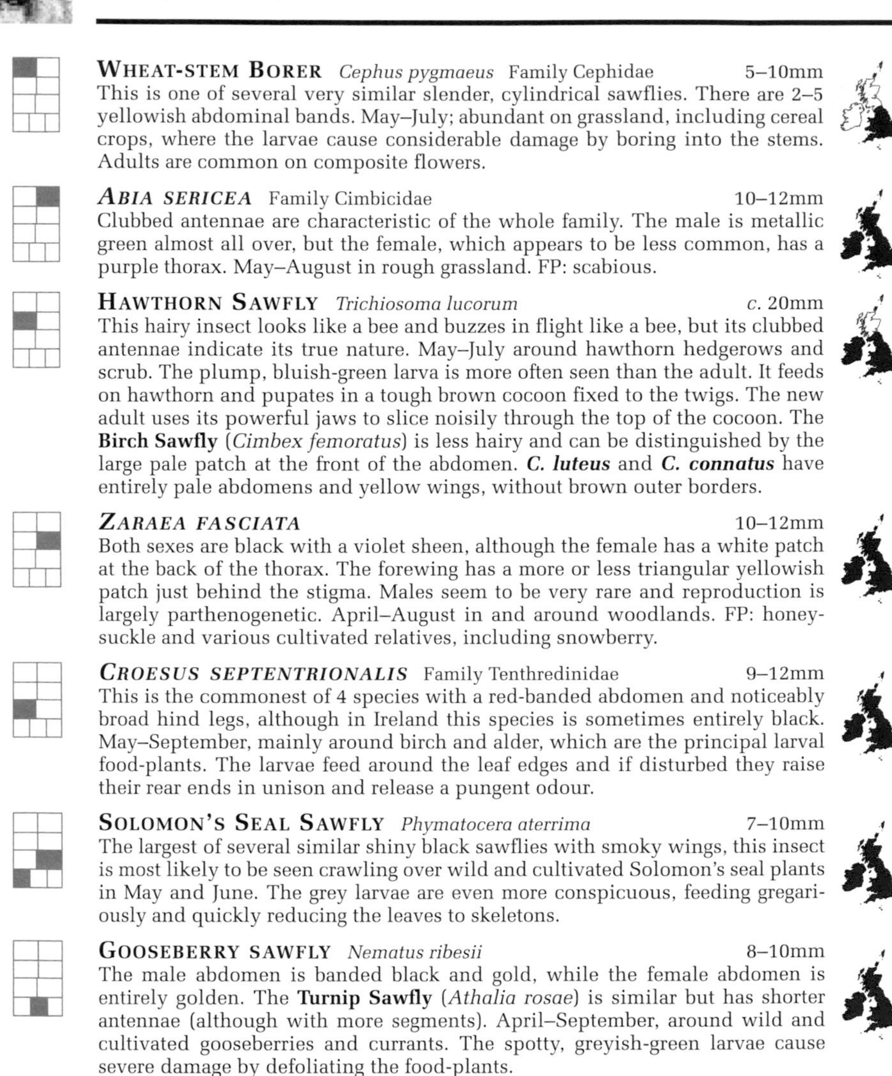

WHEAT-STEM BORER *Cephus pygmaeus* Family Cephidae 5–10mm
This is one of several very similar slender, cylindrical sawflies. There are 2–5 yellowish abdominal bands. May–July; abundant on grassland, including cereal crops, where the larvae cause considerable damage by boring into the stems. Adults are common on composite flowers.

ABIA SERICEA Family Cimbicidae 10–12mm
Clubbed antennae are characteristic of the whole family. The male is metallic green almost all over, but the female, which appears to be less common, has a purple thorax. May–August in rough grassland. FP: scabious.

HAWTHORN SAWFLY *Trichiosoma lucorum* *c.* 20mm
This hairy insect looks like a bee and buzzes in flight like a bee, but its clubbed antennae indicate its true nature. May–July around hawthorn hedgerows and scrub. The plump, bluish-green larva is more often seen than the adult. It feeds on hawthorn and pupates in a tough brown cocoon fixed to the twigs. The new adult uses its powerful jaws to slice noisily through the top of the cocoon. The **Birch Sawfly** (*Cimbex femoratus*) is less hairy and can be distinguished by the large pale patch at the front of the abdomen. ***C. luteus*** and ***C. connatus*** have entirely pale abdomens and yellow wings, without brown outer borders.

ZARAEA FASCIATA 10–12mm
Both sexes are black with a violet sheen, although the female has a white patch at the back of the thorax. The forewing has a more or less triangular yellowish patch just behind the stigma. Males seem to be very rare and reproduction is largely parthenogenetic. April–August in and around woodlands. FP: honeysuckle and various cultivated relatives, including snowberry.

CROESUS SEPTENTRIONALIS Family Tenthredinidae 9–12mm
This is the commonest of 4 species with a red-banded abdomen and noticeably broad hind legs, although in Ireland this species is sometimes entirely black. May–September, mainly around birch and alder, which are the principal larval food-plants. The larvae feed around the leaf edges and if disturbed they raise their rear ends in unison and release a pungent odour.

SOLOMON'S SEAL SAWFLY *Phymatocera aterrima* 7–10mm
The largest of several similar shiny black sawflies with smoky wings, this insect is most likely to be seen crawling over wild and cultivated Solomon's seal plants in May and June. The grey larvae are even more conspicuous, feeding gregariously and quickly reducing the leaves to skeletons.

GOOSEBERRY SAWFLY *Nematus ribesii* 8–10mm
The male abdomen is banded black and gold, while the female abdomen is entirely golden. The **Turnip Sawfly** (*Athalia rosae*) is similar but has shorter antennae (although with more segments). April–September, around wild and cultivated gooseberries and currants. The spotty, greyish-green larvae cause severe damage by defoliating the food-plants.

PONTANIA PROXIMA 5–7mm
This is one of numerous similar, black willow-feeding sawflies better known for their galls than for their adult appearance. The species can be reliably separated only by detailed study of the ovipositors and other genitalia. May–August on and around willows. The larvae inhabit bean-shaped red or yellow galls on the leaves of white and crack willow.

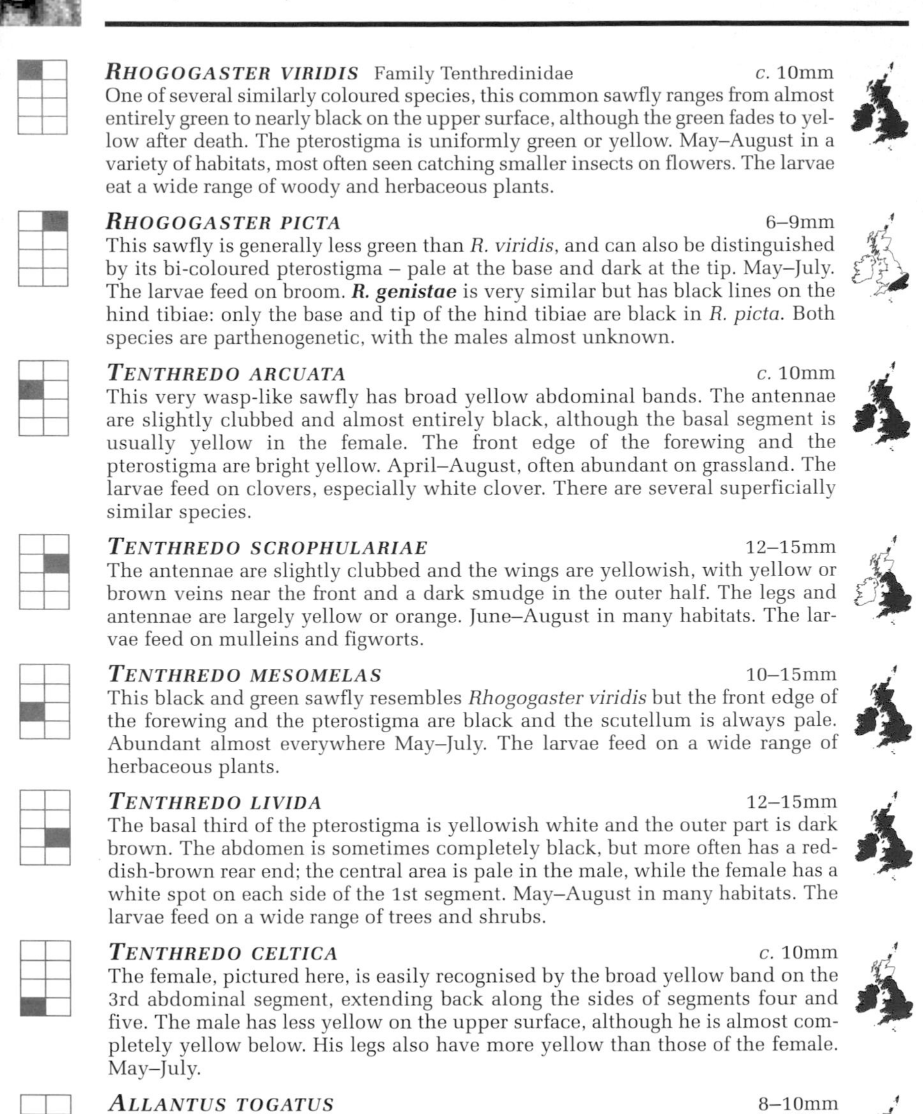

RHOGOGASTER VIRIDIS Family Tenthredinidae *c.* 10mm
One of several similarly coloured species, this common sawfly ranges from almost entirely green to nearly black on the upper surface, although the green fades to yellow after death. The pterostigma is uniformly green or yellow. May–August in a variety of habitats, most often seen catching smaller insects on flowers. The larvae eat a wide range of woody and herbaceous plants.

RHOGOGASTER PICTA 6–9mm
This sawfly is generally less green than *R. viridis*, and can also be distinguished by its bi-coloured pterostigma – pale at the base and dark at the tip. May–July. The larvae feed on broom. ***R. genistae*** is very similar but has black lines on the hind tibiae: only the base and tip of the hind tibiae are black in *R. picta*. Both species are parthenogenetic, with the males almost unknown.

TENTHREDO ARCUATA *c.* 10mm
This very wasp-like sawfly has broad yellow abdominal bands. The antennae are slightly clubbed and almost entirely black, although the basal segment is usually yellow in the female. The front edge of the forewing and the pterostigma are bright yellow. April–August, often abundant on grassland. The larvae feed on clovers, especially white clover. There are several superficially similar species.

TENTHREDO SCROPHULARIAE 12–15mm
The antennae are slightly clubbed and the wings are yellowish, with yellow or brown veins near the front and a dark smudge in the outer half. The legs and antennae are largely yellow or orange. June–August in many habitats. The larvae feed on mulleins and figworts.

TENTHREDO MESOMELAS 10–15mm
This black and green sawfly resembles *Rhogogaster viridis* but the front edge of the forewing and the pterostigma are black and the scutellum is always pale. Abundant almost everywhere May–July. The larvae feed on a wide range of herbaceous plants.

TENTHREDO LIVIDA 12–15mm
The basal third of the pterostigma is yellowish white and the outer part is dark brown. The abdomen is sometimes completely black, but more often has a reddish-brown rear end; the central area is pale in the male, while the female has a white spot on each side of the 1st segment. May–August in many habitats. The larvae feed on a wide range of trees and shrubs.

TENTHREDO CELTICA *c.* 10mm
The female, pictured here, is easily recognised by the broad yellow band on the 3rd abdominal segment, extending back along the sides of segments four and five. The male has less yellow on the upper surface, although he is almost completely yellow below. His legs also have more yellow than those of the female. May–July.

ALLANTUS TOGATUS 8–10mm
This fairly sturdy sawfly has a marked constriction at the front of the abdomen and three yellowish-white abdominal bands. The tegulae overlying the bases of the forewings are also pale yellow and the forewing has a brownish patch behind the stigma. June–September, in and around woodlands. FP: mainly oak.

BEES, WASPS AND ANTS: SUBORDER APOCRITA

This is the larger of the two suborders of the Hymenoptera, with over 6,000 British species. They vary enormously in appearance and behaviour, but all have the characteristic 'wasp waist'. Most are parasitoids (*see* p. 35) and the females use their ovipositors to lay eggs in or on their hosts. The Apocrita also include the bees, wasps and ants, in which the ovipositor has been transformed into a sting used for defence or to paralyse prey. All our social insects (*see* p. 35) belong to this group.

GALL WASPS: FAMILY CYNIPIDAE

Generally brown or black, these tiny winged or wingless insects, rarely more than 5mm long, are noticeably flattened from side to side. The insects themselves are rarely noticed, but their effects can be seen nearly everywhere for they are responsible for the development of many familiar plant galls (*see* p. 38). Some species have normal, straightforward lives but others have complex life cycles involving the alternation of sexual and agamic or parthenogenetic (female only) generations. Although the 2 generations may look alike, each gives rise to its own kind of gall and the two galls may be very different; they may even develop on different host plants. There are about 90 gall wasp species in the British Isles and 42 of these induce galls on oak trees. Other hosts include wild roses and a number of herbaceous plants. Any part of the host plant, from the roots to the flowers and fruits, may be attacked, although each gall wasp species keeps to its preferred site or sites. Not all gall wasps induce galls: some live as inquilines in the galls of other species.

MARBLE GALL *Andricus kollari*
This very common oak gall starts off green in early summer and becomes brown and woody by September. Up to 20mm across, it is home to a single agamic female, which leaves in the autumn. She lays eggs on Turkey oak buds, where the sexual generation develops in tiny egg-shaped galls.

COLA-NUT GALL *Andricus lignicolus*
This oak gall resembles the Marble Gall but is rarely more than 10mm across and has a much rougher surface. It is also very hard. The life cycle of the insect resembles that of the previous species.

ARTICHOKE GALL *Andricus fecundator*
Up to 30mm long, this oak gall is composed of enlarged bud scales surrounding a little egg-shaped inner gall in which a single gall wasp develops. The inner gall falls out in autumn, but the old galls remain on the trees for some months. Agamic adults leave the fallen inner galls in the spring and lay eggs on the oak catkins, where the sexual generation develops in tiny hairy galls.

HEDGEHOG GALL *Andricus lucidus*
This gall develops on buds and, less often, on acorn cups. It is up to 20mm across and the knobbed spines are quite sticky. Several agamic wasps develop in each gall during the autumn and leave in the spring. The sexual generation develops in galls on the catkins of Turkey oak.

COTTON WOOL GALL *Andricus quercusramuli*
This gall develops on a male oak catkin, which becomes completely enveloped in a mass of silky white hairs up to 20mm across. Hidden among the hairs, the main body of the gall consists of up to 20 woody chambers, each containing a developing male or female gall wasp. The insects mature by midsummer and the female lays her eggs in young oak buds, where the agamic generation develops in tiny egg-shaped galls.

KNOPPER GALL *Andricus quercuscalicis*
This rather sticky gall develops on acorns in the autumn and varies from green to bright red. A central opening leads to a small, egg-shaped inner gall where a single agamic wasp develops. The sexual generation develops in tiny galls on the catkins of Turkey oak.

OAK APPLE *Biorhiza pallida*
Appearing on oak twigs in the spring, this soft gall is up to about 45mm across and contains numerous male and female gall wasps. They emerge in June–July, by which time the galls have a brown and rather papery surface. The females burrow into the ground and lay their eggs on fine oak roots, where the agamic generation develops in little brown galls.

Common Spangle Gall *Neuroterus quercusbaccarum*

This disc-shaped, hairy gall, up to 5mm across, is abundant on the undersides of oak leaves in autumn. There is a slight mound in the centre but the edges are never raised. Each gall houses a single agamic gall wasp. The galls mature on the ground in the winter and the insects leave in early spring to lay their eggs in oak buds.

Currant Gall *Neuroterus quercusbaccarum*

This gall, common on both leaves and catkins in late spring, reaches 7mm in diameter and is green or purple. It is caused by the same species as the Common Spangle, but it houses the sexual generation of the species.

Cupped Spangle Gall *Neuroterus tricolor*

Much rarer than the Common Spangle, this gall can be distinguished by its distinctly raised rim. It develops on the undersides of oak leaves in the autumn. The life cycles of the two species are very similar, although the sexual galls of *tricolor* are like pale peas with scattered red or purple hairs. These galls always develop under the leaves.

Smooth Spangle Gall *Neuroterus albipes*

This gall is either flat or dish-shaped, but differs from the previous 2 species in being hairless. It often has an irregular shape. It is usually cream, often with pink or purple patches, but sometimes entirely pink. The life cycle is like that of the other spangles, the sexual gall normally being a small egg-like swelling on the edge of a leaf.

Silk Button Spangle Gall *Neuroterus numismalis*

Up to 5mm across, this easily identified gall is often abundant under oak leaves in the autumn. The galls fall from the leaves in autumn and the agamic insects, one to each gall, leave in the spring. The sexual galls are inconspicuous swellings in the oak leaves.

Cherry Gall *Cynips quercusfolii*

This striking gall, up to 20mm across, may have a smooth or warty skin. It develops on the undersides of oak leaves in the summer and each houses an agamic gall wasp. The latter matures and leaves its gall in the middle of winter. The sexual galls are like tiny purple eggs and can be found in buds on or close to the oak trunks in spring.

Robin's Pincushion Gall *Diplolepis rosae*

Also known as the Bedeguar Gall, this is very common on wild roses in the summer. It is up to 10cm across and under the branching hairs it is hard and woody and consists of many chambers, each with a single gall wasp grub. The insects mature and leave in the spring. They are almost all females and they lay their eggs in the opening buds without mating. There is no alternation of generations, so the insect causes only one kind of gall.

Spiked Pea Gall *Diplolepis nervosus*

This gall can be found on wild roses in the summer, usually under the leaves but sometimes on the flower-stalks and even on the hips. It is sometimes called a Sputnik Gall. The insect has a simple life cycle like that of *D. rosae*. The galls often lack spikes, and are then impossible to separate from those of ***D. eglanteriae***.

Liposthenus glechomae

This hairy gall, up to 20mm in diameter, develops on the leaves and stems of ground ivy in the summer. It remains green in shady places, but becomes bright red when growing in a sunlit spot. The insect has a simple life cycle, with only one kind of gall.

Diastrophus rubi

Although common, this gall is not often seen because it develops mainly on slender bramble stems deep in the vegetation. Up to 20 cm long, it matures in the autumn. Each gall contains numerous gall wasp larvae. The adults escape in spring and begin the cycle again. The life cycle is normal, with just one kind of gall.

THE ICHNEUMONS: FAMILY ICHNEUMONIDAE

The antennae in this family consist of at least 16 small segments. The front of the forewing is relatively thick as a result of the fusion of the two anterior veins, and there is always a prominent pterostigma. The insects are all parasitoids and most species attack the caterpillars of butterflies and moths. Unless otherwise stated, the larvae of the following ichneumons are all endoparasitoids of caterpillars.

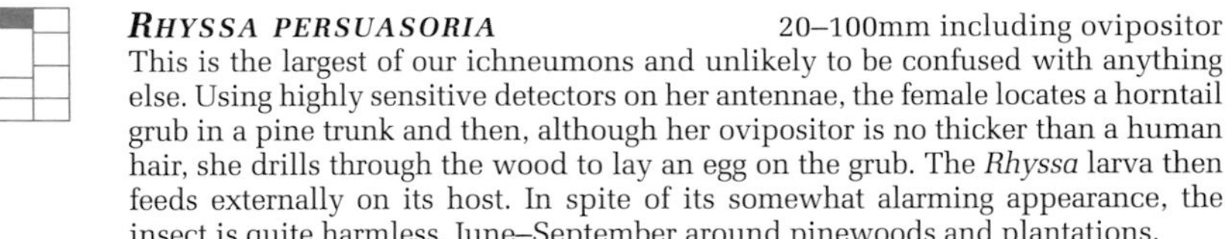

RHYSSA PERSUASORIA 20–100mm including ovipositor

This is the largest of our ichneumons and unlikely to be confused with anything else. Using highly sensitive detectors on her antennae, the female locates a horntail grub in a pine trunk and then, although her ovipositor is no thicker than a human hair, she drills through the wood to lay an egg on the grub. The *Rhyssa* larva then feeds externally on its host. In spite of its somewhat alarming appearance, the insect is quite harmless. June–September around pinewoods and plantations.

PIMPLA HYPOCHONDRIACA 20–25mm

One of several similar species, this very common parasitoid attacks both caterpillars and pupae. It has brown legs and the abdomen ranges from black to chestnut brown. Most of the year almost everywhere, but dormant in winter; often abundant in gardens. Related species are mostly a little smaller.

NETELIA TESTACEA 15–20mm

The arched abdomen is flattened from side to side and, unlike other species of *Netelia*, it has a dark tip. A tiny, more-or-less circular cell in the outer part of the forewing distinguishes *Netelia* species from the superficially similar *Ophion*. The short ovipositor is capable of piercing a finger if the insect is handled. May–September almost everywhere; markedly nocturnal. The larvae are ectoparasitoids of moth larvae.

OPHION LUTEUS 15–20mm

This is one of several closely related species resembling *Netelia* but lacking the small cell in the forewing. *Ophion* species also have a markedly triangular scutellum and the outer part of the forewing has a vein running close to and parallel to the rear edge. June–October, almost everywhere; markedly nocturnal.

AMBLYTELES ARMATORIUS 12–15mm

The cream or yellow scutellum and abdominal bands, together with the deep yellow front and middle legs, distinguish this from most other ichneumons. The hind trochanter is also yellow. All year nearly everywhere, but dormant in winter. Abundant on hogweed and similar flowers in summer.

ICHNEUMON SUSPICIOSUS 15–18mm

This is one of the commoner species in a group exhibiting a broad red or orange abdominal band and pale spots at the rear. The scutellum is pale and there is a pale section in the middle of each antenna. All year nearly everywhere, but dormant in winter.

ICHNEUMON STRAMENTARIUS 10–15mm

This species resembles *I. suspiciosus* in having a pale region in the middle of the antenna, but the legs and abdomen are banded with clear yellow. All year in many habitats, but dormant in winter.

APANTELES GLOMERATUS Family Braconidae 3–4mm

This tiny insect is better known in its pupal stage than as an adult, for it is the parasitoid that attacks and destroys the caterpillars of the Large White butterfly and then leaves their empty skins to pupate around them in clusters of fluffy yellow cocoons. May–September, mainly in cultivated regions where the host flies. There are numerous similar species, each with its own preferred host.

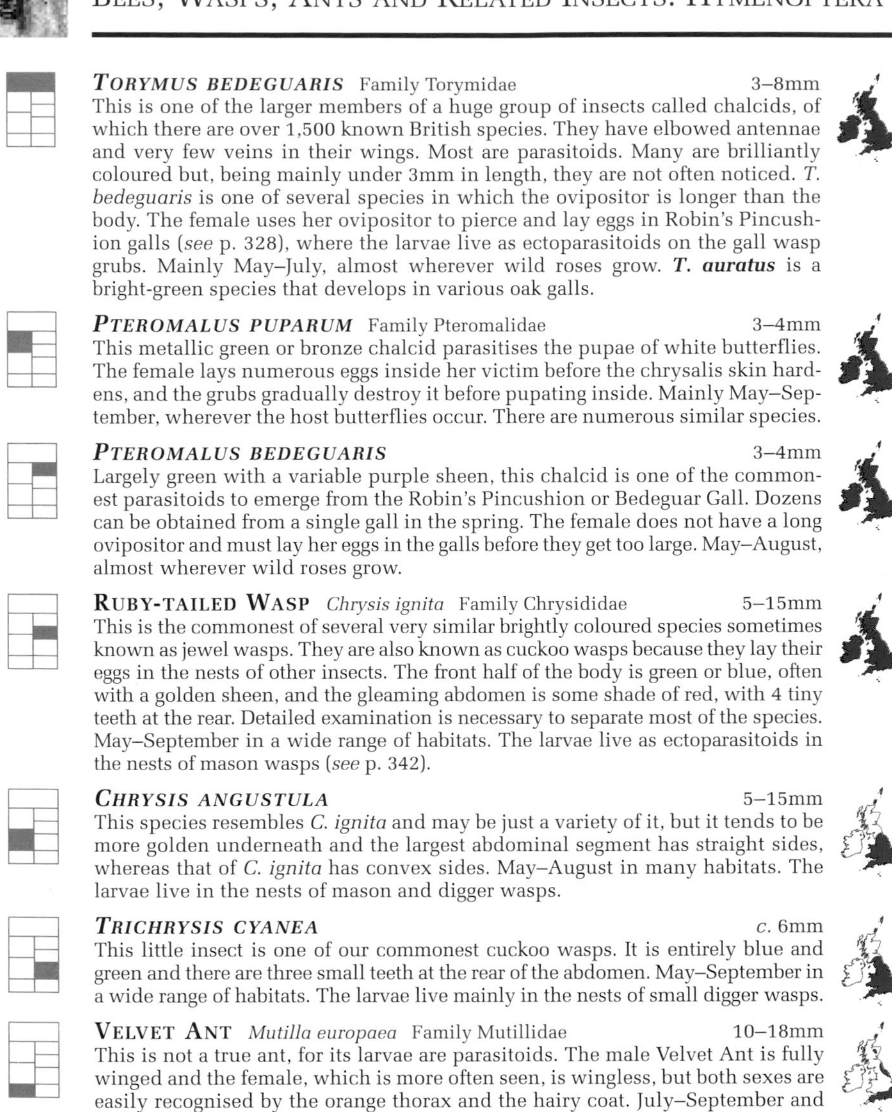

TORYMUS BEDEGUARIS Family Torymidae 3–8mm

This is one of the larger members of a huge group of insects called chalcids, of which there are over 1,500 known British species. They have elbowed antennae and very few veins in their wings. Most are parasitoids. Many are brilliantly coloured but, being mainly under 3mm in length, they are not often noticed. *T. bedeguaris* is one of several species in which the ovipositor is longer than the body. The female uses her ovipositor to pierce and lay eggs in Robin's Pincushion galls (*see* p. 328), where the larvae live as ectoparasitoids on the gall wasp grubs. Mainly May–July, almost wherever wild roses grow. ***T. auratus*** is a bright-green species that develops in various oak galls.

PTEROMALUS PUPARUM Family Pteromalidae 3–4mm

This metallic green or bronze chalcid parasitises the pupae of white butterflies. The female lays numerous eggs inside her victim before the chrysalis skin hardens, and the grubs gradually destroy it before pupating inside. Mainly May–September, wherever the host butterflies occur. There are numerous similar species.

PTEROMALUS BEDEGUARIS 3–4mm

Largely green with a variable purple sheen, this chalcid is one of the commonest parasitoids to emerge from the Robin's Pincushion or Bedeguar Gall. Dozens can be obtained from a single gall in the spring. The female does not have a long ovipositor and must lay her eggs in the galls before they get too large. May–August, almost wherever wild roses grow.

RUBY-TAILED WASP *Chrysis ignita* Family Chrysididae 5–15mm

This is the commonest of several very similar brightly coloured species sometimes known as jewel wasps. They are also known as cuckoo wasps because they lay their eggs in the nests of other insects. The front half of the body is green or blue, often with a golden sheen, and the gleaming abdomen is some shade of red, with 4 tiny teeth at the rear. Detailed examination is necessary to separate most of the species. May–September in a wide range of habitats. The larvae live as ectoparasitoids in the nests of mason wasps (*see* p. 342).

CHRYSIS ANGUSTULA 5–15mm

This species resembles *C. ignita* and may be just a variety of it, but it tends to be more golden underneath and the largest abdominal segment has straight sides, whereas that of *C. ignita* has convex sides. May–August in many habitats. The larvae live in the nests of mason and digger wasps.

TRICHRYSIS CYANEA *c.* 6mm

This little insect is one of our commonest cuckoo wasps. It is entirely blue and green and there are three small teeth at the rear of the abdomen. May–September in a wide range of habitats. The larvae live mainly in the nests of small digger wasps.

VELVET ANT *Mutilla europaea* Family Mutillidae 10–18mm

This is not a true ant, for its larvae are parasitoids. The male Velvet Ant is fully winged and the female, which is more often seen, is wingless, but both sexes are easily recognised by the orange thorax and the hairy coat. July–September and again in spring after hibernation, on heathland and in other dry habitats. The host insects are various bumble-bees.

METHOCHA ICHNEUMONIDES *Family Tiphiidae* 4–9mm

The female is shiny and wingless and, with its long reddish-brown thorax, is easily mistaken for an ant, although the antennae are not elbowed. May–September, mainly on heathland where its larvae are parasitoids of young tiger beetles. The male is winged and has a completely black body, but it is rarely seen and reproduction is probably largely parthenogenetic.

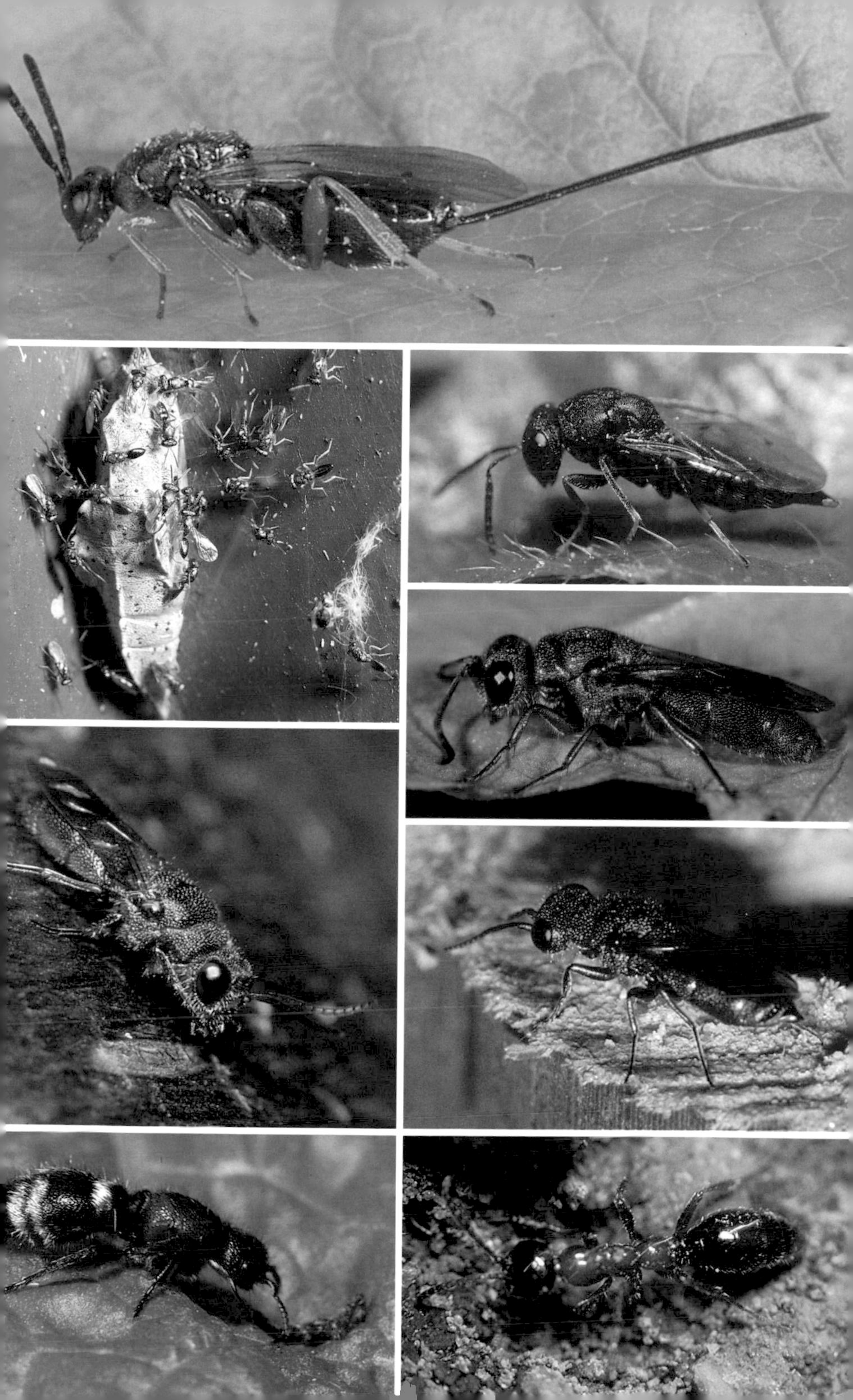

THE ANTS: FAMILY FORMICIDAE

Ants are fairly easy to recognise by their elbowed antennae, the relatively long thorax, and the very narrow waist composed of one or two segments. They are all social insects (*see* p. 35) and most of our species nest in the ground or in mounds of excavated soil. Some colonies have just one queen and some have several, but the bulk of the colony consists of smaller, wingless females called workers. Males and new queens appear in late spring or summer and, when the weather is right, they swarm from their nests and take off on their mating flights. The males soon die, but mated queens break off their wings and either return to their old nests or begin new colonies. Worker ants can be seen foraging from spring until the autumn, when they withdraw into their nests to spend the winter clustered together in a dormant state. Most of our ants are omnivorous: other insects make up the bulk of their diet, but fruit and nectar are taken when available and many ants 'milk' aphids for their sweet honeydew. There are about 50 British species, including a few established aliens, but many of them are quite rare. The lengths given here are for worker ants.

BLACK GARDEN ANT *Lasius niger* — *c.* 5mm

This uniformly dark brown ant has a single, scale-like waist segment and is by far the commonest of our small, dark, ground-nesting ants. Abundant nearly everywhere, in towns as well as in rural areas, it usually nests under stones and paths and often enters houses. It has no sting. Workers regularly climb plants in search of aphids, which yield honeydew when gently stroked by the ants' antennae. Mating swarms occur in August.

YELLOW MEADOW ANT *Lasius flavus* — *c.* 4mm

Apart from its yellowish-brown colour, this ant is very similar to the Black Garden Ant. It is abundant in well-drained grasslands, where it makes fairly solid nest mounds up to 50cm high. The mounds are quickly colonised by vegetation and they last for many years. Mating swarms usually occur in August.

WOOD ANT *Formica rufa* — *c.* 10mm

One of our largest ants, this species has a single, scale-like waist segment and the worker has a red thorax. There are several similar species but most have dark patches on the thorax. Colonies contain up to 500,000 insects and are found mainly in woodlands. The ants build large mounds thatched with leaves, especially pine needles. There is no sting, but the insects can bite fiercely and also spray formic acid from their rear ends. Mating flights take place May–June.

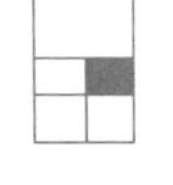

FORMICA SANGUINEA — 10–12mm

This rather rare ant resembles the Wood Ant but its thorax is a brighter red and its legs are also much redder. The species is best known for its 'slave raids' during which the workers invade the nests of related species and take away the pupae. The latter are installed in the *sanguinea* nests, and when the adults emerge they work for their new mistresses. The nests are usually established in tree stumps in open woodland or on heathland. Mating swarms occur June–July but consist mainly of males: most females remain on their nests and attract the flying males with scent.

MYRMICA RUBRA — *c.* 5mm

This is one of two very common garden ants, usually called red ants although they are actually pale brown. It has a two-segmented waist, both segments having a smoothly rounded hump. It nests under stones and logs and in tree stumps, forming colonies that rarely contain more than about 100 individuals. In common with all ants with a two-segmented waist, it has a sting. ***M. ruginodis*** is very similar but its waist segments are more angular and its colonies usually have several hundred members. Both species are abundant in many natural habitats as well as in gardens and usually have their mating flights in August.

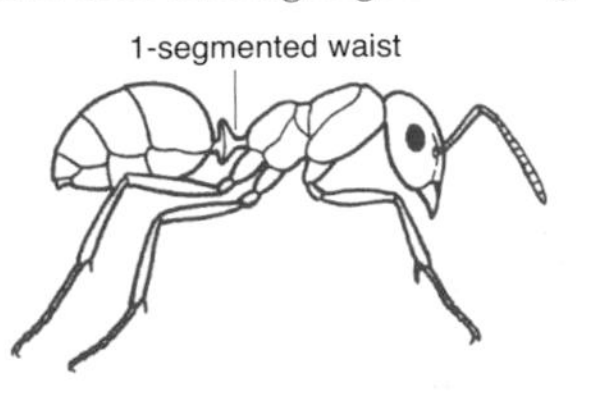

DIGGER WASPS: FAMILY SPHECIDAE

These solitary wasps get their name because most species nest in the ground. The front legs of many females are equipped with spiny 'brushes' used for digging and sweeping away the excavated soil. Each female constructs her own nest and stocks it with paralysed prey, usually other insects, although some species collect spiders. The insects rest with their wings folded flat over the body and the pronotum forms just a small collar that does not reach back to the wing bases. The submarginal cells just behind the pterostigma provide important clues to identification of the 110 or so British species. Most species have either two or three submarginal cells, although some species have just one.

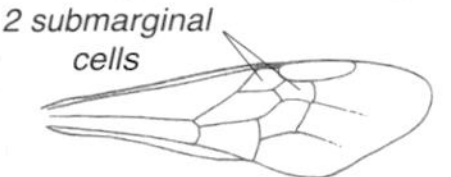

CRABRO CRIBRARIUS — 10–15mm

This is one of several similar species with a single submarginal cell, but distinguished from many because the middle part of the abdomen has yellow spots rather than complete bands. The front legs of the male are spoon-shaped. May–September, mainly in sandy areas where the nest burrows are stocked with various flies.

ECTEMNIUS CAVIFRONS — 11–17mm

There is one submarginal cell and the ocelli form a more-or-less equilateral triangle on the top of the head. The yellow bands run right across the abdomen. This species nests in rotten wood and stocks its burrows with flies, especially hover-flies, which it usually catches on flowers. June–October in many habitats; not uncommon in gardens. There are several similar species.

ECTEMNIUS CONTINUUS — 8–15mm

There is one submarginal cell and the base of the ocellar triangle (the rear edge) is longer than the other two sides. The yellow bands do not run right across the abdomen. The insect nests in rotten wood and stocks its burrows with various flies. May–September in many habitats.

CROSSOCERUS QUADRIMACULATUS — 5–10mm

There is 1 submarginal cell and the abdominal pattern is somewhat variable, although the posterior yellow bands do not run right across the abdomen; some specimens are almost black. The insect differs from most other black-and-yellow species in its equilateral ocellar triangle. June–September in many habitats, usually nesting in the ground and stocking its burrows mainly with small flies.

CROSSOCERUS MEGACEPHALUS — 6–10mm

With a single submarginal cell, this is one of several very similar black digger wasps in which the ocelli on the top of the head form an equilateral triangle. The jaws each have three teeth, although they are not easy to see, and the tip of the female abdomen curves up at the edges to form a shallow channel. Named for its relatively large head, this fast-flying insect is common in many habitats, May–September. It nests in rotten wood and stocks its burrows with various flies.

TRYPOXYLON FIGULUS — 5–12mm

The single submarginal cell, the long pointed marginal cell beyond the pterostigma, and the narrow abdomen tapering markedly towards the front distinguish *Trypoxylon* species from other black wasps. May–September, nesting in hollow stems and other narrow cavities, including woodworm holes, and stocking the burrows with small spiders. ***T. clavicerum*** has brownish front legs instead of black and ***T. attenuatum*** has eyes that converge strongly at the bottom.

PEMPHREDON LUGUBRIS — 7–12mm

There are two submarginal cells, the 1st being much larger than the 2nd, and the head and thorax are very hairy. The abdomen is connected to the thorax by a short stalk. May–September; nests are built in decaying wood and stocked with aphids. There are several very similar species and separating them is not easy. Most of them nest in hollow stems, especially bramble stems.

OXYBELUS UNIGLUMIS — 5–8mm

This is one of very few solitary wasps with a black and white or silvery coloration. The rear of the thorax carries a spine and two translucent flaps. Nest burrows are excavated in sandy soil and stocked with flies which are carried back in a very unusual way – impaled on the sting. June–September in areas with light, sandy soil. ***O. argentatus*** has light brown femora instead of black.

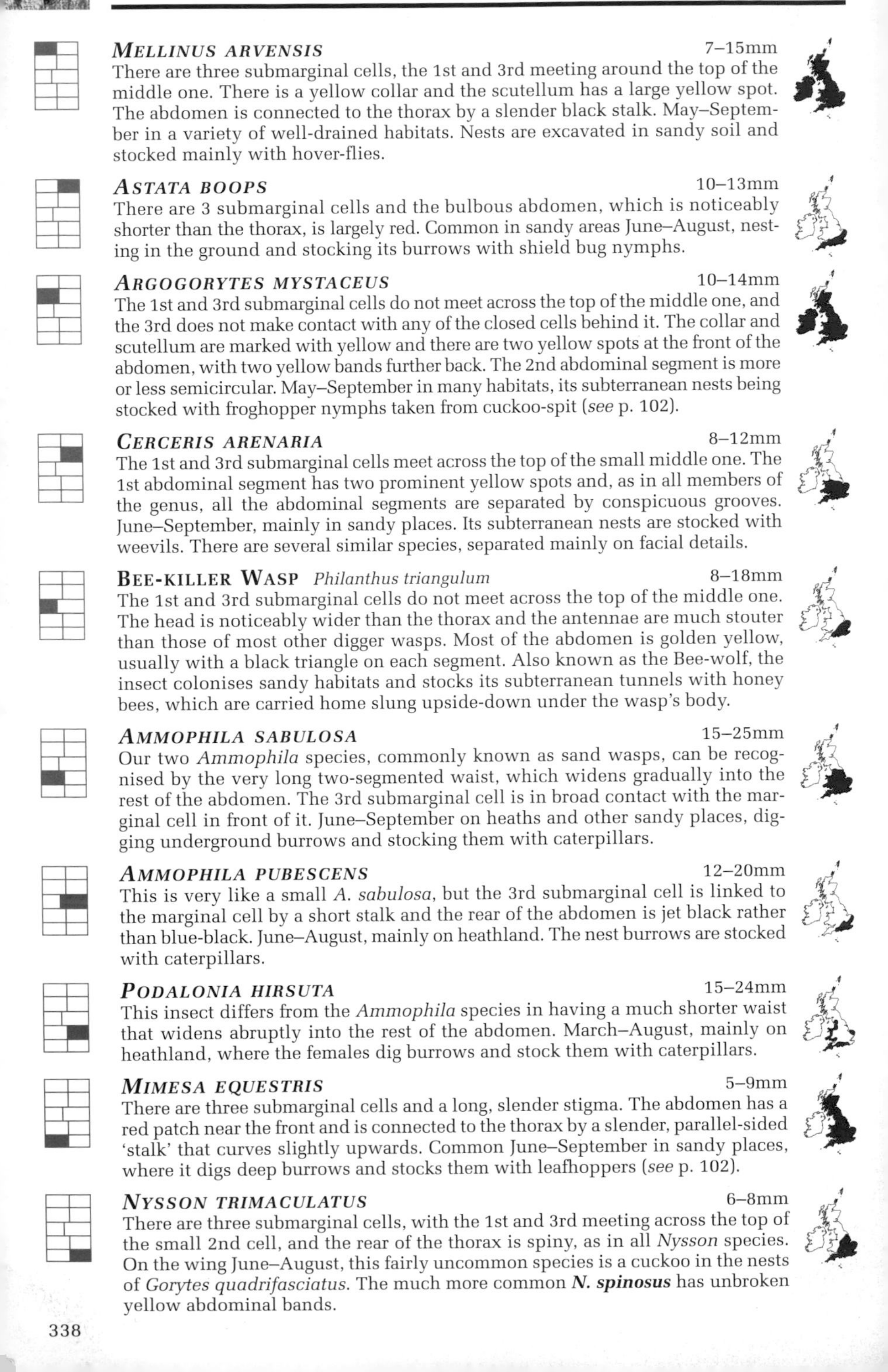

MELLINUS ARVENSIS 7–15mm

There are three submarginal cells, the 1st and 3rd meeting around the top of the middle one. There is a yellow collar and the scutellum has a large yellow spot. The abdomen is connected to the thorax by a slender black stalk. May–September in a variety of well-drained habitats. Nests are excavated in sandy soil and stocked mainly with hover-flies.

ASTATA BOOPS 10–13mm

There are 3 submarginal cells and the bulbous abdomen, which is noticeably shorter than the thorax, is largely red. Common in sandy areas June–August, nesting in the ground and stocking its burrows with shield bug nymphs.

ARGOGORYTES MYSTACEUS 10–14mm

The 1st and 3rd submarginal cells do not meet across the top of the middle one, and the 3rd does not make contact with any of the closed cells behind it. The collar and scutellum are marked with yellow and there are two yellow spots at the front of the abdomen, with two yellow bands further back. The 2nd abdominal segment is more or less semicircular. May–September in many habitats, its subterranean nests being stocked with froghopper nymphs taken from cuckoo-spit (*see* p. 102).

CERCERIS ARENARIA 8–12mm

The 1st and 3rd submarginal cells meet across the top of the small middle one. The 1st abdominal segment has two prominent yellow spots and, as in all members of the genus, all the abdominal segments are separated by conspicuous grooves. June–September, mainly in sandy places. Its subterranean nests are stocked with weevils. There are several similar species, separated mainly on facial details.

BEE-KILLER WASP *Philanthus triangulum* 8–18mm

The 1st and 3rd submarginal cells do not meet across the top of the middle one. The head is noticeably wider than the thorax and the antennae are much stouter than those of most other digger wasps. Most of the abdomen is golden yellow, usually with a black triangle on each segment. Also known as the Bee-wolf, the insect colonises sandy habitats and stocks its subterranean tunnels with honey bees, which are carried home slung upside-down under the wasp's body.

AMMOPHILA SABULOSA 15–25mm

Our two *Ammophila* species, commonly known as sand wasps, can be recognised by the very long two-segmented waist, which widens gradually into the rest of the abdomen. The 3rd submarginal cell is in broad contact with the marginal cell in front of it. June–September on heaths and other sandy places, digging underground burrows and stocking them with caterpillars.

AMMOPHILA PUBESCENS 12–20mm

This is very like a small *A. sabulosa*, but the 3rd submarginal cell is linked to the marginal cell by a short stalk and the rear of the abdomen is jet black rather than blue-black. June–August, mainly on heathland. The nest burrows are stocked with caterpillars.

PODALONIA HIRSUTA 15–24mm

This insect differs from the *Ammophila* species in having a much shorter waist that widens abruptly into the rest of the abdomen. March–August, mainly on heathland, where the females dig burrows and stock them with caterpillars.

MIMESA EQUESTRIS 5–9mm

There are three submarginal cells and a long, slender stigma. The abdomen has a red patch near the front and is connected to the thorax by a slender, parallel-sided 'stalk' that curves slightly upwards. Common June–September in sandy places, where it digs deep burrows and stocks them with leafhoppers (*see* p. 102).

NYSSON TRIMACULATUS 6–8mm

There are three submarginal cells, with the 1st and 3rd meeting across the top of the small 2nd cell, and the rear of the thorax is spiny, as in all *Nysson* species. On the wing June–August, this fairly uncommon species is a cuckoo in the nests of *Gorytes quadrifasciatus*. The much more common ***N. spinosus*** has unbroken yellow abdominal bands.

SPIDER-HUNTING WASPS: FAMILY POMPILIDAE

Mostly black or black and red, these are long-legged solitary wasps in which the hind tibia usually reaches well beyond the tip of the abdomen. The pronotum reaches back to the wing bases and most species have three submarginal cells. The antennae are proportionally longer than those of the digger wasps, almost as long as the body in some species. Females spend much of their time running rapidly over the ground, waving their wings and antennae as they search for spiders; flights tend to be short and rarely far above the ground. Males are generally smaller and, spending more time in the air, are less often noticed, although both sexes are often seen taking nectar from flowers. The females nearly all burrow in sandy ground and many have strong combs or brushes on their front legs. Each burrow is generally stocked with just a single paralysed spider, which is caught and dragged around until the wasp finds a suitable spot in which to bury it. Most digger wasps dig their burrows first and then look for their victims, which are normally carried back in flight. There are about 40 British species of spider-hunting wasps.

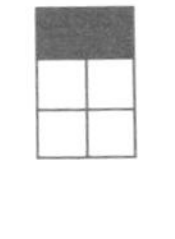

ANOPLIUS VIATICUS 7–15mm

The red bands at the front of the abdomen, together with the tuft or fan of black bristles at the rear, separate the female of this species from most other spider-hunters. The red is usually crossed by a narrow black line in the middle of each segment. Mature females occur throughout the year, although they spend the winter asleep in deep burrows. Males fly June–August, mainly on heaths and dunes. ***A. infuscatus*** is a little smaller and the red bands are not interrupted.

EPISYRON RUFIPES 5–15mm

This species is easily identified by its black and cream abdomen and reddish-brown hind tibiae. Whereas most British spider-hunters attack wolf spiders and other ground-living forms, this one goes for orb-web spiders, which it plucks deftly from their webs. May–September, mainly on heaths and dunes.

PRIOCNEMIS EXALTATA 5–12mm

The two dark bands and a pale spot near the tip of the forewing distinguish female *Priocnemis* species from most other spider-hunters, but the separation of individual species is very difficult. *P. exaltata* is one of our commonest pompilids, active from May–September in sandy areas nearly everywhere.

CRYPTOCHEILUS NOTATUS 7–15mm

The large size and the dark cloud near the tip of the forewing separate this from most other species, and the strongly curved vein at the outer end of the marginal cell should distinguish it from *Anoplius* species. Unlike most spider-hunters, it digs its burrow (or finds an existing one) and excavates several cells before going in search of spiders. June–September on heathland, but uncommon.

CEROPALES MACULATA 5–10mm

This species is easily identified by the white spots on its body and by the almost complete lack of bristles on its legs. The female has a visible sting sheath. The insect is a 'cuckoo', relying on various other pompilids to provide food and shelter for its offspring. Spotting another wasp dragging home a spider, the female nips in and lays an egg on the spider. The other wasp buries the spider as normal and lays her own egg on it, but the *Ceropales* egg hatches first and the larva eats the host egg before starting to devour the spider. June–September, wherever the host wasps occur. ***C. variegata*** is similar but the front of the abdomen is brown.

POTTERS and MASON WASPS: FAMILY EUMENIDAE

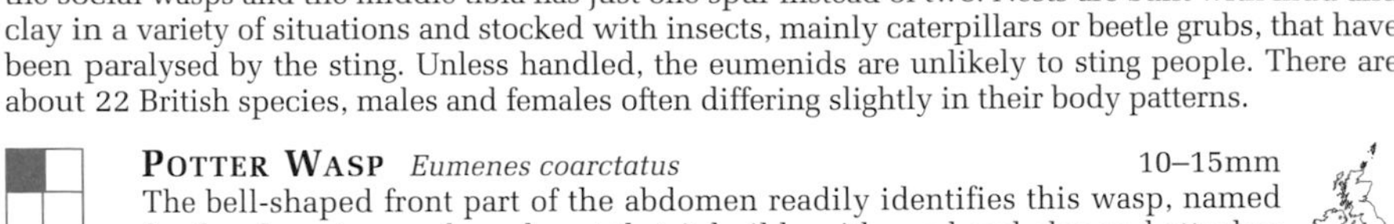

These are solitary wasps, although closely related to the social wasps (*see* p. 344) and resembling them in folding their wings lengthwise at rest and having a pronotum that reaches back to the wing bases. Nearly all are black and yellow, but the front of the abdomen tends to be more rounded than in the social wasps and the middle tibia has just one spur instead of two. Nests are built with mud and clay in a variety of situations and stocked with insects, mainly caterpillars or beetle grubs, that have been paralysed by the sting. Unless handled, the eumenids are unlikely to sting people. There are about 22 British species, males and females often differing slightly in their body patterns.

POTTER WASP *Eumenes coarctatus* 10–15mm

The bell-shaped front part of the abdomen readily identifies this wasp, named for the elegant vase-shaped nest that it builds with sand and clay and attaches to stones or low-growing vegetation (*see* p. 47). The vase is stocked with small hairless caterpillars. June–September, mainly on heathland.

ODYNERUS SPINIPES 8–12mm

The sides of the pronotum are smoothly rounded and the tegulae are rounded at the rear. There is no yellow on the rear of the thorax. Males' antennae are markedly thickened at the tips. Nests are dug in vertical sandy banks and stocked with weevil larvae. The female builds a protective 'chimney' over each nest entrance. May–July, wherever there are suitable nesting sites.

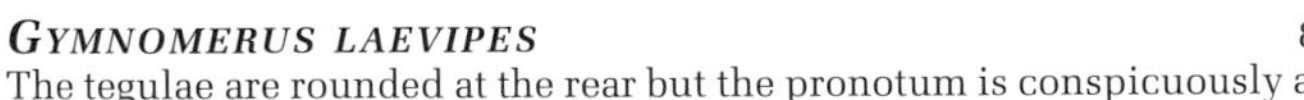

GYMNOMERUS LAEVIPES 8–11mm

The tegulae are rounded at the rear but the pronotum is conspicuously angled at the front corners. There is no yellow on the rear of the thorax. This is a rather local species, nesting May–August in hollow stems and stocking them with weevil larvae.

ANCISTROCERUS NIGRICORNIS 7–13mm

The tegulae are pointed at the rear, as in all *Ancistrocerus* species, but the reddish-brown tarsi and the conspicuously square black spot at the front of the abdomen distinguish this species from most of its relatives. The mud nests are constructed in all kinds of cavities and stocked with small caterpillars. Active April–September in a wide range of habitats; mated females hibernate and construct their nests in the spring. Often common around houses. ***A. gazella*** has a similar abdominal pattern but there is a dark smudge near the wing-tip.

ANCISTROCERUS PARIETUM 8–13mm

This common wasp has a triangular black mark on the 1st abdominal segment and there is a dark smudge near the wing-tip. It builds in a wide range of cavities May–August and provides small caterpillars for its young. There are several other, generally rarer species, but they are difficult to identify without close examination of the underside.

SYMMORPHUS CRASSICORNIS 8–15mm

This species has long pale brown hairs on the thorax and, as in all *Symmorphus* species, the 1st abdominal segment is more or less semi-circular or bell-shaped. The tegulae are pointed at the rear. There are four or five complete yellow bands on the abdomen. June–August, nesting in existing holes in timber and stocking them with the grubs of leaf beetles. ***S. mutinensis*** has less yellow on the pronotum and lacks the long brown hair.

SOCIAL WASPS: FAMILY VESPIDAE

These are the most familiar of our wasps, for they often live in and around our houses. They chew up wood and mix it with saliva to form paper with which they build their nests. In the summer, each nest may contain several thousand insects, most of which are sterile females called workers (*see* p. 35). Male wasps, recognised by their longer antennae, appear only in late summer. Only young, mated queens survive the winter, hibernating in log piles, sheds, attics and other similar places. They start new colonies in the spring. Adult wasps enjoy fruit and other sweet foods but the grubs are fed mainly on chewed-up insects, including many garden pests, so the wasps are really quite useful. The social wasps have deeply-notched eyes and they fold their wings longitudinally at rest, leaving much of the abdomen exposed. Queens and workers of the nine species living in the British Isles can be separated fairly easily by looking at their facial and thoracic patterns, but males tend to be more variable. Most species can be seen from the first warm days of spring through to the first frosts of autumn.

HORNET *Vespa crabro* 20–30mm
Easily identified by its large size and yellow and brown colour, this wasp usually nests in hollow trees and wall cavities and builds with very brittle yellowish paper. Although it has a powerful sting, it tends to be less aggressive than our other wasps.

COMMON WASP *Vespula vulgaris* 10–18mm
There are 4 large yellow spots at the rear of the thorax and the yellow stripe on each side of the pronotum is of uniform thickness. The vertical yellow band behind the eye is broken by a black spot and the face usually has a black anchor-like mark. Nests are built in holes, often in the ground but also in roof-spaces, and consist of yellowish paper with shell-like plates on the outside (*see* p. 36).

GERMAN WASP *Vespula germanica* 12–20mm
This is very like the Common Wasp but the yellow stripes on the pronotum bulge in the middle and the yellow band behind the eye is unbroken. The face usually has a triangle of three black spots. Nests are built in the same places as those of the Common Wasp but the paper is greyer.

RED WASP *Vespula rufa* 12–18mm
There are just two large yellow spots at the rear of the thorax and the front of the abdomen is usually tinged with red. The eye-notch is largely black and the face has a vertical black bar, sometimes expanded like an anchor at the bottom. Nests are usually on or under the ground and lack the shell-like plates of the previous two species. The relatively rare *V. austriaca* lives as a cuckoo in the nests of the Red Wasp and has no workers of it own. It has two large yellow spots at the rear of the thorax and the face has two or three black spots. The eye-notch is largely black and the tibiae bear long black hairs.

TREE WASP *Dolichovespula sylvestris* 12–18mm
There are two large yellow spots at the rear of the thorax and the face usually has a single black dot. The ball-shaped nests are grey and covered with smooth sheets and, although some are slung in trees and bushes, many are built on the ground.

NORWEGIAN WASP *Dolichovespula norwegica* 12–18mm
There are two large yellow spots at the rear of the thorax and the front of the abdomen usually has red patches on the sides. The face is divided by a thick vertical black bar and the eye-notch is largely black. The smooth grey, ball-shaped nests are usually built in trees and bushes. The Saxon Wasp (*D. saxonica*), only recently established in Britain, is very similar but never has red patches on the abdomen.

MEDIAN WASP *Dolichovespula media* 15–22mm
There are usually four yellow spots at the rear of the thorax and the face has a narrow black vertical bar. The eye-notch is completely yellow. The abdomen often has more black than our other wasps and the yellow is often replaced by orange in the queens. Nests are clothed with smooth grey sheets and are usually built in trees and bushes. Young nests have a tubular spout at the bottom. This species was not known in Britain until the 1980s.

BEES: FAMILY APIDAE

These hymenopterans feed primarily on flowers and the females use their stings purely for defence. The adults exist mainly on energy-rich nectar, while the young bees receive a mixture of nectar and pollen. The latter provides the proteins necessary for growth. As in the digger wasps (*see* p. 336), the pronotum does not extend back to the tegulae, but most bees can be distinguished from the wasps by their hairy legs and bodies – an adaptation to pollen-gathering. Most female bees carry the pollen back to their nests attached to specialised hair clusters known as pollen brushes or pollen baskets. The bees were once arranged in several different families, but the current trend is to put them all into a single family, the Apidae. Important features used in the identification of these insects include the submarginal cells (*see* p. 336) and the form of the tongue. The latter varies a great deal and its length is closely correlated with the kinds of flowers that are visited: only long-tongued bees can obtain nectar from deep-throated flowers. The Honey Bee and the bumble-bees are social insects (*see* p. 35), but all our other bees are solitary. Nevertheless, the females all construct nests for their young and provision them with sufficient food for their youngsters to reach maturity. There are about 250 British species.

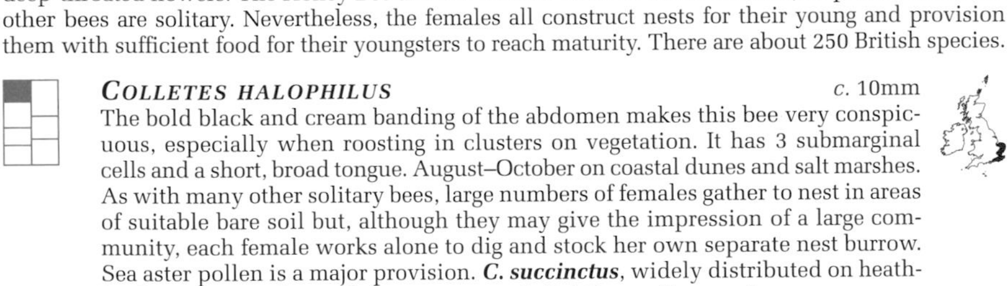

COLLETES HALOPHILUS — *c.* 10mm

The bold black and cream banding of the abdomen makes this bee very conspicuous, especially when roosting in clusters on vegetation. It has 3 submarginal cells and a short, broad tongue. August–October on coastal dunes and salt marshes. As with many other solitary bees, large numbers of females gather to nest in areas of suitable bare soil but, although they may give the impression of a large community, each female works alone to dig and stock her own separate nest burrow. Sea aster pollen is a major provision. ***C. succinctus***, widely distributed on heathland, is very similar, and there are several slightly smaller species.

ANDRENA CINERARIA — 10–13mm

Often known as mining bees because nearly all nest in the ground, *Andrena* species have fairly flat bodies and 3 submarginal cells. The tongue is short and pointed. Most species appear early in the spring and many are attracted to dandelions and sallow blossom. *A. cineraria* is easily recognised by the grey hairs on the head and thorax, although these are less obvious in the male. March–July in many open habitats, including parks and garden lawns, nesting in bare patches of ground.

ANDRENA HAEMORRHOA — 10–12mm

This bee is easily recognised by the rich brown thoracic hair and the shiny blackish-brown abdomen with its golden or reddish-brown tip. The male has brown facial hair and the female has white. March–June, on a wide range of flowers, including blackthorn. Burrows are dug in many open habitats, including parks and garden lawns.

ANDRENA FLOREA — *c.* 10mm

The first 2 abdominal segments of this uncommon little bee are largely reddish-brown. Although it visits a range of flowers for nectar, it collects pollen only from white bryony and this is where the females are most likely to be found. May–July, mainly on light soils around woodland margins and hedgerows.

TAWNY MINING BEE *Andrena fulva* — 10–13mm

The female, with her rich orange-red coat, is unlikely to be mistaken for anything else. The male is altogether more sombre and rarely noticed. The female is a common visitor to gooseberry and currant flowers April–June and frequently nests in the garden lawn, where she throws up little volcano-shaped mounds around her nest entrance.

ANDRENA FLAVIPES — 10–12mm

Named for the dense coat of yellow hair on its legs, this bee also has yellowish bands on the abdomen. March–September in a variety of open, sunny habitats, sometimes nesting in dense aggregations where the soil is suitable.

ANDRENA MARGINATA — 8–10mm

The female of this bee exists in three quite different forms. One has orange and black bands on the abdomen, as illustrated here; another has an entirely red or orange abdomen; and the third is largely black and resembles the male. July–September, feeding especially on scabious flowers on grassland and heathland and also in open woods.

HALICTUS TUMULORUM 6–8mm

Halictus species generally nest in subterranean burrows like *Andrena* species but they tend to be more slender and less flattened and can be found later in the year. Both genera have short, pointed tongues and three submarginal cells. They are best separated by looking at the basal vein near the centre of the wing: it is more or less straight in *Andrena* and gently curved in *Halictus*. In addition, *Halictus* females have a conspicuous bald patch at the tip of the abdomen. *H. tumulorum* is one of several slightly greenish species breeding in short grassland, including parks and garden lawns. Active April–July. In common with all *Halictus* species, mated females hibernate and begin nest-building in the spring.

LASIOGLOSSUM LEUCOZONIUM 8–10mm

This is one of several similar ground-nesting species with three submarginal cells and a short, pointed tongue. The shiny abdomen is banded with silvery hairs on the front of each segment, and the rear of each segment is black right to the edge. Grassland, light woodland and many other habitats in summer.

PANURGUS BANKSIANUS *c.* 8mm

This rather shiny black bee has two submarginal cells and a short, pointed tongue. The male abdomen is clothed with long black hair, while the female has brown hair and a very conspicuous golden pollen brush on each of her hind legs enabling her to carry large pollen loads back to her subterranean nest. July–September, mainly in sandy habitats.

CHELOSTOMA CAMPANULARUM 5–7mm

This is our smallest bee, slim enough to nest in old woodworm holes. It has 2 submarginal cells and a long, slender tongue, and specialises in feeding from harebells and other campanula flowers, including garden forms. The grey hairs are thin on top, but form a dense pollen brush under the female's abdomen. June–August in woods and gardens, where it may nest in old sheds and fences.

WOOL CARDER BEE *Anthidium manicatum* 10–15mm

Easily recognised by the golden spots on each side of the abdomen, this bee has 2 submarginal cells and a long, slender tongue. It is unusual in that the male is much larger than the female and also in that he maintains a territory, usually on a clump of flowers, from which he expels all but the female carder bees. The female cuts and combs fibres from the leaves of hairy plants with her large, toothed jaws and uses them to line her nest, which is usually constructed in a hollow stem or other pre-existing cavity. May–August in many habitats, including gardens.

OSMIA RUFA *c.* 10mm

This somewhat rotund bee has two submarginal cells and a long, slender tongue. The orange hair is much denser in the male. The latter has pale facial hair, while the female has black hair and also a pair of sturdy black horns just below her antennae. Often called the Red Mason Bee because of its liking for the mortar of old walls, it rakes out the mortar and constructs its nest cells in the cavity before rendering it over again. April–July nearly everywhere; very common in gardens.

OSMIA BICOLOR 10–12mm

The jet-black thorax and the bright orange abdominal hair make the female of this fairly rare bee easy to recognise. The male is yellowish grey all over. The female makes her nests in empty snail shells and covers them with dead grass stems and other plant debris when she has finished. April–July in grassland and in open woodland, mainly on lime-rich soils.

EUCERA LONGICORNIS 12–15mm

The male of this bee is easily recognised by his enormous antennae and yellow face. Both sexes have tawny hair on the thorax and a duller brown abdomen. April–August, nesting in the ground, mainly in rough grassland.

DASYPODA ALTERCATOR 12–15mm

There are two submarginal cells, but this relatively uncommon insect is easily identified by the dense yellow hair on the back legs. This is particularly long in the female. The male also has dense yellow hair on his face. June–September, mainly on heaths and coastal dunes, where it nests in the ground.

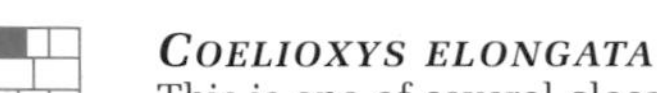

COELIOXYS ELONGATA c. 12mm

This is one of several closely related bees in which the females are all sharply pointed at the rear and the males bear a number of sharp spines. All the species are black and white and almost hairless, and they have two submarginal cells. June–September in many habitats, *C. elongata* is a cuckoo in the nests of leaf-cutter bees of the genus *Megachile*.

NOMADA FLAVA c. 12mm

This is one of many almost hairless, wasp-like bees that live as cuckoos in the nests of mining bees, mainly *Andrena* species. April–July, mainly in sandy habitats. *Nomada* species have three submarginal cells and can be distinguished from digger wasps by the fairly broad hind tarsi. The genus includes brown and yellow, brown and black, and red and black species, but most have brownish antennae. ***Sphecodes*** differs from *Nomada* in that the 1st and 3rd submarginal cells are more or less equal: in *Nomada* the 1st is much the larger. In addition, most *Sphecodes* species have black antennae.

LEAF-CUTTER BEE *Megachile centuncularis* 10–15mm

Best known for its habit of cutting neat semicircles from leaves, with its large jaws, the female can be distinguished from several related and less-common species by the bright orange pollen brush under her abdomen. There are two submarginal cells and, as in all leaf-cutter bees, the upper surface of the abdomen is scooped out at the front. May–August in many habitats; common in gardens. The leaf segments are used to make sausage-shaped nest cells in a variety of cavities.

ANTHOPHORA PLUMIPES 12–18mm

The female (A) of this fast-flying bee is jet black apart from her orange pollen brushes, while the male (B) is largely brown. He is easily mistaken for a bumble-bee, but can be recognised by his white face and the fans of long hair on his middle legs. March–June in flower-rich places, including gardens, where lungwort is a favourite flower. Nests are built in the ground and in the mortar of old walls.

MACROPIS EUROPAEA c. 10mm

There are two submarginal cells and the tongue is short and pointed. The female has pale yellow pollen brushes on her hind-legs, while the male hind-legs are enlarged and strongly curved. July–September in wetland areas, collecting pollen mainly from yellow loosestrife flowers. The nest burrows are lined with a waxy material probably derived from oils secreted by the flowers.

MELECTA ALBIFRONS c. 15mm

Named for the white hairs on the face, this bee is easily recognised by its black and white legs and the white spots around the edge of the abdomen. There are three submarginal cells. Active April–July, it is a cuckoo in the nests of *Anthophora plumipes*. It is not uncommon in gardens in the south.

XYLOCOPA VIOLACEA c. 20mm

This unmistakable bee is probably not a resident in the British Isles but it is being seen more and more often in the southern counties and may well establish itself soon. It nests in old timbers, where the cells are formed from coarse sawdust. Both sexes are active in late summer and again in spring after hibernation.

HONEY BEE *Apis mellifera* c. 12mm (workers), 16mm (drones and queens)

Apart from the bumble-bees, this is the most familiar of our bees, abundant in gardens and most other habitats from early spring until late in the autumn. As well as the numerous domestic hives, there are many wild colonies, often established in hollow trees and each containing thousands of bees. There are lots of races, with the abdomen ranging from largely orange to almost black, and they are often confused with *Andrena* species, but Honey Bees can always be identified by the very long, pointed marginal cell that reaches almost to the wing-tip. There are three submarginal cells and the insects have very long tongues. Most of the Honey Bees that we see are either workers or drones. The latter appear in the summer and have fatter bodies and longer antennae. Queens never leave the hive to forage.

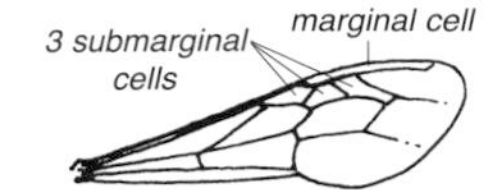

BUMBLE-BEES

The big, furry bumble-bees of the genus *Bombus* are the most familiar of our bees. They live in fairly small annual colonies, with only the young mated queens surviving the winter. These queens are most often seen in the spring, when they are seeking homes in which to rear new colonies. Most species breed in underground cavities, often making use of abandoned mouse nests. Others nest in dense vegetation on the ground or even settle in bird boxes. Bumble-bee colonies rarely contain more than a few hundred workers, and these can be seen from late spring to autumn, with those appearing early in the year noticeably smaller than later individuals. The descriptions and illustrations on this and the following pages are mostly of females (queens or workers) and most species can be distinguished by looking at the thoracic patterns and the colours of the stout hairs making up the pollen baskets on the back legs. Males, distinguished by their long antennae, may have different patterns and are usually seen only towards the end of the summer. Just 17 species inhabit the British Isles today, and only six of these are at all common and widespread. Many species have shown a marked decline in recent years, largely as a result of changing agricultural practices and loss of undisturbed, flower-rich habitats.

BOMBUS HORTORUM queen 17–22mm; worker 11–16mm
This relatively large bee has rather 'scruffy' long hair, a white tail and a yellow collar, scutellum and 1st abdominal segment. The face is distinctly longer than broad and has black hair. The pollen baskets are also black. It is one of our commonest species, occupying a wide range of habitats including gardens.

BOMBUS JONELLUS queen 15–18mm; worker 9–14mm
This species resembles *hortorum* but is smaller and 'neater' and the pollen baskets are reddish. The face is more or less rounded, about as broad as it is long. Although widely distributed, this bee is rather local and found mainly on heaths and moorland.

BOMBUS LUCORUM queen *c.* 20mm; worker 10–16mm
The lemon-yellow collar and 2nd abdominal segment, together with the white tail, readily identify this widespread and abundant species. The scutellum and 1st abdominal segment are black. Commonly known as the White-tailed Bumble-bee, it is one of the earliest bumble-bees to appear in the spring. ***B. magnus*** is difficult to separate from *B. lucorum* and some entomologists consider it to be just a subspecies of *lucorum*. It tends to be a little larger, with a wider yellow band on the abdomen and it occurs in upland areas in the north and west.

BOMBUS SOROEENSIS queen *c.* 16mm; worker 10–14mm
With its yellow collar, black scutellum and white tail, this species resembles *B. lucorum* but the yellow band on the 2nd abdominal segment is interrupted in the middle and the whole coat is rather more scruffy. Although widely distributed, this is one of our rarest bumble-bees.

BOMBUS TERRESTRIS queen *c.* 22mm; worker 10–16mm
The collar and 2nd abdominal segment are usually deep yellow or orange, much darker than in *B. lucorum*. The tail is buff or ginger in queens, but may be white in workers, which could then be confused with *lucorum*. The bee is common in gardens and many other habitats.

BOMBUS MONTICOLA queen 16–18mm; worker 10–15mm
The extensive red or orange, covering more than half of the abdomen, distinguishes this rather long-haired species from all other British bumble-bees. The collar is greyish yellow and the scutellum is either greyish yellow or black. As indicated by its scientific name, this bee is an upland species and confined mainly to areas of bilberry moor.

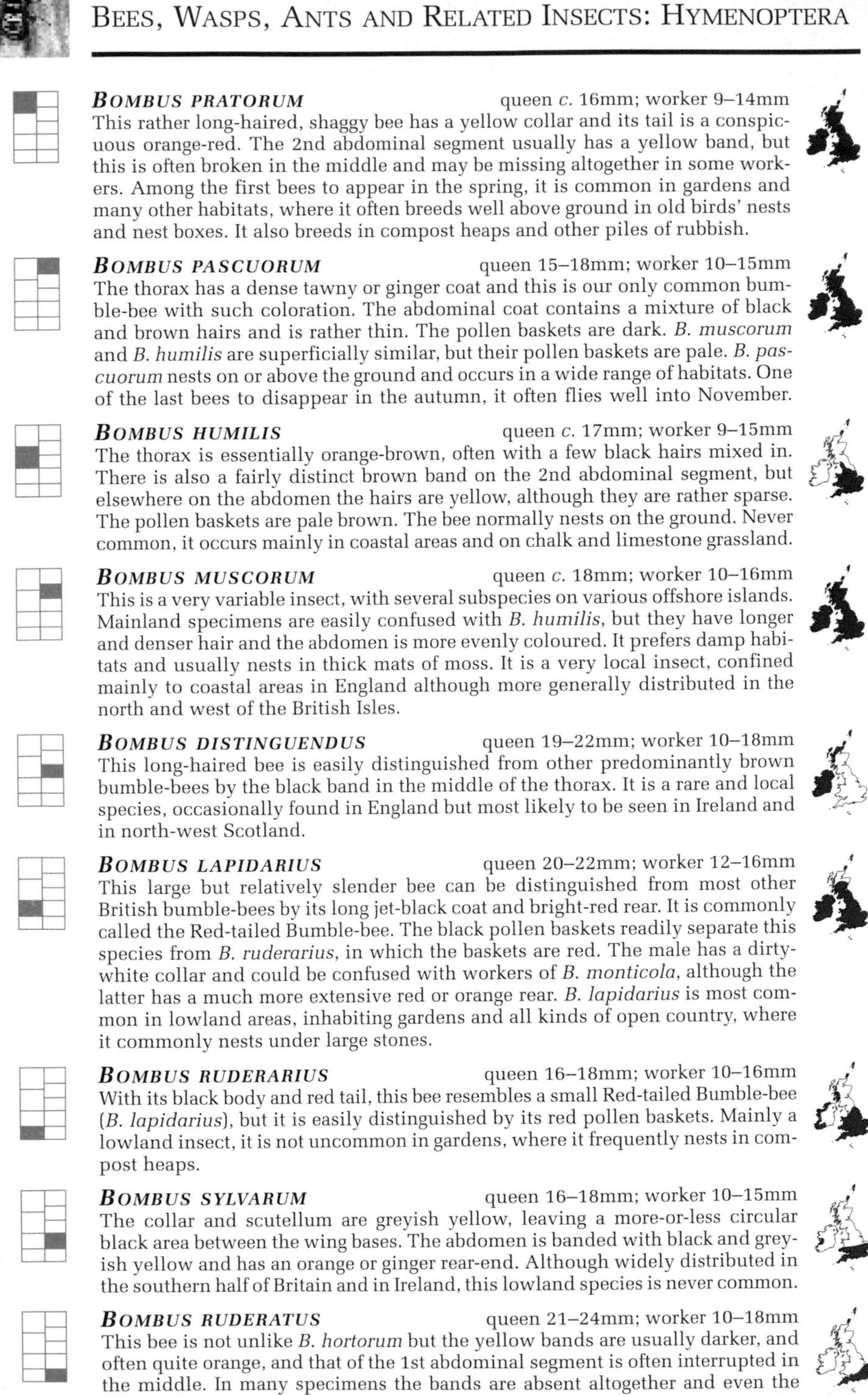

BOMBUS PRATORUM queen *c.* 16mm; worker 9–14mm
This rather long-haired, shaggy bee has a yellow collar and its tail is a conspicuous orange-red. The 2nd abdominal segment usually has a yellow band, but this is often broken in the middle and may be missing altogether in some workers. Among the first bees to appear in the spring, it is common in gardens and many other habitats, where it often breeds well above ground in old birds' nests and nest boxes. It also breeds in compost heaps and other piles of rubbish.

BOMBUS PASCUORUM queen 15–18mm; worker 10–15mm
The thorax has a dense tawny or ginger coat and this is our only common bumble-bee with such coloration. The abdominal coat contains a mixture of black and brown hairs and is rather thin. The pollen baskets are dark. *B. muscorum* and *B. humilis* are superficially similar, but their pollen baskets are pale. *B. pascuorum* nests on or above the ground and occurs in a wide range of habitats. One of the last bees to disappear in the autumn, it often flies well into November.

BOMBUS HUMILIS queen *c.* 17mm; worker 9–15mm
The thorax is essentially orange-brown, often with a few black hairs mixed in. There is also a fairly distinct brown band on the 2nd abdominal segment, but elsewhere on the abdomen the hairs are yellow, although they are rather sparse. The pollen baskets are pale brown. The bee normally nests on the ground. Never common, it occurs mainly in coastal areas and on chalk and limestone grassland.

BOMBUS MUSCORUM queen *c.* 18mm; worker 10–16mm
This is a very variable insect, with several subspecies on various offshore islands. Mainland specimens are easily confused with *B. humilis*, but they have longer and denser hair and the abdomen is more evenly coloured. It prefers damp habitats and usually nests in thick mats of moss. It is a very local insect, confined mainly to coastal areas in England although more generally distributed in the north and west of the British Isles.

BOMBUS DISTINGUENDUS queen 19–22mm; worker 10–18mm
This long-haired bee is easily distinguished from other predominantly brown bumble-bees by the black band in the middle of the thorax. It is a rare and local species, occasionally found in England but most likely to be seen in Ireland and in north-west Scotland.

BOMBUS LAPIDARIUS queen 20–22mm; worker 12–16mm
This large but relatively slender bee can be distinguished from most other British bumble-bees by its long jet-black coat and bright-red rear. It is commonly called the Red-tailed Bumble-bee. The black pollen baskets readily separate this species from *B. ruderarius*, in which the baskets are red. The male has a dirty-white collar and could be confused with workers of *B. monticola*, although the latter has a much more extensive red or orange rear. *B. lapidarius* is most common in lowland areas, inhabiting gardens and all kinds of open country, where it commonly nests under large stones.

BOMBUS RUDERARIUS queen 16–18mm; worker 10–16mm
With its black body and red tail, this bee resembles a small Red-tailed Bumble-bee (*B. lapidarius*), but it is easily distinguished by its red pollen baskets. Mainly a lowland insect, it is not uncommon in gardens, where it frequently nests in compost heaps.

BOMBUS SYLVARUM queen 16–18mm; worker 10–15mm
The collar and scutellum are greyish yellow, leaving a more-or-less circular black area between the wing bases. The abdomen is banded with black and greyish yellow and has an orange or ginger rear-end. Although widely distributed in the southern half of Britain and in Ireland, this lowland species is never common.

BOMBUS RUDERATUS queen 21–24mm; worker 10–18mm
This bee is not unlike *B. hortorum* but the yellow bands are usually darker, and often quite orange, and that of the 1st abdominal segment is often interrupted in the middle. In many specimens the bands are absent altogether and even the white tail may be inconspicuous. Once common over large areas of England, the bee is now scarce in many places and rarely seen further north than Norfolk.

CUCKOO BEES

Cuckoo bees belong to the genus *Psithyrus*, although recent work suggests that they are just another group of *Bombus* species. They get their name because, like their avian namesakes, they are social parasites: they lay their eggs in the nests of various bumble-bees, which then rear the 'cuckoo' grubs as their own. The cuckoo bees therefore have no worker caste. There are six British species, each usually sticking to a particular host species and normally resembling it quite closely, although the cuckoo bees can always be distinguished by their thinner coats through which the abdominal plates are clearly visible. In addition, the cuckoo bees have no pollen baskets and their back legs are clothed with dense hair. New females sleep through the winter but wake somewhat later than *Bombus* queens, when the host nests are already well-established. Cuckoo bees fly more slowly than bumble-bees and both sexes commonly cluster on thistles and similar flowers in the summer, although males seem to be more common than females. The latter are frequently attracted to dandelion flowers in the spring.

PSITHYRUS BARBUTELLUS — 15–19mm

With its yellow collar and scutellum and white tail, this species is very similar to its host, *Bombus hortorum*, but it can be distinguished by the yellow hairs on the top of the head as well as by the shiny abdomen. The male, pictured here, has white or grey hair on the thorax.

PSITHYRUS VESTALIS — 15–22mm

This species grows up in the nests of *Bombus terrestris*, from which it differs in having no yellow band at the front of the abdomen. In addition, *P. vestalis* has a conspicuous yellow patch on each side of its abdomen, just in front of the white tail. Although common in the south, it is rare north of the Humber and does not seem to have followed its host into Scotland or Ireland.

PSITHYRUS BOHEMICUS — 15–20mm

This cuckoo bee is very difficult to separate from *P. vestalis*, but the yellow collar is usually paler and the scutellum often has a band of pale hairs at the rear. The coat of *bohemicus* is also longer and more uneven than that of *vestalis*. Although its host is the widely distributed *Bombus lucorum*, this cuckoo is uncommon in the south and, unlike *vestalis*, it is more often found in the north and west of the British Isles.

PSITHYRUS CAMPESTRIS — 15–20mm

This species uses *B. pascuorum* as its host, although it does not look much like it. The collar is yellowish and there may be pale hairs on the scutellum. The female abdomen is mainly black, with brownish hairs towards the rear of each side, but the male abdomen is largely clothed with pale yellow hair. The species is common in most parts of the British Isles.

PSITHYRUS RUPESTRIS — 15–23mm

With its jet-black body and red tail, this species is very similar to its host *Bombus lapidarius*, but its wings are much browner and it is a good deal less common than *lapidarius*. It is probably our rarest *Psithyrus* species.

PSITHYRUS SYLVESTRIS — 13–16mm

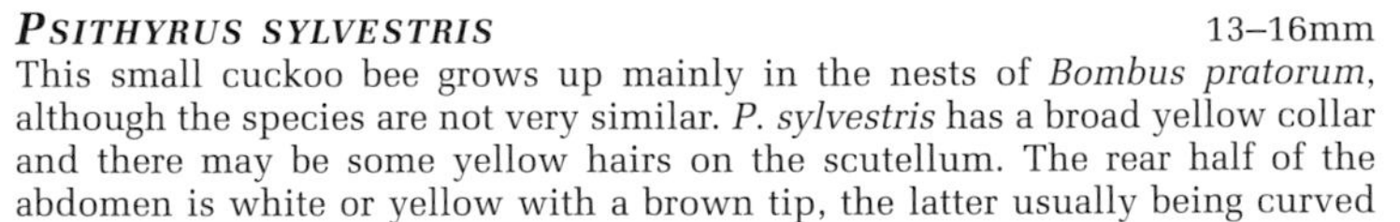

This small cuckoo bee grows up mainly in the nests of *Bombus pratorum*, although the species are not very similar. *P. sylvestris* has a broad yellow collar and there may be some yellow hairs on the scutellum. The rear half of the abdomen is white or yellow with a brown tip, the latter usually being curved under the body. The insect is common in many areas.

GLOSSARY

Technical terms have been kept to a minimum in this book and not all words listed here appear in the main text, but you may well meet some of them in other insect books referred to in the bibliography. Words printed in bold in the definitions have separate entries in the glossary.

ABDOMEN: the hindmost of the three main divisions of the insect body. It carries no legs in the adult although there may be some slender **appendages** at the rear.

ACULEATE: possessing a sting: used of certain groups of Hymenoptera.

ALAR SQUAMA: the central of three membranous lobes at or near the base of the wing in various flies.

ALITRUNK: the name given to the thorax and 1st abdominal segment of the wasp-waisted *Hymenoptera*, in which the 1st true abdominal segment is firmly fused with the **thorax**.

ALULA: the outermost of the three membranous lobes at the base of the wing in various flies.

ANNULATE: with ring-like markings.

ANTENNAE: the 'feelers': the pair of sensory appendages on the head used mainly for feeling and smelling. Some insects also use their antennae for detecting heat and sound.

ANTENODAL VEINS: the small cross-veins at the front of a dragonfly's wing, between the base and the notch or **nodus**.

APICAL: at or concerning the tip of any organ, such as a wing.

APPENDAGE: any limb or other organ, such as an antenna attached to the body by a joint.

APTEROUS: without wings.

APTERYGOTE: any member of the *Apterygota*, the suborder of insects, including the bristletails, that have never had wings during their history.

AQUATIC: living in water.

ARISTA: a bristle-like outgrowth from the **antenna** in various flies.

ARISTATE: bearing a bristle or **arista**.

AROLIUM: a small pad between the claws on an insect's foot.

BASAL: concerning the base of a structure, the part nearest to the body.

BASITARSUS: the 1st segment of the insect foot or tarsus, usually the largest segment.

BIPECTINATE: feathery, with branches sprouting from both sides of the main axis: applied mainly to **antennae**.

BRACHYPTEROUS: with short wings.

CALYPTER: the innermost of the three membranous lobes at the base of the wing in various flies. It is also known as the thoracic **squama** and it usually conceals the **haltere**.

CAMPODEIFORM LARVA: an elongated and flattened larva with well-developed legs and antennae. Many beetle larvae are of this type.

CARINA: a ridge or keel.

CASTE: any of the three or more distinct forms that make up the colonies of social insects. The usual castes are queen (fertile female), worker (sterile female), and male or drone.

CAUDAL: concerning the tail end.

CELL: an area of the wing bounded by a number of veins. Closed cells are completely surrounded by veins, while open cells are partly bounded by the wing margin. A large cell near the middle of the wing is often called *the* cell, especially in butterflies and moths.

Such a cell may also be called the **discal cell**.

CERCI: the paired appendages that spring from the tip of the abdomen in many insects. Examples include the pincers of earwigs and the outer 'tails' of mayflies.

CERVICAL: concerning the neck region, just behind the head.

CHAETAE: stiff hairs or bristles.

CHITIN: the tough, horny material making up the bulk of the insect's outer skeleton.

CILIATED: bearing minute hairs (cilia).

CLAVATE: club-shaped, applied especially to antennae.

CLAVUS: the rear part of the forewing of a heteropteran bug, the part nearest the centre when the wings are folded.

CLYPEUS: the lowest part of the insect face.

COCOON: a protective case made largely or completely from silk by a caterpillar or other insect larva in which it can pupate safely.

CONTIGUOUS: touching, used especially of eyes.

CORBICULA: the **pollen basket** on the hind-leg of many female bees, formed by stout hairs on the borders of the tibiae.

CORIUM: the main part of the wing of a heteropteran bug. It is generally horny and distinct from the membranous tip.

CORNICLE: one of the pair of small tubular outgrowths on the rear of an aphid's abdomen. They secrete waxes and other repellent substances.

COSTA: the front edge of the wing.

COXA: the basal segment of an insect leg, often immovably attached to the body.

CREMASTER: the hook or cluster of minute hooks at the rear end of the pupa of a butterfly or moth by which it clings to its support or grips the inside of its cocoon.

CROSS-VEIN: a short vein joining neighbouring longitudinal veins.

CRYPSIS: a form of camouflage or concealment in which animals simply blend in with their surroundings.

CUNEUS: a more-or-less triangular region of the forewing in certain heteropteran bugs (*see* p. 96–8). It abuts the membranous tip and is usually separated from the rest of the corium by a groove.

CURSORIAL: adapted for running.

DENTATE: toothed or sawlike.

DIAPAUSE: a period of suspended animation of regular occurrence in the lives of certain species, especially the young stages. It enables them to survive adverse conditions, such as hard winters.

DISCAL CELL: a conspicuous and often quite large cell near the middle of the wing, although not anatomically the same in all insects and may be bounded by different veins. It is sometimes simply called the **cell**, especially in butterflies and moths.

DISTAL: concerning that part of an appendage furthest from the body.

DORSAL: on or concerning the upperside or back.

DORSUM: the upper surface or back of the body. Also the rear edge of the forewing in a moth, which usually lies along the back when the insect is at rest.

ECDYSIS: the process of moulting, during which a young insect sheds its outer coat and replaces it with a new one (*see* p. 6).

ECTOPARASITE: a **parasite** living on the outside of its host. Fleas and lice are good examples.

ECTOPARASITOID: a **parasitoid** living on the outside of its host and always firmly fixed to it. Many ichneumon grubs are of this type.

Elbowed antennae: **antennae** that are distinctly angled, often near the middle. The antennae of ants are good examples.

Elytron (plural **elytra**): the tough, horny forewing of a beetle or earwig.

Emarginate: with a notched or indented margin.

Embolium: a narrow region close to the front edge of the forewing in certain heteropteran bugs and separated from the rest of the corium by a groove.

Empodium: a bristle or small pad between the claws on a fly's foot.

Endoparasite: a **parasite** that lives inside its host. The grubs of warble-flies are good examples.

Endoparasitoid: a **parasitoid** that lives inside the body of its host. Many ichneumon grubs live in this way, and so do those of many tachinid flies (*see* p. 34).

Endopterygote: any insect having a complete **metamorphosis**, with its wings developing inside the body and not visible until the pupal or chrysalis stage (*see* pp. 6–7). Butterflies and beetles are good examples.

Epiproct: an appendage arising from the midline of the last abdominal segment. It forms the central 'tail' in bristletails and some mayflies.

Eruciform: having a more-or-less cylindrical body, with stumpy legs at the rear as well as the three pairs of true legs on the thorax. Caterpillars are eruciform insects.

Exarate pupa: a pupa in which all the legs and other appendages are free and not enclosed with the rest of the body. Most beetle pupae are of this kind.

Excavate: hollowed out or concave. The coxae of many beetles have a concave surface that receives the **femur** when the legs are folded.

Exopterygote: any insect in which the wings develop gradually on the outside of the body. Examples include grasshoppers and dragonflies (*see* p. 6).

Exuvium: the cast-off skin left behind after **ecdysis** or moulting.

Family: an important category in the classification of living things, containing a number of genera related by common descent and sharing various features, although if a genus is markedly different from any others it might be given a family to itself. Family names in the animal kingdom all end in -idae, whereas those in the plant kingdom end in -aceae.

Femur: the 3rd and usually the largest segment of the insect leg.

Filiform: threadlike or hairlike.

Flabellate: with projecting flaps on one side, as in the antennae of chafer beetles (*see* p. 112).

Flagellum: the distal or outer part of an antenna, beyond the 2nd segment. It may be composed of one or many segments.

Fossorial: adapted for digging.

Frenulum: the wing-coupling mechanism found in many moths. It consists of one or more stout bristles curving forward from the base of the hindwing and held in place by a hook under the forewing.

Frons: the upper region of an insect's face, between and below the **antennae**.

Gall: an abnormal growth on or in a plant or a fungus, caused by the presence of a young insect or some other organism, such as a mite or a fungus, in the tissues. The gall-causers obtain both food and shelter from the host plant. Gall wasps (*see* p. 326), gall midges (*see* p. 300), sawflies (*see* p. 320) and aphids (*see* p. 104) are major gall-causing insects.

Gaster: the main part of the abdomen in bees, wasps, and other hymenopterans. The front part of the gaster is often narrowed to form a waist. The true 1st segment of the abdomen is fused to and functionally part of the thorax.

Gena: the cheek, that part of the face below and behind the eye.

Geniculate: bent at a sharp angle.

Genitalia: the copulatory and egg-laying apparatus, the shape and arrangement of which may be used to distinguish closely related and otherwise very similar species.

Genus (plural **genera**): a group of closely related species. The name of the genus forms part of the scientific name of every species within that genus. For example, our tiger beetles all belong to the genus *Cicindela*, with the Green Tiger Beetle being called *Cicindela campestris* and the Wood Tiger Beetle being *Cicindela sylvatica*.

Gill: a type of breathing organ possessed by many aquatic creatures, including many young insects. The gills are very delicate outgrowths or flaps containing lots of fine air-tubes, and oxygen from the water passes through the thin skin and into the air-tubes or **tracheae**.

Glabrous: without hairs.

Haltere: One of the pair of pin-shaped 'balancers' found in the true flies. They represent much-modified hindwings.

Hamuli: the minute hooks that link the front- and hind-wings together in bees and wasps and other hymenopterans. The hooks are situated on the front edge of the hind-wing and they latch on to a fold on the rear of the forewing.

Hemelytron: the forewing of a heteropteran bug, consisting of a horny basal area (the **corium**) and a membranous tip.

Hemimetabolous: having an incomplete **metamorphosis**, with no pupal stage in the life history (*see* p. 6).

Heteromerous: having unequal numbers of segments on the three pairs of legs. This term is used mainly when dealing with beetles.

Holometabolous: having a complete **metamorphosis**, with a pupal stage in the life cycle.

Holoptic: with eyes touching or almost touching on the top of the head. The term is used mainly when dealing with flies and dragonflies.

Honeydew: the sugary liquid emitted by aphids and some other sap-sucking bugs. In order to obtain sufficient proteins and other nutrients, the insects have to imbibe large quantities of sugar-rich sap, much of which is voided through the anus as honeydew.

Host: the organism, plant or animal, that is attacked by a **parasite**.

Hyaline: clear and colourless, like the wings of many dragonflies, flies and bees.

Hyperparasite: a parasitic organism that attacks another **parasite**.

Hypognathous: having a vertical head and face, with the mouthparts at the bottom: the normal arrangement in insects.

Hypopleural bristles: a curved row of bristles on the side of the thorax of blow-flies and tachinid flies (*see* p. 314), just below the **haltere** and just above the base of the hind-leg.

Imago (plural **imagines**): the adult insect.

Inquiline: an animal that shares the home of another species without having any obvious effect on that species. Many galls, for example, give food and shelter to a range of insects in addition to the rightful occupant.

Instar: a stage in an insect's development between any two moults. A 1st-instar larva, for example, is an individual that has not yet moulted. A 2nd-instar insect is between its 1st and 2nd moults, while the final instar is the adult insect.

Integument: the outer coat.

Joint: strictly speaking, a joint is the articulation between any two neighbouring parts, such as the tibia and femur of a leg, but the word is commonly used as a synonym of **segment** to mean any section of a leg or other **appendage**.

Keel: a narrow ridge, also known as a **carina**.

Larva: the name given to a young insect that is markedly different in form from the adult and has no trace of wings. Caterpillars and maggots are good examples.

Lateral: concerning the sides.

Macropterous: with fully developed wings.

Mandible: the jaw of an insect. It is usually sharp and toothed and adapted for biting and chewing solid food, as in dragonflies, grasshoppers and wasps. Mandibles are usually completely absent in butterflies and moths and in most flies, although those of mosquitoes are drawn out to form slender needles.

Mandibulate: having biting mandibles or jaws.

Marginal cell: one of the cells bordering the front edge of the wing, especially in the outer region.

Melanism: an increase in the amount of the dark pigment called melanin, causing an animal to be much darker than normal. Many moths exhibit this condition and are known as melanic individuals.

Mesonotum: the dorsal or upper surface of the **mesothorax**, the 2nd and usually the largest of the thoracic segments.

Mesoscutellum: *see* **Scutellum**.

Mesothorax: the 2nd thoracic segment.

Metamorphosis: the complete spectrum of changes that take place during an insect's life as it matures from a young animal into an adult. When these changes are gradual, as in grasshoppers and bugs (*see* p. 6), metamorphosis is said to be partial or incomplete, but when there is a sudden and major change from young to adult form, as in butterflies and moths, metamorphosis is said to be complete (*see* p. 7).

Metanotum: the dorsal or upper surface of the **metathorax**, the 3rd of the thoracic segments. It is often very small and may not be visible from above.

Metatarsus: the basal and usually the largest segment of the tarsus.

Metathorax: the 3rd and usually the smallest thoracic segment.

Moniliform antenna: an antenna composed of well-separated bead-like segments.

Nodus: the kink or notch on the front edge of a dragonfly's wing.

Nymph: the name given to the young stages of insects that undergo a partial **metamorphosis**. The nymph usually resembles the adult in shape, although it has no functional wings, and it gets more like the adult at each moult. (*see also* **larva**.)

Obtect pupa: a **pupa** in which the legs and other **appendage**s are completely enclosed with the rest of the body, although their outlines may be clearly visible.

Occiput: the hindmost region of the top of the head, just in front of the neck membrane.

OCELLUS (plural **OCELLI**): one of the simple eyes of an insect, usually carried in a triangular group of three on the top of the head, although absent in many insects.

OOTHECA: an egg-case.

OVIPAROUS: reproducing by laying eggs.

OVIPOSITOR: the egg-laying apparatus of a female insect, used to place the eggs in the appropriate position. It is concealed in many insects, but very conspicuous in some bush crickets and parasitic hymenopterans, in some of which it exceeds the length of the body (*see* p. 332).

PALP: one of a pair of short, leg-like **appendages** around the mouth. The palps play a major role in tasting food.

PARASITE: an organism that lives on or in another species, known as the **host**, taking food from it and giving nothing in return, although the host is not usually killed. **Ectoparasites** live on the outside of the host's body, while **endo-parasites** live inside the host. Most parasitic insects are ectoparasites and their hosts are generally much larger, usually birds and mammals.

PARASITOID: an organism that spends its early life living parasitically on or in another species, but differs from normal parasites in that it usually kills its host – but not until the parasitoid itself has matured and has no further need of the host. Most parasitoids are insects and their hosts are also insects. Ichneumons and chalcid wasps (*see* p. 332) are the most abundant of the parasitoids. There may be anything from one to several hun-dred parasitoids in a single **host**.

PARTHENOGENESIS: a form of reproduction in which eggs develop normally without being fertilised. It is common among gall wasps (*see* p. 326) and aphids (*see* p. 104).

PECTINATE ANTENNA: an antenna with branches arising on one or both sides of the main axis, like the teeth of a comb.

PEDICEL: the 2nd segment of the **antenna**, and also the name given to the narrow waist of an ant.

PETIOLATE: attached by a narrow stalk.

PETIOLE: the narrow waist of bees and wasps and their relatives, although usually called the **pedicel** when referring to ants.

PICTURED WING: a heavily mottled or patterned wing characteristic of certain flies (*see* p. 314).

PILOSE: densely clothed with hair.

PLUMOSE: feathery, especially applied to **antennae**.

POLLEN BASKET: the pollen-carrying region on the hind-leg of a bee. Also known as the **corbicula**, it is bordered by stout hairs that form a flexible basket. (*See also* **Scopa**.)

PORRECT ANTENNA: an **antenna** that extends horizontally forward from the head.

POSTERIOR: concerning the rear.

POSTSCUTELLUM: a small division of the mesonotum, usually very small or absent, but well-developed in tachinid flies (*see* p. 314).

PRE-APICAL: arising just before the tip. Many flies, for example, have pre-apical bristles just before the tip of the tibia.

PROBOSCIS: the name given to various kinds of sucking mouths, in which some of the components have been drawn out to form tubes.

PROGNATHOUS: having a more-or-less horizontal head, with the mouthparts at the front.

PROLEG: one of the stumpy, fleshy legs on the abdominal region of a caterpillar.

PRONOTUM: the dorsal surface of the 1st thoracic segment, often rather small, but very large in many beetles and bugs.

Propodeum: the 1st segment of the abdomen in bees, wasps and other apocritan hymenopterans (*see* p. 326). It is completely fused to the rear of the thorax and functionally part of the thorax. (*See* **Gaster**.)

Prothorax: the 1st thoracic segment.

Proximal: concerning the basal part of an **appendage**: the part nearest to the body.

Pruinose: covered with a powdery deposit, usually white or pale blue. The term is especially used in connection with dragonflies and damselflies (*see* p. 68).

Pterostigma: a small coloured cell near the wing-tip of dragonflies and various other clear-winged insects; commonly abbreviated to stigma.

Pterygote: any member of the subclass Pterygota, which includes all insects apart from the bristletails (*see* p. 11).

Pubescent: covered with short, soft hair.

Pulvillus: the little pad beneath each claw on a fly's foot.

Punctate: covered with tiny pits or depressions. Many beetle **elytra** are so described, and so are the thoraxes of many hymenopterans.

Pupa: the 3rd major stage in the life history of butterflies and moths and other insects with a complete **metamorphosis**. It is during the pupal stage, which does not feed and does not usually move about, that the larval body is broken down and rebuilt into the adult form.

Puparium: the barrel-shaped case enclosing the pupa of many true flies.

Pupate: to pupate is to turn into a pupa.

Quadrilateral: a cell near the base of a damselfly's wing, whose shape is important in separating some of the families.

Raptorial: adapted for seizing prey.

Reticulate: covered with a network pattern.

Rostrum: a beak or snout, applied mainly to the piercing mouthparts of bugs and the elongated snouts of weevils.

Scape: the 1st segment of the **antenna**, the term being used especially when this segment is longer than any of the others.

Scarabaeiform larva: a larva with a thick, soft body, a well-developed head and thoracic legs and no legs in the abdominal region. It is often permanently curved into a 'C'. The larvae of stag beetles and chafers (*see* p. 8) are of this type.

Scopa: the pollen-collecting apparatus of a bee, which may be in the form of a **pollen basket** on the hind-leg or a brush of hairs on the abdomen.

Scutellum: the usual meaning of this term is the rear section of the **mesonotum**. It is commonly triangular and often quite small, although it is very large in certain bugs (*see* p. 96–8).

Segment: one of the rings or divisions of the body, or any section of a jointed limb between two joints.

Serrate: toothed or sawlike.

Sessile: fixed to one point and unable to move about or, when referring to anatomical features, attached directly to a neighbouring part without any form of stalk.

Seta: a bristle.

Setaceous: bristle-like.

Species: the basic unit of classification, consisting of a group of individuals that all look more or less alike and that can interbreed to produce another generation of similar organisms.

Spiracle: one of the breathing pores through which insects obtain their air supplies. Spiracles occur on most segments of the body and lead into the tracheal system, which carries air to all parts of the body. Spiracles are most clearly seen on the sides of large, non-hairy caterpillars.

Spur: a large, usually movable spine, found mainly on the legs.

Spurious vein: a false vein formed by a thickening of the wing membrane and not normally connected to any of the true veins.

Squama (plural **squamae**): any of the membranous flaps that arise at or near the base of the wing in many true flies.

Stigma (plural **stigmata**): the name given to any of three spots, of various shapes, near the middle of the forewing of certain moths, especially those of the family Noctuidae (*see* p. 222). (*see also* **Pterostigma**.)

Striae: grooves running along or across the body; applied especially to the grooves on beetle **elytra**.

Stridulation: the production of sounds by rubbing 2 parts of the body together; best known in grasshoppers and their relatives (*see* pp. 15–16).

Sub-apical: situated just before the tip or apex of a structure.

Sub-costa: usually the first of the longitudinal veins behind the front edge of the wing, although often very faint. Abbreviated to Sc.

Sub-imago: found only among the mayflies, the sub-imago or dun is the winged insect that emerges from the nymphal skin (*see* p. 66).

Submarginal cell: any of the cells lying just behind the **pterostigma** in the hymenopteran forewing. Their number and arrangement are important in the classification and identification of bees and wasps.

Suture: a groove on the body surface; also the line where the two **elytra** of a beetle meet.

Synonym: one of two or more names that have been given to a single species. The earliest name usually takes precedence.

Tarsus: the insect's foot, usually consisting of several segments.

Tegmen (plural **tegmina**): the leathery forewing of a grasshopper, cricket or cockroach.

Tegula: a small lobe or scale overlying the base of the forewing like a shoulder pad.

Tergum: the dorsal surface of any body segment.

Thorax: the middle section of the insect body, bearing the legs and the wings.

Tibia: the leg segment between the femur and the **tarsus**.

Trachea: one of the minute tubes that permeate the insect body and carry air from the **spiracles** to all parts of the body.

Triangle: a triangular region near the base of a dragonfly wing, often divided into smaller cells and playing an important role in classification.

Triungulin: the active 1st-instar larva of oil beetles and a few other beetles. It appears to have three claws on each foot, although two of these are actually stout bristles.

Trochanter: a small and easily overlooked segment of the leg between the **coxa** and the **femur**.

Truncate: squared off or ending abruptly.

Tymbal: the sound-producing 'drum-skin' of a cicada.

Tympanum: the 'ear-drum', which may be situated on various parts of the body, but never on the head; in crickets it is on the front legs.

Ventral: concerning the ventral or undersurface of the body.

Vertex: the top of the head, between and behind the eyes.

Vestigial: poorly developed.

Viviparous: bringing forth active young instead of laying eggs.

FURTHER READING

Askew, R.R. (1988). *The Dragonflies of Europe*. Harley Books.

Askew, R.R. (1971). *Parasitic Insects.* Heinemann.

Balfour-Browne, F. (1940–1958). *British Water Beetles*, Vols I, II & III. Ray Society. (These books are out-of print but a CD-ROM is available from Pisces Conservation, IRC House, The Square, Pennington, Lymington, Hampshire SO41 8GN.)

Bellman, H. (1988). *A Field Guide to the Grasshoppers and Crickets of Britain and Northern Europe*. Collins.

Brooks, S., and Lewington, R. (2004). *Field Guide to the Dragonflies and Damselflies of Great Britain and Ireland*, 2nd edn. British Wildlife Publishing.

Carter, D.J., and Hargreaves, B. (2001). *A Field Guide to Caterpillars of Butterflies and Moths in Britain and Europe.* Collins.

Chinery, M. (1998). *Butterflies of Britain and Europe.* Collins. (A photographic guide in the *Wildlife Trust Guide* series.)

Chinery, M. (1993). *Insects of Britain and Northern Europe*, 3rd edn. Collins.

Csóka, G. (2003). *Leaf Mines and Leaf Miners.* Agroinform Kiadó, Hungary. (A photographic guide with text in Hungarian and English.)[1]

Csóka, G. (1997). *Plant Galls.* Agroinform Kiadó, Hungary. (A photographic guide with text in Hungarian and English.)[1]

Csóka, G., and Kovács, T. (1999). *Xylophagous Insects.* Agroinform Kiadó, Hungary. (A photographic guide with text in Hungarian and English.)[1]

d'Aguilar, J., Dommanget, J-L., and Préchac, R. (1986). *A Field Guide to the Dragonflies of Britain, Europe and North Africa.* Collins.

Gibbons, B. (1986). *Dragonflies and Damselflies of Britain and Northern Europe.* Country Life Books.

Gibbons, B. (1995). *Field Guide to Insects of Britain and Northern Europe.* The Crowood Press.

Goater, B. (1986). *British Pyralid Moths.* Harley Books.

Hammond, C.O., and Merritt, R. (1985). *The Dragonflies of Great Britain and Ireland*, 2nd edn. Harley Books.

Lewington, R. (2003). *Pocket Guide to the Butterflies of Great Britain and Ireland.* British Wildlife Publishing.

Marshall, J.A., and Haes, E.C.M. (1988). *Grasshoppers and Allied Insects of Great Britain and Ireland.* Harley Books.

Nelson, B., and Thompson, R. (2004). *The Natural History of Ireland's Dragonflies.* The National Museums and Galleries of Northern Ireland.

Redfern, M., Shirley, P., and Bloxham, M. (2002). *British Plant Galls.* Field Studies Council.

Riley, A.M., and Prior, G. (2003). *British and Irish Pug Moths.* Harley Books.

Southwood, T.R.E., and Leston, D. (1959). *Land and Water Bugs of the British Isles.* Warne. (This book is out-of-print but a CD-ROM version is available from Pisces Conservation, IRC House, The Square, Pennington, Lymington, Hampshire SO41 8GN.)

Stubbs, A.E. (2003). *Dipterists' Forum Starter Pack.* Biological Records Centre, Huntingdon. (An extensive list of references relevant to the identification of flies, available from Centre for Ecology and Hydrology, at Monks Wood, Abbots Ripton, Huntingdon, Cambs PE28 2LS.)

Stubbs, A.E., and Falk, S.J. (1983). *British Hoverflies.* British Entomological and Natural History Society.

Thomas, J. (1986). *Butterflies of the British Isles.* Country Life Books.

Tolman, T., and Lewington, R. (2004). *Field Guide to Butterflies of Britain and Europe.* Collins.

Waring, P., Townsend, M., and Lewington, R. (2003). *Field Guide to the Moths of Great Britain and Ireland.* British Wildlife Publishing.

[1] These Hungarian titles are available from Pemberley Books (**www.pembooks.demon.co.uk**)

The *Naturalists' Handbooks* published by The Richmond Publishing Co. include many titles of interest to the entomologist. The books include keys for identification and a wealth of biological information as well as helpful tips on techniques for studying the various groups. There are also good bibliographies. Titles available at present include:

Brown, V.K. *Grasshoppers*
Erzinclioğlu, Z. *Blowflies*
Forsythe, T.G. *Common Ground Beetles*
Gilbert, F.S. *Hoverflies*

Harker, J. *Mayflies*
Kirk, W.D.J. *Thrips*
Majerus, M., and Kearns, P. *Ladybirds*
Miller, P.L. *Dragonflies*
Morris, M.G. *Weevils*
Prys-Jones, O.E., and Corbett, S.A. *Bumblebees*
Redfern, M., and Askew, R.R. *Plant Galls*
Skinner, G.J., and Allen, G.W. *Ants*
Snow, K.R. *Mosquitoes*
Wheater, C.P., and Cook, P. *Studying Invertebrates*
Yeo, P.F., and Corbett, S.A. *Solitary Wasps*

The Royal Entomological Society of London publishes a series of *Handbooks for the Identification of British Insects*. Although quite technical, these provide extremely useful information on the biology and distribution of our insects as well as identification keys. A list of titles currently available can be obtained from *Royal Entomological Society*, 41 Queen's Gate, London, SW7 5HR or from **www.royensoc.co.uk**

Useful and interesting entomological journals include:

Atropos – for butterfly, moth and dragonfly enthusiasts **www.atropos.info**
The Coleopterist **www.coleopterist.org.uk**
Entomologist's Gazette **www.gempublishing.co.uk**
Entomologist's Monthly Magazine **www.gempublishing.co.uk**
The Entomologist's Record and Journal of Variation **www.entrecord.com**
British Wildlife also contains many interesting entomological articles **www.britishwildlife.com**

USEFUL ADDRESSES

The following societies all publish regular bulletins or journals. The addresses may be those of the secretaries of the organisations and may change from time to time.

Amateur Entomologists' Society, PO Box 8774, London, SW7 5ZG
www.amentsoc.org

British Dragonfly Society, c/o The Haywain, Hollywater Road, Borden, Hampshire GU35 0AD
www.dragonflysoc.org.uk

The British Entomological and Natural History Society, The Pelham-Clinton Building, Dinton Pastures Country Park, Davis Street, Hurst, Reading, Berkshire RG10 0TH
www.benhs.org.uk

British Naturalists' Association, PO Box 5682, Corby, Northants NN17 2ZW
www.bna-naturalists.org

British Plant Gall Society, c/o 2 The Dene, Nettleham, Lincolnshire LN2 2LS
www.btinternet.com/~bpgs

Butterfly Conservation, Manor Yard, East Lulworth, Wareham, Dorset BH20 5QP
This organisation also deals with moths.
www.butterfly-conservation.org

BWARS (*The Bees, Wasps and Ants Recording Society*) c/o Nightingales, Haslemere Road, Milford, Surrey GU8 5BN
www.bwars.com

Dipterists Forum, c/o British Entomological and Natural History Society, The Pelham-Clinton Building, Dinton Pastures Country Park, Davis Street, Hurst, Reading, Berkshire RG10 0TH
www.dipteristsforum.org.uk

Field Studies Council (*FSC*), Montford Bridge, Preston Montford, Shrewsbury, Shropshire SY4 1HW
Courses on a wide range of natural history and countryside topics are held at 12 centres in England and Wales. The FSC also publishes the AIDGAP identification guides.
www.field-studies-council.org

Freshwater Biological Association, The Ferry House, Far Sawrey, Ambleside, Cumbria, LA22 0LP
The Association's numerous publications include keys to many groups of freshwater insects.
www.fba.org.uk

INDEX

PHOTO CREDITS

All photographs were taken by the author with the exception of those in the list below. Each photograph is referenced by a page number followed by the species order, as it appears in the text, in parentheses.

Ted Benton 353(4), 355(5), 357(3);
Nigel Catlin 97(5), 317(8);
Robin Chittenden/FLPA 107(1);
Paul Cleary-Pugh 157(5), 159(4;5;6;9), 161(2;8), 169(11), 171(3), 175(3;5;8), 183 (6), 187(2;3;8), 189(4;5), 191(4;8;10), 195 (5), 197(3), 201(4;6;7), 203 (2;3;8), 205 (5;8;9), 211(5), 213(5;6), 221(4;5), 223(3), 227(7), 229(2), 231(6;7), 233(1;2), 237(9), 239(2;3), 241(1;3;10), 243(8), 249(5), 251(10), 253(9), 255(1;6), 261(7;8;9);
György Csóka 75(7), 77(3), 79(4a), 103(5), 105(1), 111(5), 121(2;3), 123(4;5), 129(7), 131(2), 135(8), 157(2;6), 221(10), 279(4), 281(2;10), 285(8;10;11), 287 (1;3), 289(1), 295(4;6;7), 297 (1;2;4;9), 305(2), 317(4);
Stan Dumican 69(1;5), 71(6), 73(8), 75(3a;3b;5a;5b), 77(2a;2b), 161(4), 163(3), 165(3;4;9), 167(1;2;3;4;6), 169(2;5), 171(2; 7;8;11;12), 173(1a;1b;4;10), 175(1;4), 177(2;5;6;9), 179(1;4;5;9), 181(1;3;4;6;8; 11), 183(4;5;7), 185(1;6;7;9;10), 187(1;7;9), 189(1;2;3;8;9), 191(1;3;5;6), 193(5,10), 195(1;2;7;8), 197(2;5), 199(3;10), 201(2;8; 10), 203(1;5;6), 205(1;4;7), 209(4;6), 211(4), 215(2a), 219(10), 223(2;6), 225(5; 9), 227(1;4;5;9), 229(5;6;8;9;11), 231(3;4), 233(5;7;11), 235(1;5), 237(1;3), 239(5;7;8; 9), 245(5;6;8), 251(2), 253(1;3;5;10;11), 255(4;7;8;10), 257(1;2;4;5;6;10), 259(1;2;4; 7;8;9), 261(2;3;4;6), 263(6), 265(6), 331(1);
Mike Edwards 337(4), 341(2), 343(6), 349(3), 355(7;9), 357(1;5);
David Element 85(2), 89(7), 115(4), 121(6), 123(7), 127(4;5;8), 137(5), 339(3), 345(4);
Bob Gibbons 85(4), 119(7), 310(5), 313(5), 317(6), 325(1);
Michael Hammett 67 (3;4;7;8;9a;9b);
Roger Key 8, 84(8), 93(6), 95(3), 97(8), 101(2), 107(5;6), 111(5;7;8;9;10), 113(1; 3;4;5;6;8;11), 115(1;2;3;8), 117(2;3;6), 119(6;8), 121(5;6;7), 123(10), 125(6;9), 129(3), 131(10), 135(1), 137(2; 10b), 139(6;7), 159(8), 273(5;6), 275(4), 279(2), 299(5), 301(2), 303(9), 305(3;8;9), 311(9), 321(1;5), 337(7), 339(6;10);
Iris Lane 275(5);
Roy Leverton 43(1), 157(3b), 165(11), 171(1;5;6), 173(7), 181(5;10), 185(11), 187(4), 189(10;11), 203(7), 205(6), 227(2;8), 235(8), 241 (4;6;11), 243(6), 247(5), 255(2;3;9), 259(3;6), 261(5), 287(6;7), 289(9;10), 291(2;3), 293(4;6);
Kevin McGee 303(7), 311(2), 313(2b;8), 321(2;3;4;6), 323(2;4), 325(7;8), 351(2);
Ian McLean 35, 301(3;4;5), 303(2;3;4), 307(2), 317(7), 339(8), 347(7);
Geoff Nobes 7, 13(2), 14(1), 22(1;2), 54, 57(2), 60(1), 69(4a;8), 73(4), 77(5), 97(2), 101(6;7), 111(4), 139(1a;1b;3;4;8), 155(1), 157 (1;3a), 159(2), 171(10), 173(2;5;9), 175(6), 177(8), 179(3), 183(8), 189(6), 191(2;7;11), 215(2b), 221(3), 223(7), 227(6), 229(10), 233(4), 239(1;10), 243(10), 245(2), 247(4;8), 249(7), 253(8), 255(5), 257(9), 259(5), 263(3;4;5;10), 265(9), 267(11), 273(8;9), 275(2;7;8), 277(1;3;4;5;6;7;8;9), 279(1;3;5;6;7;8;9;10), 281(4;7;11), 283(1;2;3;4;6;7;8), 285(1;9), 289(2), 291(1,11), 293(9;10;11), 295(2;5), 297(7;10), 313(4), 331(8), 333(4), 351(3a);
Richard Revels 26, 187(5), 237(10), 247(1), 273(3;7), 303(5);
Kevin Simmonds 73(7), 75(6a);
Peter Sutton 79(5), 83(1), 111(2), 119(1), 121(4);
Robert Thompson 69(4b), 71(2;3;4), 77(1b;4;6;7), 109(7), 157(5), 177(10), 179(2;6;7), 183(3), 199(7), 223(1), 235(3), 257(11), 265(5), 267(8), 289(4)
Ian Wallace 319(2;4;6;8;9);
John Walters 87(5), 91(3), 113(9;10), 115(5), 117(4;7;8), 123(2), 131(7), 135(2), 139(2;5);
PH & SL Ward 105(3), 143(2), 293(1), 319(1);
Robin Williams 69(2a), 71(5;8), 73(1b;2; 6), 75(1), 77(1a), 87 (2;3), 93(7), 95(4;6), 97(6), 98(9), 101(3;4;6), 103(7), 305(1;6), 307(8), 309 (1;2;4;5;6;7;8), 313(2b), 315(3;6;7), 317(5), 325(3), 331(5), 333(3; 5;6;7;8), 335(1;2;3;5), 337(1;4;8), 339(1;2; 4;5;7), 341(3;4;5), 343(2;3;4;5), 345(5; 6;7), 347(2;3;6), 349(1;2;7;9), 351(1;5;8a), 353(1;2;3;5;6), 355(3;4;6;8), 357(2;4;6);
Peter Wilson 127(2), 307(9), 313(8), 331(6);
Basil Yates-Smith 111(3), 141(1), 145(2; 3), 147(1;2;3;4;5;7a), 149(1;8), 151(6;7), 153(2;4;5a;10), 155(3;4;10), 161(9).

The artwork is by Denys Ovenden.

ACKNOWLEDGEMENTS

I have been bug-hunting for nearly 60 years and during that time I have been lucky enough to bump into a large number of Britain's insects. I have learned a lot from watching them, but I have learned even more from the many fine entomologists, both amateur and professional, whom I have met over the years and I am grateful to all of them for the help and encouragement that they have given to me. I cannot list them all, but I must mention the late Professor Patrick Buxton, who took me under his wing as a 12-year-old and first opened my eyes to the fact that the insect world embraced much more than the butterflies and moths that I had been chasing until then. I enjoyed many hours exploring his garden with him, often collecting fungi and examining the fungus gnats that emerged from them! I am also indebted to the numerous other entomologists whose works I have consulted in my search for knowledge and I am now happy to be able pass on some of this accumulated information to another generation of bug-hunters.

When I started to compile this book I was slightly worried by the prospect of tracking down sufficient good photographs to illustrate all the species that I wanted to include, and without the photographs there would clearly have been no book. But photographers from all over the country have readily put their excellent digital and conventional images at my disposal and various friends have scoured the countryside to find and photograph elusive species. I am deeply grateful to the following for bringing life to the book:

Ted Benton
Paul Cleary-Pugh
György Csóka
Stan Dumican
Mike Edwards
David Element
Bob Gibbons
Michael Hammett
Roger Key
Roy Leverton
Kevin McGee
Ian McLean
Geoff Nobes
Richard Revels
Kevin Simmonds
Peter Sutton
Robert Thompson
Ian Wallace
John Walters
Peter and Sondra Ward
Robin Williams
Basil Yates-Smith

While putting the book together, I have also received help from numerous friends and others who have kindly identified specimens and photographs and provided information on the distribution and scientific names of many species. In this respect, I am especially grateful to Roy Leverton and to Ian McLean, who has also worked overtime to point me in the direction of photographers and others with specialist knowledge. Peter Chandler, Mike Edwards, Mike Morris, Mike Wilson and Derek Lott have given much information on the distribution of various insects, and James O'Connor in Dublin has provided valuable information on the distribution of insects in Ireland, but responsibility for all errors and omissions remains firmly in my court.

Michael Chinery